Springer Proceedings in Mathematics & Statistics

Volume 256

Springer Proceedings in Mathematics & Statistics

This book series features volumes composed of selected contributions from workshops and conferences in all areas of current research in mathematics and statistics, including operation research and optimization. In addition to an overall evaluation of the interest, scientific quality, and timeliness of each proposal at the hands of the publisher, individual contributions are all refereed to the high quality standards of leading journals in the field. Thus, this series provides the research community with well-edited, authoritative reports on developments in the most exciting areas of mathematical and statistical research today.

More information about this series at http://www.springer.com/series/10533

Galina Filipuk · Alberto Lastra
Sławomir Michalik
Editors

Formal and Analytic Solutions of Diff. Equations

FASdiff, Alcalá de Henares, Spain, September 2017

Selected, Revised Contributions

Editors
Galina Filipuk
Institute of Mathematics
University of Warsaw
Warsaw, Poland

Alberto Lastra
Departamento de Física y Matemáticas
University of Alcalá
Alcalá de Henares, Spain

Sławomir Michalik
Faculty of Mathematics and Natural
Sciences, College of Science
Cardinal Stefan Wyszynski University
Warsaw, Poland

ISSN 2194-1009 ISSN 2194-1017 (electronic)
Springer Proceedings in Mathematics & Statistics
ISBN 978-3-030-07570-5 ISBN 978-3-319-99148-1 (eBook)
https://doi.org/10.1007/978-3-319-99148-1

Mathematics Subject Classification (2010): 34M35, 33E17, 33C45, 34M56, 35C10, 35A01, 26E05, 35C20, 34E17, 32C38, 47F05, 35Q55

This Springer imprint is published by the registered company Springer Nature Switzerland AG
The registered company address is: Gewerbestrasse 11, 6330 Cham, Switzerland

Preface

This volume aims to provide the reader with the actual state of research in the field of Diff. (differential, partial differential, difference, q-difference, q-difference-differential…) Equations. It consists of selected contributions from the conference "Formal and Analytic Solutions of Diff. Equations", held at Alcalá de Henares, Spain during September 4–8, 2017.

This volume is divided into different parts depending on the topics in which the works may be classified.

Part I is devoted to solutions of various types of nonlinear equations. It explains main notions and methods of nonlinear analysis, which allow one to obtain asymptotic expansion of solutions of nonlinear algebraic, ordinary differential, partial differential equations, and of systems of such equations.

Part II is devoted to formal and analytic solutions of partial differential equations (PDEs) and of their discrete analogues. Since formal solutions are usually divergent, one of the main problems of the theory of formal and analytic solutions is to obtain the actual analytic solution from the formal ones. For some types of formal solutions, this is possible due to the use the so-called summability methods. Such type of formal solutions for a class of linear PDEs with time-dependent coefficients is studied in the first paper of this part. In the whole part different types of equations (with constant coefficients and with time-dependent coefficients) and of their solutions are studied. In particular, existence and uniqueness of singular solutions of certain systems of PDEs, hyperasymptotic solutions, integral representation of solutions of certain types of PDEs are considered. The Stokes phenomena, which play a significant role in the theory of ordinary differential equations (ODEs), can be generalized to PDEs, and one of the papers in this part is devoted to the further development of this notion. The so-called integrable equations are special classes of equations which play a significant role in modern mathematics and mathematical physics. They appear in many areas of mathematics, and their discrete counterparts often have very interesting properties. Integrable discretization of equations is a separate and a very interesting problem. One of the papers in this part is devoted to the integrable discrete Schrödinger equation and to the study of long-time behavior of its solutions.

Part III is devoted to the development of the theory of ODEs and systems of ODEs and to their formal and analytic solutions. Nonlinear ODEs cannot be solved explicitly in general, and therefore, various methods to extract information about their solutions have to be devised. The Painlevé equations are nonlinear second-order ordinary differential equations whose solutions have no movable algebraic branch points. Their solutions are called the Painlevé transcendents, and they are often referred to as nonlinear special functions due to their appearance in many diverse areas of mathematics and mathematical physics. The six Painlevé equations possess various interesting properties, and they also appear in the result of similarity reduction of certain integrable PDEs. The first paper in this part is devoted to the derivation and study of complicated and exotic asymptotic expansions of solutions to polynomial ODEs and in particular to some of the Painlevé equations. For real differential equations, singularly perturbed by a small parameter, the so-called canard solutions can be studied. They can be generalized as overstable solutions to complex ODEs. The next paper in this part deals with such solutions by using the summability techniques. Fuchsian systems of ODEs are special types of systems with interesting properties. The so-called monodromy group can be defined. Rigid Fuchsian systems are special systems which are determined by the equivalence classes of residue matrices at singular points. An algorithm to calculate the so-called semilocal monodromy is described in one of the papers. The problem to compute the Newton polygon for linear differential systems is also considered in this part. One more paper deals with the extension of the deformation theory of linear systems with resonant irregular singularities. The deformation theory is very important for applications, in particular to Frobenius manifolds (like quantum cohomology) and the Painlevé equations.

Finally, in Part IV, various related topics, applications, and generalizations are gathered together. ODEs and PDEs are extensively used in mathematical physics. One-dimensional Schrödinger operator with complex potential and the operator of magnetic induction on a two-dimensional symmetric surface are two classes of non-self adjoint operators whose spectrum in the semiclassical limit is concentrated in the neighborhood of some curves in the complex plane. The asymptotics of their eigenvalues is calculated by using certain complex equations in the first paper in this part. In the second paper, a generalization of analytic functions for real normed vector spaces is introduced, and conditions of the uniqueness of such analytic functions are investigated. Another paper is devoted to a new functorial interpretation of asymptotics, which is related with asymptotics along a subvariety with a simple singularity. Such an algebraic study of asymptotics requires some new geometrical and combinatorial notions underlying the multi-normal deformation of a real analytic manifold and the construction of the multispecialization functor along a family of submanifolds. Finally, two papers deal with various aspects of orthogonal polynomials. Orthogonal polynomials are often solutions of differential or difference equations and have many applications in mathematics and physics. They also satisfy the so-called three-term recurrence relation with recurrence coefficients. One of the problems in the theory of orthogonal polynomials is to find explicit expressions for the recurrence coefficients (or to express them in terms of solutions of other differential or difference equations). When the orthogonal polynomials satisfy differential or difference equations, the problem of the factorization of such equations appear, and this is related to the theory of ladder operators (or creation and annihilation operators). The form of ladder operators for polynomials orthogonal with respect to a convex linear combination of discrete and continuous measures (the Laguerre–Krall polynomials) is investigated in the last paper of this volume.

The volume is aimed to graduate students and researchers in theoretical and applied mathematics, physics, and engineering seeking an overview of the recent trends in theory of formal and analytic solutions of functional equations in the complex domain.

The conference FASdiff17 brought together experts in the field of formal and analytic solutions of functional equations, such as differential, partial differential, difference, q-difference, q-difference-differential equations. It took place at the School of Architecture of the University of Alcalá, located in the city center of Alcalá de Henares, in Madrid, Spain.

This meeting was an opportunity to exchange recent results in the field and to explore different possible directions of future research. One of its main objectives was to promote both new and existing scientific collaborations of the researchers in these topics. More precisely, the topics of the conference were the following:

- Ordinary differential equations in the complex domain. Formal and analytic solutions. Stokes multipliers.
- Formal and analytic solutions of partial differential equations.
- Formal and analytic solutions of difference equations (including q-difference and differential-difference equations).
- Special functions (hypergeometric functions and others), orthogonal polynomials, continuous and discrete Painlevé equations.
- Integrable systems.
- Holomorphic vector fields. Normal forms.
- Asymptotic expansions, Borel summability.

The successful planning and organization of FASdiff17 was due to the coordinated efforts of the scientific and organizing committees, consisting of the following members.

Scientific Committee

- Stephane Malek (University of Lille, France)
- Masatake Miyake (Nagoya University, Japan)
- Jorge Mozo-Fernández (University of Valladolid, Spain)
- Javier Sanz (University of Valladolid, Spain)
- Hidetoshi Tahara (Sophia University, Japan)
- Masafumi Yoshino (Hiroshima University, Japan)

Organizing Committee

- Galina Filipuk (University of Warsaw, Poland)
- Javier Jiménez-Garrido (University of Valladolid, Spain)
- Alberto Lastra (University of Alcalá, Spain)

We refer to the conference web-page for more information: http://www3.uah.es/fasdiff17

We finally thank Ayuntamiento de Alcalá de Henares and the University of Alcalá for their support. We also acknowledge the enormous help of Yovana Rodríguez to organize the conference.

We would like to express our deep gratitude to the participants of the conference and to the authors of this volume for their contributions. All contributions have been peer reviewed by anonymous referees chosen among the experts on the subject. We also want to acknowledge the invaluable work done by them. Moreover, we thank the Springer staff for their help and valuable suggestions without whom it would not have been possible to complete this project.

We express our sincere gratitude to all for making the conference FASdiff17 so special and to all who contributed to the creation and production of this volume of proceedings.

Warsaw, Poland — Galina Filipuk
Alcalá de Henares, Spain — Alberto Lastra
Warsaw, Poland — Sławomir Michalik

Contents

Contributors

Moulay Barkatou XLIM UMR 7252 CNRS, University of Limoges, Limoges Cedex, France

Alexander D. Bruno Keldysh Institute of Applied Mathematics of RAS, Moscow, Russia

Jose Ernie C. Lope Institute of Mathematics, University of the Philippines Diliman, Quezon City, Philippines

Mark Philip F. Ona Institute of Mathematics, University of the Philippines Diliman, Quezon City, Philippines

Galina Filipuk Faculty of Mathematics, Informatics and Mechanics, University of Warsaw, Warsaw, Poland

Davide Guzzetti SISSA, Trieste, Italy

Carlos Hermoso Departamento de Física y Matemáticas, Universidad de Alcalá, Alcalá de Henares, Madrid, Spain

Naofumi Honda Department of Mathematics, Faculty of Science, Hokkaido University, Sapporo, Japan

Edmundo J. Huertas Departamento de Ingeniería Civil: Hidráulica y Ordenación del Territorio, E.T.S. de Ingeniería Civil, Universidad Politécnica de Madrid, Madrid, Spain

Kunio Ichinobe Department of Mathematics Education, Aichi University of Education, Igaya, Kariya, Aichi Prefecture, Japan

Alberto Lastra Departamento de Física y Matemáticas, Universidad de Alcalá, Alcalá de Henares, Madrid, Spain

Grzegorz Łysik Faculty of Mathematics and Natural Science, Jan Kochanowski University, Kielce, Poland

Sławomir Michalik Faculty of Mathematics and Natural Sciences, College of Science, Cardinal Stefan Wyszyński University, Warszawa, Poland

Toshio Oshima Josai University, Chiyodaku, Tokyo, Japan

P. Pavis d'Escurac UHA Mulhouse, Mulhouse, France

Luca Prelli Dipartimento di Matematica, Università degli studi di Padova, Padova, Italy

Maria das Neves Rebocho Departamento de Matemática, Universidade da Beira Interior, Covilhã, Portugal; Department of Mathematics, Centre for Mathematics, University of Coimbra, Coimbra, Portugal

Andrei Shafarevich "M.V. Lomonosov" Moscow State University, Moscow, Russia; Moscow Institute of Physics and Technology, Dolgoprudny, Russia; Institute for Problems in Mechanics of the Russian Academy of Sciences, Moscow, Russia; Russian National Scientific Centre "Kurchatov Institute", Moscow, Russia

Maria Suwińska Faculty of Mathematics and Natural Sciences, College of Science, Cardinal Stefan Wyszyński University, Warszawa, Poland

Bożena Tkacz Faculty of Mathematics and Natural Sciences, College of Science, Cardinal Stefan Wyszyński University, Warszawa, Poland

Hideshi Yamane Department of Mathematical Sciences, Kwansei Gakuin University, Hyogo, Japan

Part I
A Survey on the Elements of Nonlinear Analysis

Elements of Nonlinear Analysis

Alexander D. Bruno

Abstract We propose algorithms that allow for nonlinear equations to obtain asymptotic expansions of solutions in the form of: (a) power series with constant coefficients, (b) power series with coefficients which are power series of logarithm and (c) series of powers of exponent of a power series with coefficients which are power series as well. These algorithms are applicable to nonlinear equations (A) algebraic, (B) ordinary differential and (C) partial differential, and to systems of such equations as well. We give the description of the method for one ordinary differential equation and we enumerate some applications of these algorithms.

Keywords Expansions of solutions to ODE · Power expansions · Complicated expansions · Exponential expansions

MSC Primary 33E17 · Secondary 34E05, 41E58

1 Introduction

Tendency to solve the mathematical problems numerically increases in last time according to increasing of power of computers. And teaching mathematicians is oriented to that instead of the study Mathematics itself. I.e. Mathematics is substituted by Arithmetic. That is especially true for problems, which cannot be solved by methods of Classic Analysis and Functional Analysis. Here I will describe a set of such problems, which can be solved by methods of Nonlinear Analysis, allowing to compute asymptotic forms and asymptotic expansions of solution of different classes of equations: algebraic, ordinary differential and partial differential. And of systems of such equations as well. One-year course of lectures on Nonlinear Analysis was given at the Mathematical Department of the Lomonosov Moscow State University. In the present lecture, I will explain main notions and methods of Nonlinear Analysis

---

A. D. Bruno (✉)
Keldysh Institute of Applied Mathematics of RAS, Miusskaya sq. 4, Moscow 125047, Russia
e-mail: abruno@keldysh.ru

G. Filipuk et al. (eds.), *Formal and Analytic Solutions of Diff. Equations*, Springer Proceedings in Mathematics & Statistics 256,
https://doi.org/10.1007/978-3-319-99148-1_1

on examples of an ordinary differential equation

$$f(x, y, y', \dots, y^{(n)}) = 0,$$

where f is a polynomial of its arguments. These methods allow to obtain its solutions in the form of asymptotic expansions

$$y(x) = \sum_{k=0}^{\infty} \varphi_k(x) \tag{1}$$

when $x \to 0$ or $x \to \infty$. At the end, I will give a list of its applications.

For simplicity, here we consider the expansions with real power exponents only.

2 Selection of the Leading Terms

2.1 *Order of a Function [1]*

Let put

$$\omega = \begin{cases} -1, & \text{if } x \to 0, \\ +1, & \text{if } x \to \infty. \end{cases}$$

The number

$$p_\omega(\varphi) = \omega \limsup_{x^\omega \to \infty} \frac{\log|\varphi(x)|}{\omega \log|x|}$$

calculated for the fixed $\arg x \in [0, 2\pi)$, is called as *order* of the function $\varphi(x)$. For the power function $\varphi(x) = \text{const} \cdot x^\alpha$ the order $p(\varphi) = \operatorname{Re} \alpha$ for any ω and $\arg x$. The expansion (1) is called as *asymptotic*, if

$$\omega p(\varphi_k) > \omega p(\varphi_{k+1}), \quad k = 0, 1, 2, \dots .$$

2.2 *Truncated Sums [2, 3]*

Let x be independent and y be dependent variables, $x, y \in \mathbb{C}$. *The differential monomial* $a(x, y)$ is a product of an usual monomial $cx^{r_1}y^{r_2}$, where $c = \text{const} \in \mathbb{C}$, $(r_1, r_2) \in \mathbb{R}^2$, and a finite number or derivatives $d^l y/dx^l$, $l \in \mathbb{N}$. The sum of differential monomials

$$f(x, y) = \sum a_i(x, y) \tag{2}$$

is called as the *differential sum*. We want to select from it all such monomials $a_i(x, y)$, which have the biggest order after the substitution

$$y = \text{const}\, x^p, \quad p \in \mathbb{R} \tag{3}$$

Under the substitution

$$x^{q_1} y^{q_2} = \text{const}\, x^{q_1 + p q_2} = \text{const}\, x^{\langle P, Q \rangle},$$

where $P = (1, p) = (p_1, p_2)$, $Q = (q_1, q_2)$, $\langle P, Q \rangle = p_1 q_1 + p_2 q_2$ is the scalar product. For fixed p and ω, the biggest order will give those monomial const $x^{q_1} y^{q_2}$, for which

$$\omega \langle P, Q \rangle$$

has the maximal value.

Analogously, the differential monomial $a(x, y)$ corresponds to its (vectorial) *power exponent* $Q(a) = (q_1, q_2) \in \mathbb{R}^2$ with the following rules:

$$Q(cx^{r_1} y^{r_2}) = (r_1, r_2); \quad Q(d^l y / dx^l) = (-l, 1);$$

power exponent of a product of monomials is a vectorial sum of their exponents:

$$Q(a_1 a_2) = Q(a_1) + Q(a_2).$$

The set $\mathbf{S}(f)$ of vectorial power exponents $Q(a_i)$ of all differential monomials $a_i(x, y)$, containing in the differential sum (2), is called as the *support of the sum* $f(x, y)$. Evidently, $\mathbf{S}(f) \in \mathbb{R}^2$. The convex hull $\Gamma(f)$ of the support $\mathbf{S}(f)$ is called as the *polygon of the sum* $f(x, y)$. The boundary $\partial\Gamma(f)$ of the polygon $\Gamma(f)$ consists of vertices $\Gamma_j^{(0)}$ and edges $\Gamma_j^{(1)}$. We call them as *generalized faces* $\Gamma_j^{(d)}$, where the upper index shows the dimension of the face, and low index shows its number. Each face $\Gamma_j^{(d)}$ corresponds to the *truncated sum*

$$\hat{f}_j^{(d)}(x, y) = \sum a_i(x, y) \text{ along } Q(a_i) \in \Gamma_j^{(d)} \cap \mathbf{S}(f). \tag{4}$$

After substitution (3), all terms in (4) have the same order, which is $\langle P, Q \rangle$, if the vector $\omega P = \omega(1, p)$ is the exterior normal to the edge or vertex $\Gamma_j^{(d)}$. So the biggest value of $\omega \langle P, Q \rangle$ is achieved on $Q \in \Gamma_j^{(d)}$.

Example. Let us consider the third Painlevé equation

$$f(x, y) \stackrel{\text{def}}{=} -xyy'' + xy'^2 - yy' + ay^3 + by + cxy^4 + dx = 0, \tag{5}$$

assuming that its complex parameters $a, b, c, d \neq 0$. Here the first three differential monomials have the same power exponent $Q_1 = (-1, 2)$, then $Q_2 = (0, 3)$, $Q_3 = (0, 1)$, $Q_4 = (1, 4)$, $Q_5 = (1, 0)$. They are shown in Fig. 1 in coordinates q_1, q_2.

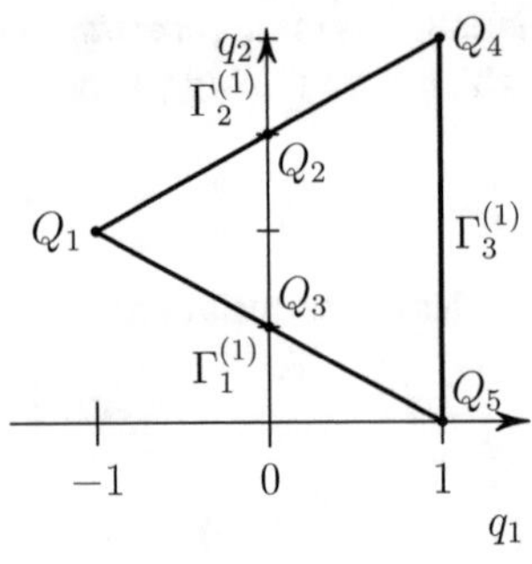

Fig. 1 Support $\mathbf{S}(f)$, polygon $\Gamma(f)$ and its edges $\Gamma_j^{(1)}$ for the third Painlevé equation (5)

Their convex hull $\Gamma(f)$ is a triangle with three vertices $\Gamma_1^{(0)} = Q_1$, $\Gamma_2^{(0)} = Q_4$, $\Gamma_3^{(0)} = Q_5$, and with three edges $\Gamma_1^{(1)}, \Gamma_2^{(1)}, \Gamma_3^{(1)}$. The vertex $\Gamma_1^{(0)} = Q_1$ corresponds to the truncated sum

$$\hat{f}_1^{(0)}(x, y) = -xyy'' + xy'^2 - yy',$$

and the edge $\Gamma_1^{(1)}$ corresponds to the truncated sum

$$\hat{f}_1^{(1)}(x, y) = \hat{f}_1^{(0)}(x, y) + by + dx. \qquad \blacksquare$$

Let the plane $\mathbb{R}_*^2$ be such conjugate to the plane $\mathbb{R}^2$, that the scalar product

$$\langle P, Q \rangle \overset{\text{def}}{=} p_1q_1 + p_2q_2$$

be defined for $P = (p_1, p_2) \in \mathbb{R}_*^2$ and $Q = (q_1, q_2) \in \mathbb{R}^2$. Each face $\Gamma_j^{(d)}$ corresponds to its own *normal cone* $\mathbf{U}_j^{(d)} \subset \mathbb{R}_*^2$. It consists of the exterior normals P to the face $\Gamma_j^{(d)}$. The normal cone $\mathbf{U}_j^{(1)}$ of the edge $\Gamma_j^{(1)}$ is a ray orthogonal to the edge $\Gamma_j^{(1)}$ and directed outside of the polygon $\Gamma(f)$. The normal cone $\mathbf{U}_j^{(0)}$ of the vertex $\Gamma_j^{(0)}$ is the open sector (angle) at the plane $\mathbb{R}_*^2$ with vertex in the origin $P = 0$ and restricted by rays, which are the normal cones of edges, adjoined to the vertex $\Gamma_j^{(0)}$. Generally

$$\mathbf{U}_j^{(d)} = \{P : \langle P, Q' \rangle = \langle P, Q'' \rangle > \langle P, Q''' \rangle,\ P', P'' \in \Gamma_j^{(d)},\ P''' \in \Gamma \backslash \Gamma_j^{(d)}\}.$$

Example. For the Eq. (5), normal cones $\mathbf{U}_j^{(d)}$ of faces $\Gamma_j^{(d)}$ are shown in Fig. 2. $\blacksquare$

So each face $\Gamma_j^{(d)}$ corresponds to the normal cone $\mathbf{U}_j^{(d)}$ in the plane $\mathbb{R}_*^2$ and to the truncated sum (4).

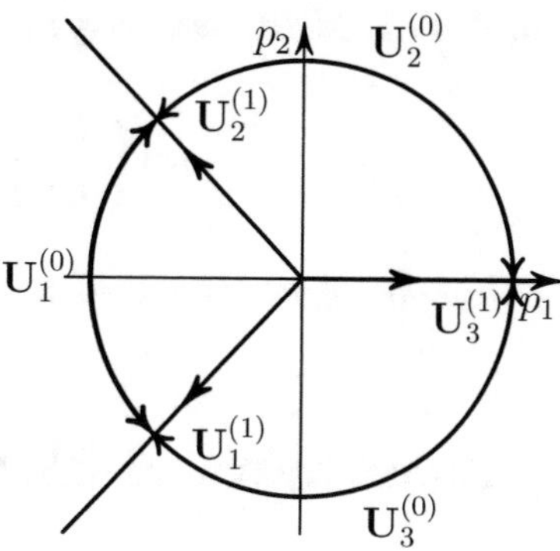

Fig. 2 Normal cones $\mathbf{U}_j^{(d)}$ to vertices and edges $\Gamma_j^{(d)}$ of the polygon of Fig. 1

2.3 Variations [3, 4]

In Classic Analysis, it is known the Taylor formula

$$f(x_0+\Delta)=\sum_{k=0}^{\infty}\frac{1}{k!}f^{(k)}(x_0)\Delta^k.$$

In the Functional Analysis, there is it analog

$$f(x,y_0+z)=\sum_{k=0}^{\infty}\frac{1}{k!}\,\frac{\delta^k}{\delta y^k}f\bigg|_{y=y_0}z^k, \tag{6}$$

where $f(x,y)$ is a differential sum, $\dfrac{\delta^k}{\delta y^k}f(x,y)$ is its *k-variation* along y (derivative of Frechet or Gateaux). It is taken on the function $y=y_0(x)$ and is an operator which is applied to the k-power of the small addendum z^k. All that is doing in the infinity-dimensional spaces. If $f(x,y)$ is an usual polynomial (without derivatives), then $\dfrac{\delta^k}{\delta y^k}f=\dfrac{\partial^k}{\partial y^k}f$. But variations are defined for differential polynomials containing derivatives.

Example. If $f=\dfrac{\partial^k y}{\partial x^k}$, then $\dfrac{\delta f}{\delta y}=\dfrac{\partial^k}{\partial x^k}$. ■

Theorem 1 *Let* $p(y_0)=p_0$, $p\left(y_0^{(k)}\right)=p_0-k$, $k=1,2,\ldots$, $p(z)=p_1$, *along the curves* $y=\text{const}\,x^p$ *the order* $p(f)=\tilde{p}$ *and* $\omega(p_1-p_0)<0$, *then expansion* (6) *is asymptotic, where*

$$\omega p\left(\frac{\delta^k}{\delta y^k}f\bigg|_{y=y_0}z^k\right)\leqslant\omega\tilde{p}+k\omega(p_1-p_0),\quad k=0,1,2,\ldots$$

Corollary 1 *In the situation of the Theorem 1*

$$\omega p \left(\frac{\delta}{\delta y} f \Big|_{y=y_0} z \right) < \omega p \left(f(y_0) \right).$$

i.e. the order of the first variation is less than the order of $f(y_0)$.

3 Power Expansions of Solutions [3, 5]

3.1 Statement of the Problem

Problem 1 Let we have the ordinary differential equation

$$f(x, y) = 0, \tag{7}$$

where $f(x, y)$ is a differential sum. For solutions $y = \varphi(x)$ of the Eq. (7) for $x \to 0$ and $x \to \infty$ to find all expansions of the form

$$y = c_r x^r + \sum c_s x^s, \quad c_r, c_s = \text{const} \in \mathbb{C}, \quad c_r \neq 0, \tag{8}$$

where the power exponents $r, s \in \mathbb{R}$,

$$\omega r > \omega s.$$

■

Computation of the expansions (8) consists of two steps: computation of the first term

$$y = c_r x^r, \quad c_r \neq 0$$

and computation of other terms in (8).

Theorem 2 *If the expansion* (8) *satisfies Eq.* (7) *and* $\omega(1, r) \in \mathbf{U}_j^{(d)}$, *then the truncation* $y = c_r x^r$ *of the solution* (8) *is a solution of the truncated equation* $\hat{f}_j^{(d)}(x, y) = 0$.

So, to find all truncated solutions $y = c_r x^r$ of the Eq. (7), we must compute: the support $\mathbf{S}(f)$, polygon $\Gamma(f)$, all its faces $\Gamma_j^{(d)}$ and their normal cones $\mathbf{U}_j^{(d)}$. Then for each truncated equation $\hat{f}_j^{(d)}(x, y) = 0$, we must find all such its power solutions $y = c_r x^r$, for which one of two vectors $\pm(1, r)$ is in the normal cone $\mathbf{U}_j^{(d)}$.

3.2 Solving a Truncated Equation

The vertex $\Gamma_j^{(0)} = \{Q\}$ corresponds to the truncated equation $\hat{f}_j^{(0)}(x, y) = 0$ with the point support $Q = (q_1, q_2)$. Let put $g(x, y) = x^{-q_1} y^{-q_2} \hat{f}_j^{(0)}(x, y)$, then $g(x, cx^r)$ does not depend from x and c and is a polynomial of r. Hence, the power exponent r of the solution $y = c_r x^r$ to the equation $\hat{f}_j^{(0)}(x, y) = 0$ is a root of the *characteristic equation*

$$\chi(r) \stackrel{\text{def}}{=} g(x, x^r) = 0, \tag{9}$$

and the coefficient c_r is arbitrary. Among real roots r of the Eq. (9), we must take only such, for which the vector $\omega(1, r)$ is in the normal cone $\mathbf{U}_j^{(0)}$ of the vertex $\Gamma_j^{(0)}$.

Example. In Eq. (5), the vertex $\Gamma_1^{(0)} = Q_1 = (-1, 2)$ corresponds to the truncated equation

$$\hat{f}_1^{(0)}(x, y) \stackrel{\text{def}}{=} -xyy'' + xy'^2 - yy' = 0, \tag{10}$$

and $\hat{f}_1^{(0)}(x, x^r) = x^{2r-1}[-r(r-1) + r^2 - r] \equiv 0$, i.e. any expression $y = cx^r$ is a solution of the Eq. (10). Here $\omega = -1$ and we are interested only in such these solutions, for which the vector $-(1, r) \in \mathbf{U}_1^{(0)}$. According to Fig. 2, it means that $r \in (-1, 1)$. So the vertex $\Gamma_1(0)$ corresponds to two-parameter family of power asymptotic forms of solutions

$$y = cx^r, \text{ arbitrary } c \neq 0, \quad r \in (-1, 1). \tag{11}$$

■

The edge $\Gamma_j^{(1)}$ corresponds to the truncated equation $\hat{f}_j^{(1)}(x, y) = 0$. Its normal cone $\mathbf{U}_j^{(1)}$ is the ray $\{P = \lambda\omega'(1, r'), \ \lambda > 0\}$. Inclusion $\omega(1, r) \in \mathbf{U}_j^{(1)}$ means equalities $\omega = \omega'$ and $r = r'$. They determine exponent r of the truncated solution $y = c_r x^r$ and value ω. To find the coefficient c_r, we must substitute the expression $y = c_r x^r$ into the truncated equation $\hat{f}_j^{(1)}(x, y) = 0$. After cancellation of some power of x, we obtain the algebraic *determining equation* for the coefficient c_r

$$\tilde{\tilde{f}}(c_r) \stackrel{\text{def}}{=} x^{-s} \hat{f}_j^{(1)}(x, c_r x^r) = 0.$$

Each its root $c_r \neq 0$ corresponds to its own asymptotic form $y = c_r x^r$.

Example. In Eq. (5), the edge $\Gamma_1^{(1)}$ corresponds to the truncated equation

$$\hat{f}_1^{(1)}(x, y) \stackrel{\text{def}}{=} -xyy'' + xy'^2 - yy' + by + dx = 0. \tag{12}$$

As $\mathbf{U}_1^{(1)} = \{P = -\lambda(1, 1), \ \lambda > 0\}$, then $\omega = -1$ and $r = 1$. After substitution $y = c_1 x$ in the truncated Eq. (12) and cancel by x, we obtain for c_1 equation $bc_1 + d = 0$. Hence, $c_1 = -d/b$. So, the edge $\Gamma_1^{(1)}$ corresponds to unique power asymptotic form of solution

$$y = -(d/b)x, \quad x \to 0. \tag{13}$$

■

3.3 Critical Numbers of the Truncated Solution

If the truncated solution $y = c_r x^r$ is found, then the change $y = c_r x^r + z$ brings the equation $f(x, y) = 0$ to the form

$$f(x, cx^r + z) \stackrel{\text{def}}{=} \tilde{f}(x, z) \stackrel{\text{def}}{=} \mathscr{L}(x)z + h(x, z) = 0,$$

where $\mathscr{L}(x)$ is a liner differential operator and the support $\mathbf{S}(\mathscr{L}z)$ consists of one point $(\tilde{\nu}, 1)$, which is a vertex $\tilde{\Gamma}_1^{(0)}$ of the polygon $\Gamma(\tilde{f})$, and the support $\mathbf{S}(h)$ has not the point $(\tilde{\nu}, 1)$. The operator $\mathscr{L}(x)$ is the first variation $\delta \hat{f}_j^{(d)}/\delta y$ on the curve $y = c_r x^r$. Let $\nu(k)$ be the characteristic polynomial of the differential sum $\mathscr{L}(x)z$, i.e.

$$\nu(k) = x^{-\tilde{\nu}-k} \mathscr{L}(x) x^k.$$

The real roots $k_1, \ldots, k_\varkappa$ of the polynomial $\nu(k)$, which satisfy the inequality $\omega r > \omega k_i$, are called as *critical numbers of the truncated solution* $y = c_r x^r$.

Example. The first variation for the truncated Eq. (10) is

$$\frac{\delta \hat{f}_1^{(0)}}{\delta y} = -xy'' - xy\frac{d^2}{dx^2} + 2xy'\frac{d}{dx} - y' - y\frac{d}{dx}.$$

At the curve $y = c_r x^r$, that variation gives the operator

$$\mathscr{L}(x) = c_r x^{r-1}\left[-r(r-1) - x^2\frac{d^2}{dx^2} + 2rx\frac{d}{dx} - r - x\frac{d}{dx}\right].$$

The characteristic polynomial of the sum $\mathscr{L}(x)z$, i.e. $\mathscr{L}(x)x^k$, is

$$\nu(k) = c_r[-r(r-1) - k(k-1) + 2rk - r - k] = -c_r(k-r)^2.$$

It has one double root $k_1 = r$, which is not a critical number, because it does not satisfy the inequality $\omega r > \omega k_1$. Hence, the truncated solution (11) has not critical numbers.

The first variation for the truncated Eq. (12) is

$$\frac{\delta \hat{f}_1^{(1)}}{\delta y} = \frac{\delta \hat{f}_1^{(0)}}{\delta y} + b.$$

At the curve (13), i.e. $y = c_1 x$, $c_1 = -d/b$, the variation gives the operator

$$\mathscr{L}(x) = c_1 \left[-x^2 \frac{d^2}{dx^2} + 2x \frac{d}{dx} - 1 - x \frac{d}{dx} - \frac{b^2}{d} \right]$$

and the characteristic polynomial

$$\nu(k) = -c_1 \left[k^2 - 2k + 1 + \frac{b^2}{d} \right].$$

Its roots are $k_{1,2} = 1 \pm b/\sqrt{-d}$. If Im $(b/\sqrt{-d}) \neq 0$, then the critical numbers are absent. If Im $(b/\sqrt{-d}) = 0$, then only one root $k_1 = 1 + |b/\sqrt{-d}|$ satisfies the inequality $\omega r > \omega k_i$, and it is the unique critical number of the power asymptotic form (13). ■

3.4 *Computation of the Power Expansion of a Solution [3, §3]*

Let $\Gamma_j^{(0)}$ be a vertex of the polygon $\Gamma(f)$ and vectors M_1 and M_2 are directed from the vertex along the adjoint edges, and all points of the shifted support $\mathbf{S}(f) - \Gamma_j^{(0)}$ have the form $l_1 M_1 + l_2 M_2$ with integers $l_1, l_2 \geqslant 0$. Then the set

$$\mathbf{K}_j^{(0)}(r) \stackrel{\text{def}}{=} \{s = r + l_1 r_1 + l_2 r_2, l_i \geqslant 0, l_1 + l_2 > 0\},$$

where $r_i = \langle (1, r), M_i \rangle$, $i = 1, 2$.

Let $\Gamma_j^{(1)}$ be an edge of the polygon $\Gamma(f)$ with vertexes $\Gamma_k^{(0)}$, $\Gamma_l^{(0)}$ and with the normal $\omega(1, r)$. Then

$$\mathbf{K}_j^{(1)} \stackrel{\text{def}}{=} \mathbf{K}_k^{(0)}(r) \cap \mathbf{K}_l^{(0)}(r).$$

Theorem 3 *If the truncated solution $y = c_r x^r$ corresponds to the vertex $\Gamma_j^{(0)}$ with $\omega(1, r) \subset \mathbf{U}_j^{(0)}$ or to the edge $\Gamma_j^{(1)}$ with $\omega(1, r) \subset \mathbf{U}_j^{(1)}$ and all critical numbers of the truncated solution does not lie in the set $\mathbf{K} = \mathbf{K}_j^{(0)}(r)$ or $\mathbf{K}_j^{(1)}$, then the initial equation has a solution in the form of expansion* (8), *where s runs the set $\mathbf{K}_j^{(0)}(r)$ or $\mathbf{K}_j^{(1)}$ correspondingly.*

Proof is based on the asymptotic expansions

$$f(x, y) = \hat{f}_j^{(d)}(x, y) + \hat{\hat{f}}(x, y) + \cdots, \quad y = c_r x^r + c_s x^s + \cdots$$

Substituting one into another and using Corollary 1, we obtain the equation

$$f(x, c_r x^r + c_s x^s + \cdots) = \hat{f}_j^{(d)}(x, c_r x^r) + \left.\frac{\delta \hat{f}_j^{(d)}}{\delta y}\right|_{y=c_r x^r} \cdot c_s x^s + \\ + \hat{\hat{f}}(x, c_r x^r) + \cdots = 0.$$

But here $\hat{f}_j^{(d)}(x, c_r x^r) = 0$ and the leading terms are next two. Hence, the equation

$$\left.\frac{\delta \hat{f}_j^{(d)}}{\delta y}\right|_{y=c_r x^r} \cdot c_s x^s + \hat{\hat{f}}(x, c_r x^r) = 0,$$

must be satisfied. It gives the equation of the form

$$\nu(s)c_s + b_s = 0, \quad b_s = \text{const} \in \mathbb{C}.$$

As $s \in \mathbf{K}$ and according to condition of the Theorem $\nu(s) \neq 0$, then moving along $s \in \mathbf{K}$ with decreasing ωs we successfully compute coefficients c_s of expansion (8).

Example. The vertex $\Gamma_1^{(0)} = Q_1$ for the Eq. (5) corresponds to vectors $M_1 = (1, 1)$, $M_2 = (1, -1)$, so $r_1 = 1 + r$, $r_2 = 1 - r$, where $|r| < 1$ and the set

$$\mathbf{K}_1^{(0)}(r) = \{s = r + l_1(1 + r) + l_2(1 - r), l_1, l_2 \geqslant 0, l_1 + l_2 > 0\}. \tag{14}$$

As there are no critical numbers, then according to Theorem 3, each truncated solution (11) corresponds to the solution (8) with $s \in \mathbf{K}_1^{(0)}(r)$.

The edge $\Gamma_1^{(1)}$ has two vertices Q_1 and $Q_5 = (1, 0)$, $r = 1$. Here, according to (14), $\mathbf{K}_1^{(0)}(1) = \{1 + 2l_1\}$ for the vertex Q_1. For the vertex $\Gamma_3^{(0)} = Q_5$, we have $M_1 = (-1, 1)$, $M_2 = (0, 2)$. So $r_1 = 0$, $r_2 = 2$, and $\mathbf{K}_3^{(0)}(1) = \mathbf{K}_1^{(0)}(1) = \{1 + 2l_1, \text{ integral } l_1 > 0\}$. If $\operatorname{Im}(b/\sqrt{-d}) \neq 0$, then the truncated solution (13) has no critical numbers and, in the expansion (8) all power exponents s are odd integral numbers more then 1, and coefficients c_s are unique constants. If $\operatorname{Im}(b/\sqrt{-d}) = 0$, then there is only one critical number $k_1 = 1 + |b/\sqrt{-d}|$. Hence, if the number k_1 is not odd, then there is the expansion (8). ■

4 Complicated Expansions of Solutions [3, § 5], [6, 7]

Truncated equations can have nonpower solutions, which can be continued into asymptotic expansions. Here we will look for solutions of the full equation $f(x, y) = 0$ in the form of the *complicated asymptotic expansions*

$$y = \varphi_r(\log x)x^r + \sum \varphi_s(\log x)x^s, \quad \omega s < \omega r, \tag{15}$$

where $\varphi_r(\log x)$ and $\varphi_s(\log x)$ are series on decreasing powers of logarithm.

Theorem 4 *If the series* (15) *is a solution of the full Eq.* (7) *and* $\omega(1,r) \subset \mathbf{U}_j^{(d)}$, *then* $y = \varphi_r x^r$ *is a solution of the corresponding truncated equation* $\hat{f}_j^{(d)}(x, y) = 0$.

A truncated equation, corresponding to a vertex, has a nonpower solution only in very degenerate cases [3, § 5]. So, here we will consider only truncated equations, corresponding to edges $\Gamma_j^{(1)}$.

4.1 *Case of the Vertical Edge* $\boldsymbol{\Gamma_j^{(1)}}$

If the edge $\Gamma_j^{(1)}$ is vertical, then its normal cone is

$$\mathbf{U}_j^{(1)} = \lambda\omega(1, 0), \quad \lambda > 0,$$

and all points $Q = (q_1, q_2) \in \Gamma_j^{(1)}$ have the same coordinate q_1. Let put

$$g(x, y) = x^{-q_1} \hat{f}_j^{(d)}(x, y),$$

then the support $\mathbf{S}(g)$ lies at the coordinate axis $q_1 = 0$. For the truncated equation, all power solutions with $\omega(1, r) \in \mathbf{U}_j^{(1)}$ are constants $y = y^0 = \text{const}$, where y^0 is a root of the determining equation

$$\tilde{g}(y) \stackrel{\text{def}}{=} g(0, y) = 0.$$

To find nonpower solutions of the equation $g(x, y) = 0$ we make the *logarithmic transformation*

$$\xi \stackrel{\text{def}}{=} \log x. \tag{16}$$

According to Theorem 2.4 from [2, Ch. VI], here the differential sum $g(x, y)$ comes to the differential sum $h(\xi, y) \stackrel{\text{def}}{=} g(x, y)$ and the equation $g = 0$ takes the form

$$h(\xi, y) = 0. \tag{17}$$

From (16), we see that $\xi \to \infty$ as $x \to 0$ and as $x \to \infty$, because ξ and x are complex, i.e. for the Eq. (17) we obtain the problem with

$$p \geqslant 0.$$

Applying the described above technique to the Eq. (17), we select truncated equations $\hat{h}_l^{(d)}(\xi, y) = 0$ with $\omega = 1$ and find their power solutions $y = c_\rho \xi^\rho$. Each of them corresponds to its characteristic polynomial $\nu^*(k^*)$, its own critical numbers k_j^* and its own set $\mathbf{K}^*$. Under conditions of Theorem 3 on k_j^* and $\mathbf{K}^*$, we find the power expansion of solution to equation $h(\xi, y) = 0$ in the form

$$y = c_\rho \xi^\rho + \sum c_\sigma \xi^\sigma, \quad \sigma \in \mathbf{K}^*, \quad \sigma < \rho, \quad c_\rho, c_\sigma = \text{const} \in \mathbb{C}.$$

Besides, the solution $y = c\xi^\rho$ to the truncated equation $\hat{h}_l^{(d)}(\xi, y) = 0$ corresponds to its own *complicated characteristic equation* $\mu(\varkappa) = 0$. It is formed by the following way. We have the variation

$$\frac{\delta \hat{h}_l^{(d)}}{\delta y} = \sum_{i=1}^{M} b_i(\xi, y)\mu_i\left(\frac{d}{d\xi}\right),$$

where b_i are differential monomials and μ_i are differential operators with constant coefficients

$$\mu_i\left(\frac{d}{d\xi}\right) = \sum_{k=0}^{l_i} \alpha_{ik}\frac{d^k}{d\xi^k}, \quad \alpha_{ik} = \text{const} \in \mathbb{C}.$$

Among all monomials $b_i(\xi, y)$, we select such, which give the maximal power of ξ after the substitution $y = \xi^\rho$: $b_i = \beta_i \xi^n + \cdots, i = 0, \ldots, M$, where n is the maximal power of ξ in all b_i and $\beta_i = 0$ or const. *Polynomial*

$$\mu(\varkappa) = \sum_{i=0}^{M} \beta_i \mu_i(\varkappa),$$

where $d^k/d\xi^k$ are changed by $\varkappa^k$, is called as *complicated-characteristic* for the double truncated solution $y = c_\rho \xi^\rho$.

Theorem 5 *If roots of polynomials $\nu^*(k^*)$ and $\mu(\varkappa)$ for a vertical edge donot ly in sets $\mathbf{K}^*$ and $\mathbf{K}$ correspondingly, then the double truncated solution $y = c_\rho \xi^\rho$ corresponds to a solution to the full equation in the form of complicated expansion* (15).

Proof is similar to the proof of Theorem 3.

4.2 Inclined Edge

Theorem 6 *The power transformation*

$$y = x^\alpha z \tag{18}$$

transforms the differential sum $f(x, y)$ into the differential sum $g(x, z) = f(x, y)$. Here their supports and normal cones are connected by the affine transformations

$$\mathbf{S}(g) = \mathbf{S}(f)A, \quad \mathbf{U}_g = A^{*-1}\mathbf{U}_f,$$

where matrices are $A = \begin{pmatrix} 1 & 0 \\ \alpha & 1 \end{pmatrix}$, $A^{*-1} = \begin{pmatrix} 1 & -\alpha \\ 0 & 1 \end{pmatrix}$.

The case of the inclined edge $\Gamma_j^{(1)}$ with the normal vector $(1, r)$ is reduced to the case of the vertical edge $\widetilde{\Gamma}_j^{(1)}$ by means of the power transformation (18) with $\alpha = r$. After computation of Sects. 3 and 4.1 for the transformed equation, we obtain a double truncated solution $z = c_\rho \xi^\rho$ together with characteristic polynomial $\nu^*(k^*)$ and complicated-characteristic polynomial $\mu(\varkappa)$. From Theorems 5 and 6 we obtain

Corollary 2 *For an inclined edge with normal* $(1, r)$*, if roots of polynomials* $\nu^*(k^*)$ *and* $\mu(\varkappa)$ *do not belong to sets* $\mathbf{K}^*$ *and* $\mathbf{K} - r$ *correspondingly, then the double truncated solution* $z = c_\rho \xi^\rho$ *corresponds to a solution to the full equation in the form of complicated expansion.*

Example. In the truncated Eq. (12), corresponding to the edge $\Gamma_1^{(1)}$ with normal $-(1, 1)$, we make the power transformation $y = xz$, with the matrix

$$A = \begin{pmatrix} 1 & 0 \\ 1 & 1 \end{pmatrix}.$$

As $y' = xz' + z$, $y'' = xz'' + 2z'$, then Eq. (12) after canceling by x and grouping takes the form

$$\hat{g}(x, z) \stackrel{\text{def}}{=} -x^2 zz'' + x^2 z'^2 - xzz' + bz + d = 0. \tag{19}$$

Its support consists of three points $\tilde{Q}_2 = (0, 2)$, $\tilde{Q}_4 = (0, 1)$, $\tilde{Q}_1 = 0$, lying at the axis $q_1 = 0$. Now we make the logarithmic transformation $\xi = \ln x$. As $z' = \dot{z}/x$, $z'' = (\ddot{z} - \dot{z})/x^2$, where $\dot{} = d/d\xi$, then Eq. (19) after grouping takes the form

$$h \stackrel{\text{def}}{=} -z\ddot{z} + \dot{z}^2 + bz + d = 0. \tag{20}$$

Its support and polygon are shown in Fig. 3 in the case $bd \neq 0$.

Let us consider case $b \neq 0$. The edge $\tilde{\Gamma}_1^{(1)}$ of Fig. 3 corresponds to the truncated equation

$$\hat{h}_1^{(1)} \stackrel{\text{def}}{=} -z\ddot{z} + \dot{z}^2 + bz = 0. \tag{21}$$

It has power solution $z = -b\xi^2/2$. The edge $\widetilde{\Gamma}_1^{(1)}$ has 2 vertices $(-2, 2)$ and $(0, 1)$. For the vertex $(-2, 2)$, vectors $M_1 = (2, -1)$ and $M_2 = (2, -2)$. Here $r = 2$. So $r_1 = 0$, $r_2 = -2$, $\mathbf{K}^* = \{2 - 2l_1\}$. For the vertex $(0, 1)$, vectors $M_1 = (-2, 1)$ and $M_2 = (0, -1)$. So $r_1 = 0$, $r_2 = -2$, $\mathbf{K}^* = \{2 - 2l_1\}$, integral $l_1 > 0$. The characteristic polynomial of the solution $z = -b\xi^2/2$ is $\nu^*(k^*) = (b/2)\left(k^{*2} - 5k^* + 4\right) = (b/2)(k^* - 1)(k^* - 4)$. As here $r^* = 2$, then there is only one critical number

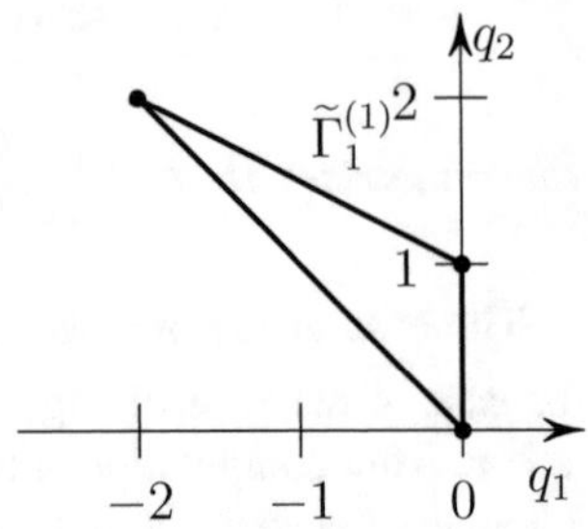

Fig. 3 Support and polygon of Eq. (20)

$k_1^* = 1 < r^*$. As it does not belong to the set $\mathbf{K}^*$, then according to Theorem 3, the Eq. (20) has a solution of the form

$$z = -\frac{b}{2}\xi^2 + \sum_{k=0}^{\infty} c_{-2k}\xi^{-2k}. \tag{22}$$

Indeed solutions to Eq. (20) have the form

$$z = -\frac{b}{2}(\xi + \tilde{c})^2 - \frac{d}{2b}, \tag{23}$$

where $\tilde{c}$ is arbitrary constant. The solution (22) corresponds to the case $\tilde{c} = 0$. According to (21), the first variation is

$$\frac{\delta \hat{h}_1^{(1)}}{\delta z} = -\ddot{z} - z\frac{d^2}{d\xi^2} + 2\dot{z}\frac{d}{d\xi} + b.$$

So

$$\frac{\delta \hat{h}_1^{(1)}}{\delta z} = b_1\mu_1\left(\frac{d}{d\xi}\right) + b_2\mu_2\left(\frac{d}{d\xi}\right) + b_3\mu_3,$$

where $b_1 = -z$, $b_2 = 2\dot{z}$, $b_3 = -\ddot{z} + b$, $\mu_1 = \dfrac{d^2}{d\xi^2}$, $\mu_2 = \dfrac{d}{d\xi}$, $\mu_3 = 1$. As $\rho = 2$, then the leading term is $b_1\mu_1$. It gives the characteristic polynomial $\mu_1(\varkappa) = \varkappa^2$ without nonzero roots. Hence, there are no complicated critical roots, and we can apply Theorem 5. After the power transformation $y = xz$ and cancel by x, the full Eq. (5) takes the form

$$g \stackrel{\text{def}}{=} -x^2 z z'' + x^2 z'^2 - xzz' + bz + d + ax^2 z^3 + cx^4 z^4 = 0. \tag{24}$$

The set **K** consists of all even natural numbers. According to Theorem 5, solution to (24) has the form

$$z = \varphi_0(\xi) + \sum_{k=1}^{\infty} \varphi_{2k}(\xi) x^{2k},$$

where φ_0 is given by (23) and $x \to 0$. ■

5 Exponential Expansions of Solutions [8, 9]

Let the truncated equation $\hat{f}_j^{(1)}(x, y) = 0$ correspond to the horizontal edge $\Gamma_j^{(1)}$ of the polygon $\Gamma(f)$. Hence, at the edge $q_2 = m \in \mathbb{N}$. According to [3, § 5], we make the logarithmic transformation

$$\zeta = d \log y / dx\,, \tag{25}$$

and from the truncated equation $\hat{f}_j^{(1)} = 0$, we obtain the equation

$$h(x, \zeta) y^m \stackrel{\text{def}}{=} \hat{f}_j^{(1)}(x, y) = 0$$

where $h(x, \zeta)$ is a differential sum [2, Ch. VI]. Let $\Gamma(h) = \tilde{\Gamma}$ be its polygon and $\tilde{\Gamma}_i^{(1)}$ is its edge with outside normal $\tilde{N} = (1, \rho)$, lying in the cone of the problem $\mathcal{K}_\omega = \{\tilde{P} = (\tilde{p}_1, \tilde{p}_2) : \tilde{p}_1 + \tilde{p}_2 > 0,\ \operatorname{sgn} \tilde{p}_1 = \omega\}$. That determines the sign of ω and direction of tendency of x (to zero or to infinity). The edge $\tilde{\Gamma}_i^{(1)}$ corresponds to the truncated equation $\hat{h}_i^{(1)}(x, \zeta) = 0$, which is algebraic and has several power solutions $\zeta = \gamma^* x^\rho$, where $\gamma = \gamma^* = \text{const}$ is one of the roots of the determining equation $\hat{h}_i^{(1)}(1, \gamma) = 0$. Each power solution $\zeta = \gamma^* x^\rho$ to the truncated equation $\hat{h}_i^{(1)}(x, \zeta) = 0$ is continued by the unique manner into power expansion

$$\zeta = \gamma^* x^\rho + \sum \gamma_\sigma x^\sigma \stackrel{\text{def}}{=} \varphi'(x) \tag{26}$$

of a solution to the full equation $h(x, \zeta) = 0$. The first variation can be written as

$$\frac{\delta \hat{f}_j^{(1)}}{\delta y} = y^{m-1} g\left(x, \zeta, \frac{d}{dx}\right),$$

where g is a polynomial of its arguments, if $\left(\frac{d}{dx}\right)^l$ means $\frac{d^l}{dx^l}$. Its order in $\zeta, \zeta', \ldots, \zeta^{(n-1)}$ is less than m. Now in the operator g, we change $\frac{d^l}{dx^l}$ by $k^l \zeta^l$ and ζ by $\gamma^* x^\rho$. Then we select the leading term $\lambda(\gamma^*, k) x^\tau$ in x. Coefficient $\lambda(\gamma^*, k)$ is the *exponential characteristic polynomial*, corresponding to the truncated solution $\zeta = \gamma^* x^\rho$.

If the equation $h(x, \zeta) = 0$ has a solution of form (26), then the truncated equation $\hat{f}_j^{(1)}(x, y) = 0$ has the family of solutions

$$y = c \exp \varphi(x), \tag{27}$$

where c is arbitrary constant and $\varphi(x)$ is an integral of the power expansion (26).

Now we come to the full equation $f(x, y) = 0$. Let the set Σ be the projection of the support $\mathbf{S}(f)$ on axis q_2 parallel to axis q_1. Let put $\Sigma' = \Sigma - m$, i.e. Σ' is a shifted on m set Σ. Finally, Σ'_+ is a set of all possible sums of numbers of the set Σ'.

Theorem 7 *Let $\hat{f}_j^{(d)}(x, y) = 0$ be a truncated equation of $f(x, y) = 0$, corresponding to a horizontal edge of height m. If no one of numbers $k \in \Sigma'_+ + 1$, $k \neq 1$ is not a root of the exponential characteristic polynomial $\lambda(\gamma^*, k)$, then solutions* (27) *of the truncated equation $\hat{f}_j^{(d)}(x, y) = 0$ are continued in the form of the exponential expansions*

$$y = c \exp \varphi(x) + \sum b_k(x) c^k \exp(k\varphi(x)) \quad k \in \Sigma'_+ + 1, \quad k \neq 1, \tag{28}$$

of solutions to the full equation $f(x, y) = 0$, where $b_k(x)$ are power expansions.

Let $\varphi(x) = \alpha x^\beta + \cdots$, where α and $\beta = \text{const} \in \mathbb{C}$. Then for $x^\beta \to \infty$

$$\exp \varphi(x) \to \begin{cases} 0, \text{ if } \operatorname{Re}\,(\alpha x^\beta) < 0, \\ \infty, \text{ if } \operatorname{Re}\,(\alpha x^\beta) > 0 \end{cases}$$

If $\Gamma_j^{(1)}$ is the lower horizontal edge, then its normal cone is $\mathbf{U}_j^{(1)} = \{P = (0, -\lambda),\ \lambda > 0\}$. Hence, the corresponding truncated equation $\hat{f}_j^{(1)} = 0$ is an approximation of the full equation $f = 0$ only for $y \to 0$, and solutions $y = \exp \varphi(x)$ of the truncated equation $\hat{f}_j^{(1)} = 0$ can be asymptotic forms of solutions to the full equation $f = 0$ only in those domains of the complex variable x, where $\exp \varphi(x) \to 0$, i.e. for $\operatorname{Re}\,\alpha x^\beta < 0$. Thus, expansion (28) gives only parts of solutions for sectors of complex plane x with $\operatorname{Re}\,\alpha x^\beta < 0$ and it does not give information about solutions outside these sectors.

If $\Gamma_1^{(1)}$ is the upper edge, then expansion (28) gives only parts of solutions in sectors with $\operatorname{Re}\,\alpha x^\beta > 0$. Then the exponential expansion is

$$y = c \exp \varphi(x) \sum_{k=0}^{\infty} b_{-k}(x) [c \exp \varphi]^{-k}.$$

Example. Let us consider the fourth Painlevé equation

$$f(x, y) \stackrel{\text{def}}{=} -2yy'' + y'^2 + 3y^4 + 8xy^3 + 4(x^2 - a)y^2 + 2b = 0, \tag{29}$$

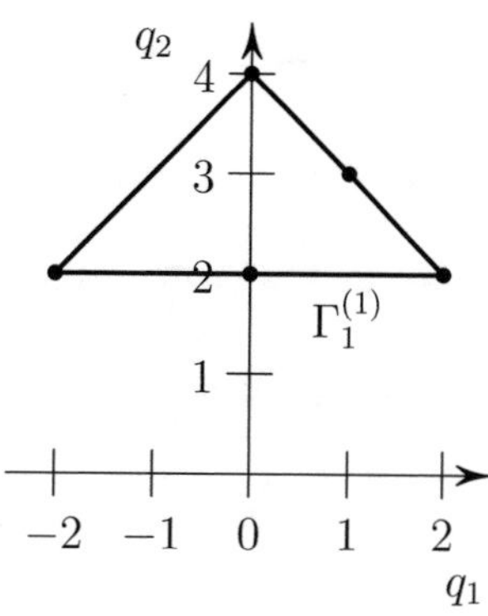

Fig. 4 Support and polygon of the Eq. (29) with $b = 0$

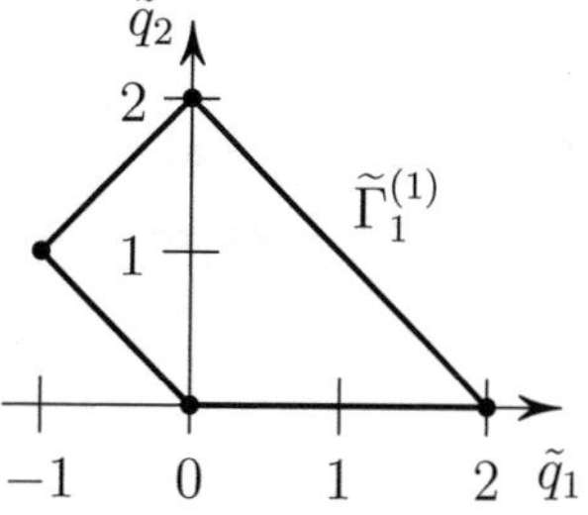

Fig. 5 Support and polygon for $h(x, \zeta)$

where a and b are complex parameters. If $b = 0$, its polygon $\Gamma(f)$ has a horizontal edge $\Gamma_1^{(1)}$ of height $m = 2$ (Fig. 4), which corresponds to the truncated equation

$$\hat{f}_1^{(1)} \stackrel{\text{def}}{=} -2yy'' + (y')^2 + 4\left(x^2 - a\right)y^2 = 0.$$

After the logarithmic transformation (25), we obtain $y' = \zeta y$, $y'' = y\left(\zeta' + \zeta^2\right)$ and

$$h(x, \zeta) = -2(\zeta' + \zeta^2) + \zeta^2 + 4x^2 - 4a.$$

Support $\mathbf{S}(h)$ and polygon $\Gamma(h)$ are shown in Fig. 5.

Polygon $\Gamma(h)$ has the inclined edge $\widetilde{\Gamma}_1^{(1)}$ corresponding to $\omega = 1$ with the truncated equation

$$\hat{h}_1^{(1)}(x, \zeta) \stackrel{\text{def}}{=} -\zeta^2 + 4x^2 = 0.$$

Hence, $\zeta = \pm 2x$, i.e. $\gamma^* = \pm 2$ and $\rho = 1$. According to Theorem 3, equation $h(x, \zeta) = 0$ has two solutions

$$\zeta_i = (-1)^i 2x + \alpha_i x^{-1} + \beta_i x^{-3} + x^{-1} \sum_{l=2}^{\infty} c_{i,l} x^{-2l} \stackrel{\text{def}}{=} \varphi_i'(x), \quad i = 1, 2,$$

$$\alpha_i = (-1)^i a - 1, \quad \beta_i = (-1)^{i+1}(a^2 + 3) + 4a.$$

If one of these numbers α_i, β_i is zero, then the corresponding expansion $\varphi_i'(x)$ is finite.

Let us compute the exponentially characteristic polynomial

$$\frac{\delta \hat{f}_1^{(1)}}{\delta y} = -2y'' - 2y \frac{d^2}{dx^2} + 2y' \frac{d}{dx} + 8(x^2 - a)y =$$
$$= y\left[-2(\zeta' + \zeta^2) - 2\frac{d^2}{dx^2} + 2\zeta \frac{d}{dx} + 8(x^2 - a)\right].$$

We change $\dfrac{d^2}{dx^2}$ and $\dfrac{d}{dx}$ by $k^2\zeta^2$ and $k\zeta$ correspondingly and ζ by $\gamma^* x$. Then the leading term for $x \to \infty$ is

$$-2\zeta^2 - 2k^2\zeta^2 + 2k\zeta^2 + 8x^2 = -2\zeta^2(k^2 - k).$$

Hence, the exponentially characteristic polynomial is $\lambda(\gamma^*, k) = -2(k^2 - k)$ for both values $\gamma^* = \pm 2$. The set Σ consists of numbers $2, 3, 4$; so $\Sigma' = \Sigma - 2 = \{0, 1, 2\}$ and the set Σ_+' consists of all nonnegative integral numbers, but $\Sigma_+' + 1$ is the set of all natural numbers. Roots of the polynomial $\lambda(\gamma^*, k)$ are $k = 0$ and $k = 1$. As the root $k = 0$ does not ly in the set $\Sigma_+' + 1$ and $k = 1$ was excluded, then according to Theorem 7, for $x \to \infty$ solutions to Eq. (29) with $b = 0$ are expanded in series

$$y = c \exp \varphi_i(x) + \sum_{k=2}^{\infty} b_{ik}(x) c^k \exp k\varphi_i(x), \quad i = 1, 2, \tag{30}$$

where

$$\varphi_i = (-1)^i x^2 + \alpha_i \ln x - \beta_i x^{-2}/2 - \sum_{l=2}^{\infty} c_{i,l} x^{-2l}/(2l), \quad i = 1, 2. \tag{31}$$

Here $\Gamma_1^{(1)}$ is the lower edge and $x \to \infty$, i.e. $\omega = 1$. So, expansions (30) describe families of solutions for $(-1)^i \operatorname{Re} x^2 < 0, i = 1, 2$. In the complex plane x, equality $\operatorname{Re} x^2 = 0$ corresponds to two bissectrices $\operatorname{Re} x = \pm \operatorname{Im} x$, dividing the plane into 4 domains $\mathscr{D}_1$, $\mathscr{D}_2$, $\mathscr{D}_3$, $\mathscr{D}_4$ (Fig. 6). So, the expansion (30) with $i = 1$ represents two families of solutions in domains $\mathscr{D}_1$ and $\mathscr{D}_3$, and the expansions (30) with $i = 2$ represents two families of solutions in domains $\mathscr{D}_2$ and $\mathscr{D}_4$. Series (31) diverge, but they are summable in some sectors of the complex plane [10].

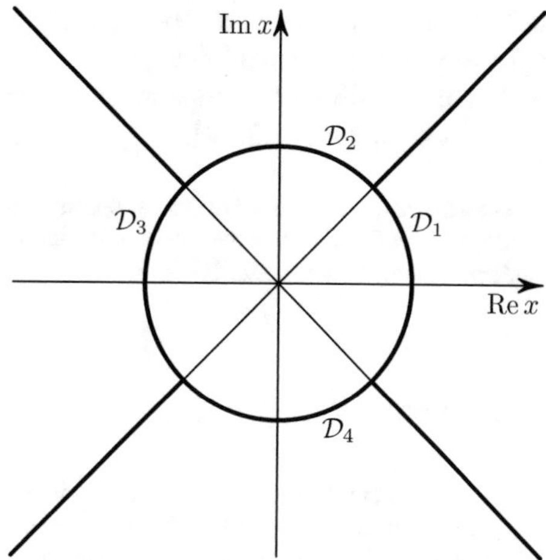

Fig. 6 Domains, where expansion (30) describe solutions of initial Eq. (29) with $b = 0$

More complicated examples of computation of domains of existence of solutions, described by expansions of type (30), see in [11].

Exponential expansions were proposed in [12].

6 Generalizations

1. The technique was used for algebraic equations [1, 2, 13, 14], for equations in partial derivations [1, 2, 15] and for systems [1, 2, 16].
2. Solutions in the form of power expansions with complex power exponents were studied in [3, 17]. Then we have the new type of expansions: exotic [5, 18].
3. We have studied asymptotic expansions of such solutions, for which difference of orders of two next derivatives is different from one [1, 19–21].
4. In Theorem 3 one can reject condition that critical numbers do not lie in the set **K**. Then there are the power-logarithmic expansions [3], or Dulac series. Similarly, in Theorem 5 we obtain expansions with multiple logarithm [7].
5. Comparison with other approaches [22].

7 Applications

- Solutions to the Painleve equations [5, 15, 19–21, 23, 24].
- The Beletsky equation [25, 26].
- The Euler-Poisson equations [27].
- The restricted three-body problem [28–30].
- Integrability of an ODE system [31, 32].

- The boundary layer on a needle [1, 15, 16, 23].
- Evolution of a turbulent flow [15].
- Sets of stability of a multiparameter ODE system [13].
- Waves on water [2, Ch. V].

Acknowledgements The work was supported by RFBR, grant No. 18-01-00422A and by the Program of the Presidium of the Russian Academy of Sciences 01 "Fundamental Mathematics and its Applications" under grant PRAS-18-01.

References

1. Bruno, A.D.: Asymptotic solving essentially nonlinear problems. Math. Stat. **4**(1), 27–39 (2016). https://doi.org/10.13189/ms.2016.040104
2. Bruno, A.D.: Power Geometry in Algebraic and Differential Equations, 288 p. Fizmatlit, Moscow (1998) (Russian). Power Geometry in Algebraic and Differential Equations, 385 p. Elsevier Science (North-Holland), Amsterdam (2000) (English)
3. Bruno, A.D.: Asymptotics and expansions of solutions to an ordinary differential equation. Uspekhi Matem. Nauk **59**(3), 31–80 (2004) (Russian). Russ. Math. Surv. **59**(3), 429–480 (2004) (English)
4. Tikhomirov, V.M.: Fréchet differential (https://www.encyclopediaofmath.org/index.php/Fr'echet_differential). Encyclopedia of Mathematics. Springer, Berlin (2001)
5. Bruno, A.D., Goryuchkina, I.V.: Asymptotic expansions of solutions of the sixth Painlevé equation. Trudy Mosk. Mat. Obs. **71**, 6–118 (2010) (Russian). Trans. Moscow Math. Soc. **71**, 1–104 (2010) (English)
6. Bruno, A.D.: Complicated expansions of solutions to an ordinary differential equation. Doklady Akademii Nauk **406**(6), 730–733 (2006) (Russian). Dokl. Math. **73**(1), 117–120 (2006) (English)
7. Bruno, A.D.: On complicated expansions of solutions to ODE. Keldysh Institute Preprints, No. 15, Moscow, 2011, 26 p. (Russian), http://library.keldysh.ru/preprint.asp?id=2011-15
8. Bruno, A.D.: Exponential expansions of solutions to ODE. Keldysh Institute Preprints, No. 36, Moscow, 2011, 16 p. (Russian), http://library.keldysh.ru/preprint.asp?id=2011-36
9. Bruno, A.D.: Exponential expansions of solutions to an ordinary differential equation. Doklady Akademii Nauk **443**(5), 539–544 (2012) (Russian). Dokl. Math. **85**(2), 259–264 (2012) (English)
10. Sibuya, Y.: Linear Differential Equations in the Complex Domain: Problems of Analytic Continuation. AMS, Providence (1985)
11. Bruno, A.D., Kudryashov, N.A.: Expansions of solutions to the equation P_1^2 by algorithms of power geometry. Ukr. Math. Bull. **6**(3), 311–337 (2009)
12. Varin, V.P.: Flat expansions of solutions to ODE at singularities. Keldysh Institute Preprints, No. 64, Moscow, 2010 (Russian), http://library.keldysh.ru/preprint.asp?id=2010-64
13. Batkhin, A.B., Bruno, A.D., Varin, V.P.: Sets of stability of multiparameter Hamiltonian systems. Prikladnaja Matematika i Mekhanika **76**(1), 80–133 (2012) (Russian). J. Appl. Math. Mech. **76**(1), 56–92 (2012) (English)
14. Bruno, A.D.: Solving the polynomial equations by algorithms of power geometry. Keldysh Institute Preprints, No. 34, Moscow, 2017, 28 p. (Russian). http://library.keldysh.ru/preprint.asp?id=2017-34
15. Bruno, A.D.: Power geometry in differential equations. Contemporary problems of mathematics and mechanic. Mathematics. Dynamical Systems. MSU, Moscow **4**(2), 24–54 (2009) (Russian)

16. Bruno, A.D., Shadrina, T.V.: Axisymmetric boundary layer on a needle. Trudy Mosk. Mat. Obsch. **68**, 224–287 (2007) (Russian). Trans. Moscow Math. Soc. **68**, 201–259 (2007) (English)
17. Bruno, A.D., Goryuchkina, I.V.: Convergence of power expansions of solutions to an ODE. Keldysh Institute Preprints, No 94, Moscow, 2013, 16 p. (Russian), http://library.keldysh.ru/preprint.asp?id=2013-94
18. Bruno, A.D.: Exotic expansions of solutions to an ordinary differential equation. Doklady Akademii Nauk **416**(5), 583–587 (2007) (Russian). Dokl. Math. **76**(2), 714–718 (2007) (English)
19. Bruno, A.D.: Power-elliptic expansions of solutions to an ordinary differential equation. Zurnal Vychislitel'noi Matematiki i Matematicheskoi Fiziki **51**(12), 2206–2218 (2012) (Russian). Comput. Math. Math. Phys. **52**(12), 1650–1661 (2012) (English)
20. Bruno, A.D.: Power geometry and elliptic expansions of solutions to the Painlevé equations. Int. J. Differ. Equ. V.2015, 13 p. (2015). https://doi.org/10.1155/2015/340715. Article ID 340715
21. Bruno, A.D.: Regular asymptotic expansions of solutions to one ODE and $P_1 - P_5$. In: Bruno, A.D., Batkhin, A.B. (eds.) Painlevé Equations and Related Topics, pp. 67–82. De Gruyter, Berlin/Boston (2012)
22. Bruno, A.D.: On some geometric methods in differential equations. Trans. J. Math. Anal. Appl. **4**(1–2), 37–47 (2016)
23. Bruno, A.D., Parusnikova, A.V.: Local expansions of solutions to the fifth Painlevé equation. Doklady Akademii Nauk **438**(4), 439–443 (2011) (Russian). Dokl. Math. **83**(3), 348–352 (2011) (English)
24. Bruno, A.D., Parusnikova, A.V.: Expansions and asymptotic forms of solutions to the fifth Painlevé equation near infinity. Keldysh Institute Preprints, No 61, Moscow, 2012, 32 p. (Russian). http://library.keldysh.ru/preprint.asp?id=2012-61
25. Bruno, A.D.: Families of periodic solutions to the Beletsky equation. Kosmicheskie Issledovanija **40**(3), 295–316 (2002) (Russian). Cosm. Res. **40**(3), 274–295 (2002) (English)
26. Bruno, A.D., Varin, V.P.: Classes of families of generalized periodic solutions to the Beletsky equation. Celest. Mech. Dyn. Astron. **88**(4), 325–341 (2004)
27. Bruno, A.D.: Analysis of the Euler–Poisson equations by methods of power geometry and normal form. Prikladnaja Matem. Mekhan. **71**(2), 192–227 (2007) (Russian). J. Appl. Math. Mech. **71**(2), 168–199 (2007) (English)
28. Bruno, A.D.: The Restricted Three-Body Problem: Plane Periodic Orbits, 296 p. Nauka, Moscow (1990) (Russian). The Restricted 3-Body Problem: Plane Periodic Orbits, 362 p. Walter de Gruyter, Berlin-New York (1994) (English)
29. Bruno, A.D., Varin, V.P.: Periodic solutions of the restricted three-body problem for small mass ratio. Prikladnaja Matem. Mekhan. **71**(6), 1034–1066 (2007) (Russian). J. Appl. Math. Mech. **71**(6), 933–960 (2007) (English)
30. Bruno, A.D., Varin, V.P.: Periodic solutions of the restricted three body problem for small μ and the motion of small bodies of the solar system. Astron. Astrophys. Trans. (AApTr) **27**(3), 479–488 (2012)
31. Bruno, A.D., Edneral, V.F.: Algorithmic analysis of local integrability. Doklady Akademii Nauk **424**(3), 299–303 (2009) (Russian). Dokl. Math. **79**(1), 48–52 (2009) (English)
32. Bruno, A.D., Edneral, V.F., Romanovski, V.G.: On new integrals of the Algaba–Gamero–Garcia system. In: Gerdt, V.P. et al. (eds.) Proceedings CASC 2017. LNCS, vol. 10490, pp. 40–50. Springer, Berlin (2017)

Part II
Summability of Divergent Solutions of PDEs

On the k-Summability of Formal Solutions for a Class of Higher Order Partial Differential Equations with Time-Dependent Coefficients

Kunio Ichinobe

Abstract We study the k-summability of divergent formal solutions to the Cauchy problem of a class of higher order linear partial differential equations with time-dependent coefficients. The problem is reduced to the k-summability of formal solutions for linear singular ordinary differential equations associated with the original Cauchy problem by means of successive approximation method.

Keywords k-summability · Formal solutions · Higher order partial differential equations

MSC Primary 35C10 · Secondary 35K35

1 Result

Let us consider the following partial differential operator L with time-dependent coefficients

$$L = \partial_t^M - P^M(t, \partial_t, \partial_x), \quad P^M(t, \partial_t, \partial_x) = \sum_{\substack{1 \le j \le M,\\ 0 \le \alpha \le \overline{\alpha}}} a_{\alpha j}(t)\partial_t^{M-j}\partial_x^{\alpha}, \tag{1}$$

where $t, x \in \mathbb{C}$, $M \ge 1$, $\overline{\alpha} \in \mathbb{N} = \{0, 1, 2, \ldots\}$ and $a_{\alpha j}(t) \in \mathbb{C}[t]$ for all j and α.

We consider the following Cauchy problem

$$\begin{cases} LU(t, x) = 0, \\ \partial_t^n U(0, x) = 0 \quad (0 \le n \le M - 2), \\ \partial_t^{M-1} U(0, x) = \varphi(x) \in \mathcal{O}_x, \end{cases} \tag{CP}$$

where $\mathcal{O}_x$ denotes the set of holomorphic functions at $x = 0$. This Cauchy problem has a unique formal power series solution of the form

K. Ichinobe (✉)
Department of Mathematics Education, Aichi University of Education,
1 Hirosawa, Igaya, Kariya City, Aichi Prefecture 448-8542, Japan
e-mail: ichinobe@auecc.aichi-edu.ac.jp

G. Filipuk et al. (eds.), *Formal and Analytic Solutions of Diff. Equations*, Springer Proceedings in Mathematics & Statistics 256,
https://doi.org/10.1007/978-3-319-99148-1_2

$$\hat{U}(t,x) = \sum\nolimits_{n \geq M-1} U_n(x) t^n / n!, \quad U_{M-1}(x) = \varphi(x). \tag{2}$$

Throughout this paper, we assume

$$\max\{\alpha - j; a_{\alpha j}(t) \not\equiv 0\} > 0. \tag{A-1}$$

This assumption is called a non-Kowalevskian condition, which means that the above formal solution is divergent in general under this assumption. Therefore we shall ask the k-summability condition of the divergent solution. We shall explain the conditions by using the Newton polygon, which is defined as follows.

Let $i(\alpha, j) = O(a_{\alpha j}(t))$ denote the order of zero of $a_{\alpha j}(t)$ at $t = 0$. Then we define a domain $\mathrm{N}(\alpha, j)$ by

$$\mathrm{N}(\alpha, j) := \{(x, y);\ x \leq M - j + \alpha,\ y \geq i(\alpha, j) - M + j\}$$

for $a_{\alpha j}(t) \not\equiv 0$, and $\mathrm{N}(\alpha, j) := \phi$ for $a_{\alpha j}(t) \equiv 0$. Then the Newton polygon $\mathrm{N}(L)$ is defined by

$$\mathrm{N}(L) := \mathrm{Ch}\left\{\mathrm{N}(0,0) \cup \bigcup\nolimits_{\alpha, j} \mathrm{N}(\alpha, j)\right\}, \tag{3}$$

where $\mathrm{Ch}\{\cdots\}$ denotes the convex hull of points in $\mathrm{N}(0,0) \cup \cup_{\alpha, j} \mathrm{N}(\alpha, j)$. Here $\mathrm{N}(0,0) := \{(x, y);\ x \leq M,\ y \geq -M\}$.

We assume that the Newton polygon $\mathrm{N}(L)$ is given by the following figure.

(A-2) There exists a unique pair (α_*, j_*) such that the Newton polygon $\mathrm{N}(L)$ has only one side of a positive slope with end points $(M, -M)$ and $(M - j_* + \alpha_*, i_* - M + j_*)$, $(i_* = i(\alpha_*, j_*))$.

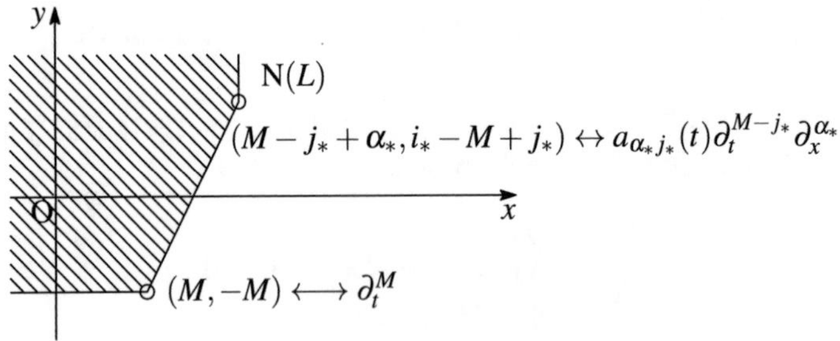

Moreover, we assume that for any (α, j) with $a_{\alpha j}(t) \not\equiv 0$ we have

$$\frac{\alpha_*}{i_* + j_*} = \max_{\alpha, j} \frac{\alpha}{i(\alpha, j) + j}, \tag{A-3}$$

whose value is called the modified order of the operator L, and we assume

$$\frac{\alpha_*}{i_* + j_*} = 1. \tag{A-4}$$

This notion was introduced by M. Miyake [12] when $M = 1$ in 1974/75 and by K. Kitagawa and T. Sadamatsu [8] when M is general in 1975/76 for a construction of the Cauchy data $\varphi(x) \in \mathscr{O}_x$ such that the formal solution to the Cauchy problem of non-Kowalevski type equations diverges.

Finally, we assume that

$$a_{\alpha j}(t) = \sum_i a_i^{(\alpha,j)} t^i, \tag{A-5}$$

where the sum is taken over i with the following conditions.

$$\begin{array}{ll} i(\alpha, j) \le i \le \overline{i_{j1}} := (i_*/j_*)j, & \text{when } j \le j_* \\ i(\alpha, j) \le i < i_*, & \text{when } j > j_*. \end{array}$$

Especially, we have

$$a_{\alpha_* j_*}(t) = a_{i_*}^{(\alpha_*, j_*)} t^{i_*}. \tag{4}$$

In order to state our result, we shall define a characteristic equation for L. Let

$$k = \frac{i_* + j_*}{\alpha_* - j_*}, \tag{5}$$

which gives the non trivial slope of $\mathrm{N}(L)$, and we put

$$L_0(t, \partial_t, \partial_x) := \partial_t^M - \sum_{(i,j,\alpha)} a_i^{(\alpha,j)} t^i \partial_t^{M-j} \partial_x^\alpha, \tag{6}$$

where the sum is taken over (i, j, α) with conditions $\alpha = i + j, i + j = k(\alpha - j)$ and $1 \le j \le j_*$. Then L_0 is the sum of the operators on the side with non trivial slope of $\mathrm{N}(L)$. In this case, we define a characteristic equation for L with respect to z by

$$L_0(1, 1, z) = 0 \tag{7}$$

whose degree with respect to z is α_*. Let $\{z_n\}_{n=1}^{\alpha_*}$ be the roots of the characteristic equation. Moreover, we give the notation $S(d, \beta, \rho)$. For $d \in \mathbb{R}$, $\beta > 0$ and ρ with $0 < \rho \le \infty$, we define a sector $S = S(d, \beta, \rho)$ by

$$S(d, \beta, \rho) := \left\{ t \in \mathbb{C}; |d - \arg t| < \frac{\beta}{2}, 0 < |t| < \rho \right\}.$$

We write $S(d, \beta, \infty) = S(d, \beta)$ for short.

Under these preparations, our result is stated as follows.

Theorem 1 *Let* $k = (i_* + j_*)/(\alpha_* - j_*)$. *For a fixed* $d \in \mathbb{R}$, *let* $d_n := d - \arg z_n$ *for* $1 \le n \le \alpha_*$. *We assume that the Cauchy data* $\varphi(x)$ *can be analytically continued in* $\bigcup_{n=1}^{\alpha_*} S(d_n, \varepsilon)$ *for some* $\varepsilon > 0$, *and has an exponential growth estimate of order at most k there, that is,*

$$|\varphi(x)| \le C \exp\left(\delta|x|^k\right) \tag{8}$$

by some positive constants C and δ. *Then under the assumptions (A-1)–(A-5), the formal solution* $\hat{U}(t, x)$ *of (CP) is k-summable in d direction.*

The author has been studying the k-summability of divergent solutions of non-Kowalevskian equations with time-dependent coefficients. In the papers [3, 4], we treated the first and higher order equations with respect to t, whose coefficients are monomials of t, respectively. In the paper [7], we treated the first order equation with respect to t, whose coefficients are polynomials of t and modified order is equal to 1. Therefore this result is a generalization as a higher order version of [3, 7]. In the paper [6], we treated the first order equations with respect to t, whose coefficients are polynomials of t and modified order is general.

Moreover, in the paper [5], we treated higher order equations with respect to t, whose coefficients are polynomials of t and modified order is general. However, we treated a special case of $j_* = M$, where j_* is given by the assumption (A-2). Therefore this result is a generalization of results in [4, 5].

Hence in the following, we assume

$$M \ge 2, \quad j_* < M. \tag{9}$$

We note that by the assumption (A-4), we have $k = \alpha_*/i_* = (i_* + j_*)/i_* > 1$.

The paper consists of the following contents. We give a brief review of k-summability in Sect. 2. In Sect. 3, we give a construction of formal solution of Cauchy problem (CP) by employing the method of successive approximation. We give a result of Gevrey order of formal solution and its simple proof in Sect. 4. In Sect. 5, we introduce formal series associated with the formal solutions obtained in Sect. 3 and an important result for the k-summability of the series is given, and in Sect. 6 we will give a simple proof of Theorem 1. Section 7 is devoted to a proof of the result for the k-summability of the series given in Sect. 5. Especially, we remark that the construction of the successive approximation of solutions for the associated convolution equations is a slight different from the one in [5].

2 Review of k-Summability

In this section, we give some notation and definitions (cf. W. Balser [1] for detail).

Let $k > 0$, $S = S(d, \beta)$ and $B(\sigma) := \{x \in \mathbb{C}; |x| \le \sigma\}$. Let $v(t, x) \in \mathcal{O}(S \times B(\sigma))$ which means that $v(t, x)$ is holomorphic in $S \times B(\sigma)$. Then we define that

$v(t,x) \in \mathrm{Exp}_t^k(S \times B(\sigma))$ if, for any closed subsector S' of S, there exist some positive constants C and δ such that

$$\max_{|x|\le\sigma} |v(t,x)| \le Ce^{\delta|t|^k}, \quad t \in S'. \tag{10}$$

For $k > 0$, we define that $\hat{v}(t,x) = \sum_{n=0}^{\infty} v_n(x)t^n \in \mathscr{O}_x[[t]]_{1/k}$ (we say $\hat{v}(t,x)$ is a formal power series of Gevrey order $1/k$) if $v_n(x)$ are holomorphic on a common closed disk $B(\sigma)$ for some $\sigma > 0$ and there exist some positive constants C and K such that for any n,

$$\max_{|x|\le\sigma} |v_n(x)| \le CK^n\Gamma\left(1+\frac{n}{k}\right). \tag{11}$$

Here when $v_n(x) \equiv v_n$ (constants) for all n, we use the notation $\mathbb{C}[[t]]_{1/k}$. In the following, we use the similar usage of notations.

Let $k > 0$, $\hat{v}(t,x) = \sum_{n=0}^{\infty} v_n(x)t^n \in \mathscr{O}_x[[t]]_{1/k}$ and $v(t,x)$ be an analytic function on $S(d,\beta,\rho) \times B(\sigma)$. Then we define the k-asymptotic expansion that

$$v(t,x) \cong_k \hat{v}(t,x) \quad \text{in } S = S(d,\beta,\rho), \tag{12}$$

if for any closed subsector S' of S, there exist some positive constants C and K such that for any $N \ge 1$, we have

$$\max_{|x|\le\sigma} \left| v(t,x) - \sum_{n=0}^{N-1} v_n(x)t^n \right| \le CK^N|t|^N\Gamma\left(1+\frac{N}{k}\right), \quad t \in S'. \tag{13}$$

For $k > 0$, $d \in \mathbb{R}$ and $\hat{v}(t,x) \in \mathscr{O}_x[[t]]_{1/k}$, we say that $\hat{v}(t,x)$ is *k-summable* in d direction, and denote it by $\hat{v}(t,x) \in \mathscr{O}_x\{t\}_{k,d}$, if there exist a sector $S = S(d,\beta,\rho)$ with $\beta > \pi/k$ and an analytic function $v(t,x)$ on $S \times B(\sigma)$ such that $v(t,x) \cong_k \hat{v}(t,x)$ in S.

We remark that the function $v(t,x)$ above for a *k-summable* $\hat{v}(t,x)$ is unique if it exists. Therefore such a function $v(t,x)$ is called the *k-sum* of $\hat{v}(t,x)$ in d direction.

3 Construction of the Formal Solution of (CP)

3.1 Decomposition of P^M

In this subsection, we give a decomposition of the operator P^M.

For a given $\ell \ge 0$, we define

$$K_\ell := \left\{(i, j, \alpha)\,;\ \ell = i + j - \alpha,\ a_i^{(\alpha,j)} \neq 0\right\}$$

and we put $P_\ell^M(t, \partial_t, \partial_x) := \sum_{(i,j,\alpha)\in K_\ell} a_i^{(\alpha,j)} t^i \partial_t^{M-j} \partial_x^\alpha$. When $j \le j_*$, we have $\ell = i + j - \alpha \le \frac{i_*}{j_*} j + j \le i_* + j_* = \alpha_*$, and when $j > j_*$, we have $\ell = i + j - \alpha < i_* + M = \alpha_* + M - j_*$. We put $\ell_* := \alpha_* + M - j_* - 1 (\ge \alpha_*)$. In this case, we have

$$P^M(t, \partial_t, \partial_x) = \sum_{0\le\ell\le\ell_*} P_\ell^M(t, \partial_t, \partial_x). \tag{14}$$

We put

$$K_{\ell 1} := \left\{(i, j, \alpha) \in K_\ell;\ 1 \le j \le j_*,\ 0 \le i \le \overline{i_{j1}} = i_* j / j_*,\ 0 \le \alpha = i + j - \ell\right\},$$
$$K_{\ell 2} := \{(i, j, \alpha) \in K_\ell;\ j_* < j \le M,\ 0 \le i < i_*,\ 0 \le \alpha = i + j - \ell\},$$

where $a_i^{(\alpha,j)} = 0$ if $i < i(\alpha, j)$.

3.2 A Sequence of Cauchy Problems

By employing the decomposition of P^M, we consider a sequence of the following Cauchy problems for $\nu \ge 0$.

$$\begin{cases} \partial_t^M u_\nu(t, x) = \sum_{\ell=0}^{\min\{\ell_*,\nu\}} P_\ell^M u_{\nu-\ell}(t, x), \\ \partial_t^n u_\nu(0, x) = 0\ (0 \le n \le M - 2), \\ \partial_t^{M-1} u_\nu(0, x) = \begin{cases} \varphi(x), & \nu = 0, \\ 0, & \nu \ge 1. \end{cases} \end{cases} \tag{E_ν}$$

For each ν, the Cauchy problem (E_ν) has a unique formal solution

$$\hat{u}_\nu(t, x) = \sum_{n\ge0} u_{\nu,n}(x) t^n / n!. \tag{15}$$

Then $\hat{U}(t, x) = \sum_{\nu\ge0} \hat{u}_\nu(t, x)$ is the formal solution of the original Cauchy problem (CP).

The formal solution $\hat{u}_\nu(t, x)$ is obtained in the following form inductively.

Lemma 1 *For each ν, we have*

$$u_{\nu,n}(x) = A_\nu(n) \varphi^{(n-M+1-\nu)}(x) \quad (n \ge M - 1 + \nu)$$

and $u_{\nu,n}(x) \equiv 0$ $(n < M - 1 + \nu)$, where $\{A_\nu(n)\}$ satisfy the following recurrence formula.

$$\begin{cases} A_\nu(n+M) = \sum_{\ell=0}^{\min\{\ell_*,\nu\}} \sum_{K_\ell} a_i^{(\alpha,j)} [n]_i A_{\nu-\ell}(n+M-i-j) \quad (n \geq 0), \\ A_\nu(n) = 0 \quad (n < M-1) \\ A_\nu(M-1) = \begin{cases} 1, & \nu = 0, \\ 0, & \nu \geq 1. \end{cases} \end{cases} \tag{R_ν}$$

Here the notation $[n]_i$ is defined by

$$[n]_i = \begin{cases} n(n-1)\cdots(n-i+1), & i \geq 1, \\ 1, & i = 0. \end{cases}$$

4 Gevrey Order of the Formal Solution $\hat{U}$

We give the Gevrey order of formal solution $\hat{U}(t,x)$ by employing a result of Gevrey order of formal solutions $\hat{u}_\nu(t,x)$ without proof (see [3, 4, 7] for detail).

Proposition 1 *Let $\varphi(x) \in \mathcal{O}_x$ and $k = (i_* + j_*)/(\alpha_* - j_*)$. Then for each ν, we have $\hat{u}_\nu(t,x) \in \mathcal{O}_x[[t]]_{1/k}$. More exactly, we have*

$$\max_{|x| \leq \sigma} \left| \frac{u_{\nu,n}(x)}{n!} \right| \leq \frac{AB^{\nu+n}}{\nu!} \Gamma\left(1 + \frac{n}{k}\right) \tag{16}$$

with some positive constants A, B and σ for any $n \geq M - 1 + \nu$.

Proposition 1 implies that $\hat{U}(t,x) \in \mathcal{O}_x[[t]]_{1/k}$. In fact, we have

$$\hat{U}(t,x) = \sum_{\nu=0}^{\infty} \hat{u}_\nu(t,x) = \sum_\nu \sum_n \frac{u_{\nu,n}(x)}{n!} t^n = \sum_n \frac{\sum_\nu^{\text{finite}} u_{\nu,n}(x)}{n!} t^n = \sum_n \frac{U_n(x)}{n!} t^n.$$

Then we have

$$\begin{aligned} \max_{|x| \leq \sigma} \left| \frac{U_n(x)}{n!} \right| &\leq \sum_\nu^{\text{finite}} \max_{|x| \leq \sigma} \left| \frac{u_{\nu,n}(x)}{n!} \right| \leq \sum_\nu^{\infty} \frac{AB^{\nu+n}}{\nu!} \Gamma\left(1 + \frac{n}{k}\right) \\ &= AB^n \Gamma\left(1 + \frac{n}{k}\right) \sum_\nu^{\infty} \frac{B^\nu}{\nu!} = Ae^B B^n \Gamma\left(1 + \frac{n}{k}\right). \end{aligned}$$

5 Preliminaries for the Proof of Theorem 1

In this section, we prepare some results for the proof of Theorem 1.

First, we give an important lemma for the summability theory (cf. [1, 9]).

Lemma 2 *Let $k > 0$, $d \in \mathbb{R}$ and $\hat{v}(t,x) = \sum_{n=0}^{\infty} v_n(x)t^n \in \mathcal{O}_x[[t]]_{1/k}$. Then the following statements are equivalent:*

(i) $\hat{v}(t,x) \in \mathcal{O}_x\{t\}_{k,d}$.

(ii) Let

$$v_B(s,x) = (\hat{\mathscr{B}}_k\hat{v})(s,x) := \sum_{n=0}^{\infty} \frac{v_n(x)}{\Gamma(1+n/k)} s^n \in \mathcal{O}_{(s,x)} \tag{17}$$

be the formal k-Borel transformation of $\hat{v}(t,x)$. Then $v_B(s,x) \in \mathrm{Exp}_s^k(S(d,\varepsilon) \times B(\sigma))$ for some $\varepsilon > 0$ and $\sigma > 0$.

Next, we introduce formal series associated with the formal solutions $\hat{u}_\nu$. We define

$$\hat{f}_\nu(t) = \sum_{n\geq 0, n-M+1-\nu\in\mathbb{N}} A_\nu(n)t^n =: \sum\nolimits_{n\geq 0}^{(\nu)} A_\nu(n)t^n, \tag{18}$$

which is the generating functions of $\{A_\nu(n)\}$ and

$$\hat{g}_\nu(t) := \sum\nolimits_{n\geq 0}^{(\nu)} A_\nu(n) \frac{(n-M+1-\nu)!}{n!} t^n, \tag{19}$$

which is called the moment series of $\hat{f}_\nu$. Then we can find $\hat{g}_\nu$ in $\hat{u}_\nu(t,x)$ by formal use of the Cauchy integral formula.

$$\begin{aligned}\hat{u}_\nu(t,x) &= \sum\nolimits_{n\geq 0}^{(\nu)} A_\nu(n)\varphi^{(n-M+1-\nu)}(x)\frac{t^n}{n!} \\ &= \frac{1}{2\pi i}\oint \varphi(x+\zeta)\zeta^{M-1+\nu-1}\hat{g}_\nu\left(\frac{t}{\zeta}\right)d\zeta.\end{aligned}$$

Moreover, we have a formal relationship between $\hat{f}_\nu$ and $\hat{g}_\nu$ by

$$\hat{g}_\nu(t) = \frac{1}{\Gamma(\nu+M-1)}\int_0^1 \tau^{-M+1-\nu}(1-\tau)^{\nu+M-2}\hat{f}_\nu(\tau t)d\tau.$$

Finally, we prepare a lemma for the k-summability of $\hat{f}_\nu$, whose proof will be given in Sect. 7.

Lemma 3 *Let $k = (i_* + j_*)/(\alpha_* - j_*)$, $f_{\nu B}(s) = (\hat{\mathscr{B}}_k\hat{f}_\nu)(s)$ and $\{z_n\}_{n=1}^{\alpha_*}$ be the roots of the characteristic equation (7). Then we obtain $f_{\nu B}(s) \in \mathrm{Exp}_s^k(S(\theta,\varepsilon_0))$ for θ satisfying*

$$\theta \not\equiv \arg z_n \pmod{2\pi} \quad (n = 1, 2, \ldots, \alpha_*) \tag{20}$$

and $\varepsilon_0 > 0$. Especially, we obtain the following estimates

$$|f_{\nu B}(s)| \leq CK^{\nu}|s|^{\nu+M-1}\exp(\delta|s|^k), \quad s \in S(\theta, \varepsilon_0), \tag{21}$$

where C, K and δ are independent of ν.

Lemma 3 means that $\hat{f}_\nu(t) \in \mathbb{C}\{t\}_{k,\theta}$ for θ satisfying (20).

6 Proof of Theorem 1

By employing Lemmas 2 and 3, we obtain the following results, which means that $\hat{u}_\nu(t,x) \in \mathscr{O}_x\{t\}_{k,d}$. We omit the proof (cf. [7, 9–11, 13]).

Proposition 2 *Let d be a fixed real number and define $u_{\nu B}(s,x) = (\hat{\mathscr{B}}_k\hat{u}_\nu)(s,x)$. We assume that the Cauchy data $\varphi(x)$ satisfies the same assumptions as in Theorem 1. Then we have*

$$\max_{|x|\leq\sigma}|u_{\nu B}(s,x)| \leq CK^{\nu}\frac{|s|^{\nu+M-1}}{(\nu+M-1)!}\exp\left(\delta|s|^k\right), \quad s \in S(d,\varepsilon) \tag{22}$$

by some positive constants C, K, δ and σ independent of ν.

We can prove Theorem 1 under these preparations.

Proof of Theorem 1. Let $\hat{U}(t,x) = \sum_{\nu\geq 0}\hat{u}_\nu(t,x)$ be the formal solution of the Cauchy problem (CP). We finish the proof of Theorem 1 by showing that $U_B(s,x) = (\hat{\mathscr{B}}_k\hat{U})(s,x) = \sum_{\nu\geq 0}u_{\nu B}(s,x) \in \mathrm{Exp}_s^k(S(d,\varepsilon)\times B(\sigma))$.

By using Proposition 2, we obtain the desired estimate of $U_B(s,x)$ for $s \in S(d,\varepsilon)$.

$$\begin{aligned}\max_{|x|\leq\sigma}|U_B(s,x)| &\leq \sum_{\nu\geq 0}\max_{|x|\leq\sigma}|u_{\nu B}(s,x)| \leq CK^{-M+1}\exp(\delta|s|^k)\sum_{\nu\geq 0}\frac{(K|s|)^{\nu+M-1}}{(\nu+M-1)!}\\ &\leq CK^{-M+1}\exp(\delta|s|^k)\cdot\exp(K|s|) \leq \widetilde{C}\exp(\tilde{\delta}|s|^k)\end{aligned}$$

by some positive constants $\widetilde{C} > C$ and $\tilde{\delta} > \delta$ because of $k > 1$. □

7 Proof of Lemma 3

We shall give the proof of Lemma 3. For the purpose, we will obtain the convolution equations of $f_{\nu B}$. After that, we will prove Lemma 3 by employing the method of successive approximation of solutions for the convolution equations.

7.1 A Canonical Form for Differential Equation of $\hat{f}_\nu$

We reduce the differential equation of $\hat{f}_\nu(t)$ to a certain canonical form.

By multiplying the both sides of the recurrence formula (R_ν) of $\{A_\nu(n)\}$ by t^{n+M} and taking sum $\sum_{n\geq 0}'$ satisfying $n+1-\nu \in \mathbb{N}$, we get

$$\sum_{n\geq 0}' A_\nu(n+M)t^{n+M} = \sum_\ell \sum_{K_\ell} a_i^{(\alpha,j)} \sum_{n\geq 0}' [n]_i A_{\nu-\ell}(n+M-i-j)t^{n+M}.$$

By noticing $\ell = i+j-\alpha$ and $A_{\nu-\ell}(n) = 0$ if $n-M+1-(\nu-\ell) \notin \mathbb{N}$ or $n < M-1$, we have $\sum_{n\geq 0}' A_{\nu-\ell}(n+M-i-j) = \sum_{n\geq 0}^{(\nu-\ell)} A_{\nu-\ell}(n)$. Therefore we obtain a differential equation of $\hat{f}_\nu$.

$$\hat{f}_\nu(t) - c_\nu t^{M-1} = \sum_\ell \sum_{K_\ell} a_i^{(\alpha,j)} t^{i+j} [\delta_t - M + i + j]_i \hat{f}_{\nu-\ell}(t), \tag{23}$$

where $\delta_t = t(d/dt)$ denotes the Euler operator and $c_\nu = 1\ (\nu = 0),\ = 0\ (\nu \geq 1)$.

Lemma 4 *For $a, b \in \mathbb{R}$ and $k > 0$, we have*

$$[a\delta_t + b + n]_n = \sum_{m=0}^{n} d_{n,m}^{[a,b]} t^{-km} (t^k \delta_t)^m, \tag{24}$$

where $d_{0,0}^{[a,b]} = 1$ and for $n \geq 1$

$$d_{n,m}^{[a,b]} = a d_{n-1,m-1}^{[a,b]} + (b+n-akm) d_{n-1,m}^{[a,b]}, \quad 0 \leq m \leq n,$$

with $d_{n-1,-1}^{[a,b]} = d_{n-1,n}^{[a,b]} = 0$. Especially we have $d_{n,n}^{[a,b]} = a^n$ and $d_{n,0}^{[a,b]} = [b+n]_n$.

Hereafter we fix $k = (i_* + j_*)/(\alpha_* - j_*)$.

By using this lemma, we have

$$\hat{f}_\nu(t) - c_\nu t^{M-1} = \sum_{\ell, K_\ell} a_i^{(\alpha,j)} t^{i+j} \sum_{m=0}^{i} d_{i,m}^{[1,-M+j]} t^{-km} (t^k \delta_t)^m \hat{f}_{\nu-\ell}(t).$$

We exchange the order of the summations $\Sigma := \sum_\ell \sum_{K_\ell} \sum_m$ in the form

$$\Sigma = \sum_{\ell=0}^{\min\{\ell_*,\nu\}} \left\{ \sum_{K_{\ell 1}} + \sum_{K_{\ell 2}} \right\} \sum_{m=0}^{i} =: \Sigma_1 + \Sigma_2,$$

where

$$\Sigma_1 = \sum_\ell \sum_{1\leq j\leq j_*} \sum_{0\leq i\leq \overline{i_{j1}}} \sum_{0\leq m\leq i} = \sum_\ell \sum_{1\leq j\leq j_*} \sum_{0\leq m\leq \overline{i_{j1}}} \sum_{m\leq i\leq \overline{i_{j1}}},$$

$$\Sigma_2 = \sum_\ell \sum_{j_*<j\leq M} \sum_{0\leq i<i_*} \sum_{0\leq m\leq i} = \sum_\ell \sum_{j_*<j\leq M} \sum_{0\leq m<i_*} \sum_{m\leq i<i_*}.$$

We put

$$\sum_{m\le i\le \overline{i_{j1}},\alpha=i+j-\ell} a_i^{(\alpha,j)} d_{i,m}^{[1,-M+j]} t^{i+j-km} =: A_{1jm}^{[\ell]}(t),$$
$$\sum_{m\le i< i_*,\alpha=i+j-\ell} a_i^{(\alpha,j)} d_{i,m}^{[1,-M+j]} t^{i+j-km} =: A_{2jm}^{[\ell]}(t).$$

Then we have $O(A_{1jm}^{[\ell]}(t)) \ge 0$ on Σ_1 and $O(A_{2jm}^{[\ell]}(t)) > 1$ on Σ_2, since $i + j - km \ge m + j - km = \frac{j_*}{i_*}(\frac{i_*}{j_*} j - m) = j - \frac{j_*}{i_*} m$. Especially, we have $A_{1j\overline{i_{j1}}}^{[\ell]}(t) \equiv a_{\frac{i_*}{j_*}j}^{(\frac{\alpha_*}{j_*}j-\ell,j)} =: A_{[j]}^{[\ell]}$ (constant). Therefore we have

$$\left[1-\sum_J A_{[j]}^{[0]}(t^k\delta_t)^{\overline{i_{j1}}}\right]\hat{f}_\nu(t) \;=\; c_\nu t^{M-1} + \sum_{\ell=1}^{\min\{\ell_*,\nu\}} A_{[j]}^{[\ell]}(t^k\delta_t)^{\overline{i_{j1}}}\hat{f}_{\nu-\ell}(t)$$
$$+\sum_{\ell=0}^{\min\{\ell_*,\nu\}}\left\{\sum_{1\le j\le j_*}\sum_{0\le m<\overline{i_{j1}}} A_{1jm}^{[\ell]}(t) + \sum_{j_*<j\le M}\sum_{0\le m<i_*} A_{2jm}^{[\ell]}(t)\right\}(t^k\delta_t)^m \hat{f}_{\nu-\ell}(t),$$

where $J = \{1 \le j \le j_*; i_*j/j_*, \alpha_*j/j_* \in \mathbb{N}\}$. Finally, by exchanging the order of summation over j and m, we obtain a canonical differential equation of $\hat{f}_\nu$.

$$\left[1-\sum_J A_{[j]}^{[0]}(t^k\delta_t)^{\overline{i_{j1}}}\right]\hat{f}_\nu(t) \;=\; c_\nu t^{M-1} + \sum_{\ell=1}^{\min\{\ell_*,\nu\}}\sum_J A_{[j]}^{[\ell]}(t^k\delta_t)^{\overline{i_{j1}}}\hat{f}_{\nu-\ell}(t) \quad (25)$$
$$+\sum_{\ell=0}^{\min\{\ell_*,\nu\}}\sum_{0\le m<i_*}\left\{\mathscr{A}_{1m}^{[\ell]}(t)+\mathscr{A}_{2m}^{[\ell]}(t)\right\}(t^k\delta_t)^m\hat{f}_{\nu-\ell}(t),$$

where

$$\mathscr{A}_{1m}^{[\ell]}(t) := \sum_{\max\{0,\frac{j_*}{i_*}m\}<j\le j_*} A_{1jm}^{[\ell]}(t), \qquad \mathscr{A}_{2m}^{[\ell]}(t) := \sum_{j_*<j\le M} A_{2jm}^{[\ell]}(t).$$

7.2 Convolution Equations

We obtain the convolution equations by operating the Borel transform to the canonical differential equations obtained in the previous subsection.

After operating the formal k-Borel transformation to the equation (25) and differentiating the both sides, we put $w_\nu(s) = D_s f_{\nu B}(s)$ or $f_{\nu B}(s) = D_s^{-1} w_\nu(s)$, where $D_s = d/ds$ and $D_s^{-1} = \int_0^s$. Then the convolution equation for $w_\nu(s)$ is obtained by the following expression

$$\left[1-\sum_J A_{[j]}^{[0]}(ks^k)^{\overline{i_{j1}}}\right] w_\nu(s) = \tilde{c}_\nu s^{M-2} + \sum_{\ell=1}^{\min\{\ell_*,\nu\}}\sum_J A_{[j]}^{[\ell]}(ks^k)^{\overline{i_{j1}}} w_{\nu-\ell}(s) \quad (26)$$
$$+D_s\left[\sum_{p=1,2}\sum_{\ell=0}^{\min\{\ell_*,\nu\}}\sum_{0\le m<i_*}\mathscr{A}_{pmB}^{[\ell]}(s) *_k D_s^{-1}(ks^k)^m w_{\nu-\ell}(s)\right],$$

where $\mathscr{A}^{[\ell]}_{pmB}(s) = (\hat{\mathscr{B}}_k \mathscr{A}^{[\ell]}_{pm})(s)$ for $p = 1, 2$ and $\tilde{c}_\nu = kc_\nu/\Gamma((M-1)/k)$.

Here the k-convolution $a(s) *_k b(s)$ with $a(0) = b(0) = 0$ is defined by the following integral

$$(a *_k b)(s) = \int_0^s a\left((s^k - u^k)^{1/k}\right) \frac{d}{du} b(u) du. \tag{27}$$

We remark that if $a(0) = b(0) = 0$, the convolution is commutative. Note that this formula is same with that in [1, Sec. 5.3] although the expression is a little different from it.

We put

$$A_*(s) := 1 - \sum_J A^{[0]}_{[j]}(ks^k)^{\overline{i_{j1}}},$$

which is the characteristic polynomial given (7) with $z = k^{1/k}s$ and

$$T_\nu(w_\nu)(s) := \frac{1}{A_*(s)}\left(\text{the right hand side of (26)}\right).$$

Then we remark that for all ν

$$T_\nu : \mathbb{C}[[s]] \to \mathbb{C}[[s]].$$

Therefore for each ν, the function $w_\nu(s) = D_s f_{\nu B}(s) = D_s \sum^{(\nu)}_{n \geq 0} A_\nu(n) s^n / \Gamma(1 + n/k)$ is a unique holomorphic solution of the convolution equation $w_\nu = T_\nu(w_\nu)$ in $\{|s| \leq 2\sigma_*\}$, where σ_* is sufficiently small (see [4, 6] for detail). We have to remark that for each ν, the solution w_ν may be analytically continued into $S(\theta, \varepsilon_0)$ with $\theta \not\equiv \arg z_n \pmod{2\pi}$ $(n = 1, 2, \ldots, \alpha_*)$, because the singular points of convolution equation $w_\nu = T_\nu(w_\nu)$ are only the roots of $A_*(s)$, which is proved in the next subsection.

7.3 *Proof of Lemma 3*

In this subsection, we shall prove that $w_\nu(s)$ has the exponential growth estimate of order at most k in a sector with infinite radius. Therefore $f_{\nu B}(s) = D_s^{-1} w_\nu(s)$ also has the exponential growth estimate of the same order as that of $w_\nu(s)$.

For obtaining the desired estimate of w_ν, we consider the convolution equation $w_\nu = T_\nu(w_\nu)$ on $S_1 := \{s \in S(\theta, \varepsilon_0); |s| \geq \sigma_*\}$ (cf. [2]).

We take $s_0 \in S_1$ with $|s_0| = \sigma_*$ arbitrary and fix it. We put $S_0 = S(\theta, \varepsilon_0)$ and $S_* := \{s \in \mathbb{C}; (s^k - s_0^k)^{1/k} \in S_0\} \subset S_1$. We modify the operator T_ν by $\tilde{T}_\nu$ on S_* by replacing the convolutions $a *_k b$ by $a \,\tilde{*}\, b$ defined by

$$(a\,\tilde{*}\,b)(s) := \int_{s_0}^{s} a\left((s^k - u^k)^{1/k}\right)\frac{d}{du}b(u)du, \quad s \in S_*$$

where $(s^k - u^k)^{1/k} \in S_0$.

Then we obtain a convolution equation $w_\nu = \tilde{T}_\nu(w_\nu)$ on S_* with

$$\begin{aligned}\tilde{T}_\nu(w_\nu) &= F_\nu(s) + \frac{1}{A_*(s)}\sum_{\ell=1}^{\min\{\ell_*,\nu\}}\sum_J A_{[j]}^{[\ell]}(ks^k)^{\overline{i_{j1}}}w_{\nu-\ell}(s)\\ &+\frac{1}{A_*(s)}D_s\left[\sum_{p=1,2}\sum_{\ell=0}^{\min\{\ell_*,\nu\}}\sum_{0\le m<i_*}\left(\mathscr{A}_{pmB}^{[\ell]}\,\tilde{*}\,D_s^{-1}(ks^k)^m w_{\nu-\ell}\right)(s)\right],\end{aligned}$$

$$F_\nu(s) = \frac{1}{A_*(s)}\left[\tilde{c}_\nu s^{M-2} + D_s\sum_{p=1,2}\sum_{\ell=0}^{\min\{\ell_*,\nu\}}\sum_{0\le m<i_*}\int_0^{s_0}\mathscr{A}_{pmB}^{[\ell]}\left((s^k-u^k)^{\frac{1}{k}}\right)(ku^k)^m w_{\nu-\ell}(u)du\right].$$

We remark that since $\{w_\nu(u)\}$ are already given in $|u| \le \sigma_*$, $\{F_\nu(s)\}$ are also given functions in $\{|s| \le \sigma_*\} \cup S_0$.

By employing the method of successive approximation to the convolution equation $w_\nu = \tilde{T}_\nu(w_\nu)$ on S_*, we shall obtain the desired exponential estimate for w_ν.

The case $\nu = 0$. We define the functions $\{w_{0,n}(s)\}$ by the following.

$$\begin{cases} w_{0,0}(s) = F_0(s), \\ w_{0,n}(s) = \dfrac{1}{A_*(s)}D_s\left[\sum_{0\le m<i_*}\left(\mathscr{A}_{1mB}^{[0]}\,\tilde{*}\,D_s^{-1}(ks^k)^m w_{0,n-1}\right)(s)\right. \\ \qquad\qquad \left. +\sum_{0\le m<i_*}\left(\mathscr{A}_{2mB}^{[0]}\,\tilde{*}\,D_s^{-1}(ks^k)^m w_{0,n-(M-j_*)}\right)(s)\right] \quad (n \ge 1). \end{cases}$$

The case $\nu \ge 1$. We define the functions $\{w_{\nu,n}(s)\}$ by the following

$$\begin{cases} w_{\nu,0}(s) = F_\nu(s) + \dfrac{1}{A_*(s)}\sum_{\ell=1}^{\min\{\ell_*,\nu\}}\sum_J A_{[j]}^{[\ell]}(ks^k)^{\overline{i_{j1}}}w_{\nu-\ell,0}(s), \\ w_{\nu,n}(s) = \dfrac{1}{A_*(s)}\left[\sum_{\ell=1}^{\min\{\ell_*,\nu\}}\sum_J A_{[j]}^{[\ell]}(ks^k)^{\overline{i_{j1}}}w_{\nu-\ell,n}(s)\right. \\ \quad +D_s\sum_{\ell=0}^{\min\{\ell_*,\nu\}}\sum_{0\le m<i_*}\left(\mathscr{A}_{1mB}^{[\ell]}\,\tilde{*}\,D_s^{-1}(ks^k)^m w_{\nu-\ell,n-1}\right)(s) \\ \quad \left.+D_s\sum_{\ell=0}^{\min\{\ell_*,\nu\}}\sum_{0\le m<i_*}\left(\mathscr{A}_{2mB}^{[\ell]}\,\tilde{*}\,D_s^{-1}(ks^k)^m w_{\nu-\ell,n-(M-j_*)}\right)(s)\right] \quad (n \ge 1). \end{cases} \tag{28}$$

Here we interpret $w_{\nu,n}(s) \equiv 0$ if $n < 0$.

We remark that from the above construction for each ν, $W_\nu(s) := \sum_{n\ge 0} w_{\nu,n}(s)$ is a formal power series solution of the convolution equation $w_\nu(s) = \tilde{T}_\nu(w_\nu)(s)$ in a neighborhood of the origin, and therefore $W_\nu(s)$ coincides with $w_\nu(s) = D_s\sum_{n\ge 0}^{(\nu)} A_\nu(n)s^n/\Gamma(1+n/k)$ in a neighborhood of the origin.

We assume that for $s \in S_0$,

$$\left|\frac{1}{A_*(s)}\right| \le \frac{B_1}{1+|s|^{\alpha_*}}, \quad \left|D_s \mathscr{A}_{1mB}^{[\ell]}(s)\right| \le B_2|s|^{\alpha_*-km-1},$$
$$\left|D_s \mathscr{A}_{2mB}^{[\ell]}(s)\right| \le B_2|s|^{\alpha_*+(M-j_*)-km-1} \quad \frac{|s|^{\alpha_*}}{1+|s|^{\alpha_*}} \le B_3, \tag{29}$$
$$\frac{|s|^{\alpha_*-1}}{1+|s|^{\alpha_*}} \le B_3 \quad \frac{1}{1+|s|^{\alpha_*}}\left|\sum\nolimits_J A_{[j]}^{[\ell]}(ks^k)^{\overline{i_{j1}}}\right| \le B_2B_3$$

with some positive constants B_1, B_2 and B_3 for all ℓ and m, and for $s \in S_*$, we assume $|F_\nu(s)| \le B_4|s|^{M-2}$ for all ν with $B_4 > B_1$.

Then we have the following proposition.

Proposition 3 *For each ν, we have*

$$|w_{\nu,n}(s)| \le C_\nu K^n \frac{|s|^{M-2+n+\nu}}{\Gamma\left(\frac{n+1}{k}\right)}, \quad s \in S_*, \tag{30}$$

where $C_0 = 2B_4\Gamma(1/k)$ and $C_\nu = \beta^\nu C_0$ with $\beta \ge 4B_2B_3B_4\sigma_*^{-\ell_*} \ge 4$ and K is a positive constant independent of n (cf. (31) *and Lemma 5, (1) and (2)).*

Proof of Lemma 3 This proposition means that $w_\nu(s) = \sum_n w_{\nu,n}(s)$ has the exponential growth estimate of order at most k in S_*. Since s_0 is arbitrary, the estimate of the same order holds in $S_0 = S(\theta, \varepsilon_0)$. Therefore the estimate of $f_{\nu B}(s)$ in Lemma 3 follows from this proposition immediately. □

7.4 Proof of Proposition 3

We shall prove Proposition 3 by induction on ν and n.

Proof The case $\nu = 0$.

When $n = 0$, it is trivial by the estimate $|F_0(s)| \le B_4|s|^{M-2}$ for $s \in S_*$.

When $n \ge 1$, we have for $s \in S_*$

$$\begin{aligned}|w_{0,n}(s)| \le& \left|\frac{1}{A_*(s)}\right|\Bigg[\sum\nolimits_{0\le m<i_*}\left|D_s\left(\mathscr{A}_{1mB}^{[0]} \tilde{*} D_s^{-1}(ks^k)^m w_{0,n-1}\right)(s)\right| \\ &+ \sum\nolimits_{0\le m<i_*}\left|D_s\left(\mathscr{A}_{2mB}^{[0]} \tilde{*} D_s^{-1}(ks^k)^m w_{0,n-(M-j_*)}\right)(s)\right|\Bigg] \\ =:& \left|\frac{1}{A_*(s)}\right|\sum\nolimits_{0\le m<i_*}(I_{1m}^n(s) + I_{2m}^n(s)).\end{aligned}$$

Since

$$I_{1m}^n(s) = \left| k^m s^{k-1} \int_{s_0}^{s} D_s \mathscr{A}_{1mB}^{[0]}((s^k - u^k)^{\frac{1}{k}})(s^k - u^k)^{\frac{1}{k}-1} u^{km} w_{0,n-1}(u)) du \right|,$$

we get by the estimates of coefficients (29) and the assumption of induction

$$\begin{aligned} I_{1m}^n(s) &\le k^m B_2 C_0 \frac{K^{n-1}}{\Gamma(n/k)} |s|^{k-1} \left| \int_{s_0}^{s} |s^k - u^k|^{\frac{\alpha_* - km - 1}{k} + \frac{1}{k} - 1} |u|^{km + (M-2+n-1)} |du| \right| \\ &\le k^m B_2 C_0 \frac{K^{n-1}}{\Gamma(n/k)} |s|^{M-2+n+\alpha_* - 1} \int_0^1 (1 - t^k)^{\frac{\alpha_*}{k} - m - 1} t^{km + M - 2 + n - 1} dt. \end{aligned}$$

Here, we use the following formula. For $a, b > 0$ and $k > 0$, we have

$$\int_0^1 (1 - t^k)^{a-1} t^{b-1} dt = \frac{1}{k} \frac{\Gamma(a)\Gamma(b/k)}{\Gamma(a + b/k)}.$$

By using this formula, we get

$$I_{1m}^n(s) \le k^{m-1} B_2 C_0 \frac{K^{n-1}}{\Gamma(n/k)} |s|^{M-2+n+\alpha_*-1} \frac{\Gamma\left(\frac{\alpha_*}{k} - m\right) \Gamma\left(m + \frac{M-2+n}{k}\right)}{\Gamma\left(\frac{\alpha_* + M - 2 + n}{k}\right)}.$$

Similarly, by putting $n_* = M - j_*$, we get

$$I_{2m}^n(s) \le k^{m-1} B_2 C_0 \frac{K^{n-n_*}}{\Gamma\left(\frac{n-n_*+1}{k}\right)} |s|^{M-2+n+\alpha_*} \frac{\Gamma\left(\frac{\alpha_*+n_*}{k} - m\right) \Gamma\left(m + \frac{M-2+n-n_*+1}{k}\right)}{\Gamma\left(\frac{\alpha_*+M-2+n+1}{k}\right)}.$$

We use the following lemma which will be proved after the proof of Proposition 3 will be completed

Lemma 5 *Let $n_*, A \in \mathbb{N}$ with $n_* \ge 1$ and $0 \le m < i_*$. Let $k > 1$. Then there exists a $\hat{B} > 0$ such that we have*

$$(1)\ \frac{1}{\Gamma(n/k)} \frac{\Gamma\left(m + \frac{A+n}{k}\right)}{\Gamma\left(i_* + \frac{A+n}{k}\right)} \le \frac{\hat{B}}{\Gamma\left(\frac{n+1}{k}\right)}, \quad (2)\ \frac{1}{\Gamma\left(\frac{n-n_*+1}{k}\right)} \frac{\Gamma\left(m + \frac{A+n-n_*+1}{k}\right)}{\Gamma\left(i_* + \frac{A+n+1}{k}\right)} \le \frac{\hat{B}}{\Gamma\left(\frac{n+1}{k}\right)}.$$

By admitting this lemma and noticing $k = \alpha_*/i_*$, we have

$$\begin{aligned} I_{1m}^n(s) &\le k^{m-1} \hat{B} B_2 C_0 \frac{K^{n-1}}{\Gamma\left(\frac{n+1}{k}\right)} |s|^{M-2+n+\alpha_*-1}, \\ I_{2m}^n(s) &\le k^{m-1} \hat{B} B_2 C_0 \frac{K^{n-n_*}}{\Gamma\left(\frac{n+1}{k}\right)} |s|^{M-2+n+\alpha_*}. \end{aligned}$$

We take a positive constant K such that

$$\hat{B}B_2B_3B_4 \sum_{0\le m<i_*} k^{m-1}\{\Gamma(i_*-m)+K^{1-n_*}\Gamma\left(\tfrac{\alpha_*+n_*}{k}-m\right)\} \le K/(2\sigma_*^{-\ell_*}) < K. \tag{31}$$

Then we obtain the desired estimate of $w_{0,n}(s)$.

$$\begin{aligned}|w_{0,n}(s)| &\le \hat{B}B_2B_3B_4C_0\frac{K^{n-1}}{\Gamma\left(\frac{n+1}{k}\right)}|s|^{M-2+n}\\ &\quad\times\sum\nolimits_m k^{m-1}\left\{\Gamma(i_*-m)+K^{1-n_*}\Gamma\left(\tfrac{\alpha_*+n_*}{k}-m\right)\right\}\\ &\le C_0\frac{K^n}{\Gamma\left(\frac{n+1}{k}\right)}|s|^{M-2+n}.\end{aligned}$$

The case $\nu \ge 1$

We assume that the estimates (30) hold up to $\nu-1$.

When $n=0$, we have for $s \in S_*$

$$\begin{aligned}|w_{\nu,0}(s)| &\le B_4|s|^{M-2}+B_1B_2B_3\sum\nolimits_{\ell=1}^{\min\{\ell_*,\nu\}} C_{\nu-\ell}\frac{|s|^{M-2+\nu-\ell}}{\Gamma(1/k)}\\ &\le \frac{|s|^{M-2+\nu}}{\Gamma(1/k)}\left\{\frac{C_0}{2}\beta^\nu + B_2B_3B_4\sigma_*^{-\ell_*}\sum\nolimits_{\ell=1}^{\nu} C_{\nu-\ell}\right\}.\end{aligned}$$

By $C_{\nu-\ell}=C_\nu/\beta^\ell$, we have $\sum_{\ell=1}^\nu C_{\nu-\ell}=(C_\nu/\beta)\sum_{\ell=1}^\nu 1/\beta^{\ell-1}$. By $\beta \ge 4B_2B_3B_4\sigma_*^{-\ell_*} \ge 4$, we have

$$B_2B_3B_4\sigma_*^{-\ell_*}\sum\nolimits_{\ell=1}^{\nu} C_{\nu-\ell} \le \frac{\beta}{4}\cdot\frac{C_\nu}{\beta}\cdot\frac{\beta}{\beta-1} \le \frac{C_\nu}{3}.$$

Therefore we obtain the desired estimate of $w_{\nu,0}(s)$.

$$|w_{\nu,0}(s)| \le C_\nu\frac{|s|^{M-2+\nu}}{\Gamma(1/k)}.$$

When $n \ge 1$, from the recurrence formula (28) we put $w_{\nu,n}(s) =: I_1^n(s)+I_2^n(s)+I_3^n(s)$.

Similarly with the estimate of $w_{\nu,0}(s)$, we have

$$|I_1^n(s)| \le B_2B_3B_4\sigma_*^{-\ell_*}\frac{K^n}{\Gamma\left(\frac{n+1}{k}\right)}|s|^{M-2+n+\nu}\sum\nolimits_{\ell=1}^{\min\{\ell_*,\nu\}} C_{\nu-\ell} \le \frac{C_\nu}{3}K^n\frac{|s|^{M-2+n+\nu}}{\Gamma\left(\frac{n+1}{k}\right)}.$$

Similarly with the estimates of $w_{0,n}(s)$ and $w_{\nu,0}(s)$, we have

$$|I_2^n(s)| \le B_1 B_2 B_3 \sigma_*^{-\ell_*} \frac{K^{n-1}}{\Gamma(n/k)} |s|^{M-2+n+\nu}$$
$$\times \sum_{\ell=0}^{\min\{\ell_*,\nu\}} C_{\nu-\ell} \sum_{0\le m<i_*} k^{m-1} \frac{\Gamma\left(\frac{\alpha_*}{k}-m\right)\Gamma\left(\frac{M-2+n+\nu-\ell}{k}+m\right)}{\Gamma\left(\frac{\alpha_*+M-2+n+\nu-\ell+1}{k}\right)}$$
$$\le \frac{4}{3} C_\nu \frac{K^{n-1}|s|^{M-2+n+\nu}}{\Gamma\left(\frac{n+1}{k}\right)} \times \hat{B} B_2 B_3 B_4 \sigma_*^{-\ell_*} \sum_m k^{m-1}\Gamma(i_*-m).$$

$$|I_3^n(s)| \le B_1 B_2 B_3 \sigma_*^{-\ell_*} \frac{K^{n-n_*}}{\Gamma\left(\frac{n-n_*+1}{k}\right)} |s|^{M-2+n+\nu}$$
$$\times \sum_{\ell=0}^{\nu} C_{\nu-\ell} \sum_{0\le m<i_*} k^{m-1} \frac{\Gamma\left(\frac{\alpha_*+n_*}{k}-m\right)\Gamma\left(\frac{M-2+n-n_*+\nu-\ell+1}{k}+m\right)}{\Gamma\left(\frac{\alpha_*+M-2+n+\nu-\ell+1}{k}\right)}$$
$$\le \frac{4}{3} C_\nu \frac{K^{n-1}|s|^{M-2+n+\nu}}{\Gamma\left(\frac{n+1}{k}\right)} \times \hat{B} B_2 B_3 B_4 \sigma_*^{-\ell_*} \sum_m k^{m-1}\Gamma\left(\frac{\alpha_*+n_*}{k}-m\right) K^{1-n_*}.$$

Therefore by the condition (31) of K, we obtain the desired estimate of $w_{\nu,n}(s)$.

$$|w_{\nu,n}(s)| \le \frac{1}{3} C_\nu K^n \frac{|s|^{M-2+n+\nu}}{\Gamma\left(\frac{n+1}{k}\right)} + \frac{4}{3} C_\nu K^{n-1} \frac{|s|^{M-2+n+\nu}}{\Gamma\left(\frac{n+1}{k}\right)}$$
$$\times \hat{B} B_2 B_3 B_4 \sigma_*^{-\ell_*} \sum_{0\le m<i_*} k^{m-1}\left\{\Gamma(i_*-m) + K^{1-n_*}\Gamma\left(\frac{\alpha_*+n_*}{k}-m\right)\right\}$$
$$\le C_\nu K^n \frac{|s|^{M-2+n+\nu}}{\Gamma\left(\frac{n+1}{k}\right)}.$$

Thus the proof of Proposition 3 is completed. □

Finally, we prove Lemma 5.
Proof of Lemma 5 *(1)* Since $k > 1$, we have

$$\frac{\Gamma(n/k)\Gamma(1/k)}{\Gamma\left(\frac{n+1}{k}\right)} = \int_0^1 t^{n/k-1}(1-t)^{1/k-1}dt \ge \int_0^1 t^{n/k-1}dt = 1/O(n) \quad (n\to\infty).$$

Since $0 \le m < i_*$, we have

$$I_m(n) := \frac{\Gamma\left(m+\frac{A+n}{k}\right)}{\Gamma\left(i_*+\frac{A+n}{k}\right)} = \frac{1}{O(n^{i_*-m})} \le \frac{1}{O(n)}.$$

Therefore we see that the inequality (1) holds from the above inequalities. □

Proof of Lemma 5 *(2)* (i) The case $(1 \le) n_* \le k$. In this case, since $n_*/k \le 1$, we have

$$\frac{\Gamma\left(\frac{n-n_*+1}{k}\right)\Gamma(n_*/k)}{\Gamma\left(\frac{n+1}{k}\right)} = \int_0^1 t^{\frac{n-n_*+1}{k}-1}(1-t)^{\frac{n_*}{k}-1}dt \ge \frac{1}{O(n)} \ (n \to \infty).$$

Hence, we have

$$\frac{1}{\Gamma\left(\frac{n-n_*+1}{k}\right)} \le \frac{O(n)}{\Gamma\left(\frac{n+1}{k}\right)}. \tag{32}$$

We put

$$J_m(n) := \frac{\Gamma\left(m + \frac{A+n-n_*+1}{k}\right)}{\Gamma\left(i_* + \frac{A+n+1}{k}\right)}.$$

Then we have

$$\begin{aligned} J_m(n) &= \frac{1}{O(n^{i_*-m-1})} \frac{\Gamma\left(m + \frac{A+n-n_*+1}{k}\right)\Gamma\left(1+\frac{n_*}{k}\right)}{\Gamma\left(m+1+\frac{A+n+1}{k}\right)} \\ &= \frac{1}{O(n^{i_*-m-1})}\int_0^1 t^{m+\frac{A+n-n_*+1}{k}-1}(1-t)^{\frac{n_*}{k}}dt \le \frac{1}{O(n^{i_*-m})}. \end{aligned} \tag{33}$$

From the inequalities (32) and (33), we obtain the desired inequality (2). □

(ii) The case $k < n_*$.

(ii)-(1) We consider the case when $n_* = qk$ by some positive integer $q \ge 2$. In this case, since we have

$$\frac{1}{\Gamma\left(\frac{n-n_*+1}{k}\right)} = \frac{1}{\Gamma\left(\frac{n+1}{k} - q\right)} = \frac{O(n^q)}{\Gamma\left(\frac{n+1}{k}\right)},$$

$$\Gamma\left(m + \frac{A+n-n_*+1}{k}\right) = \Gamma\left(m + \frac{A+n+1}{k} - q\right) = \frac{\Gamma\left(m+\frac{A+n+1}{k}\right)}{O(n^q)},$$

we obtain the desired inequality (2). □

(ii)-(2) We consider the case where there exists a $q \in \mathbb{N}$ such that $qk < n_* < (q+1)k$. In this case, we put

$$\begin{aligned} n_* &= qk + r, \quad 0 < r < k, \\ n_* &= (q-1)k + \tilde{r}, \quad k < \tilde{r} < 2k. \end{aligned}$$

By noticing $r/k < 1$, we get from the inequality (32)

$$\frac{1}{\Gamma\left(\frac{n-n_*+1}{k}\right)} = \frac{1}{\Gamma\left(\frac{n-r+1}{k} - q\right)} = \frac{O(n^q)}{\Gamma\left(\frac{n-r+1}{k}\right)} \le \frac{O(n^{q+1})}{\Gamma\left(\frac{n+1}{k}\right)}. \tag{34}$$

Similarly we get

$$\begin{aligned} J_m(n) &= \frac{\Gamma\left(m+\frac{A+n-\tilde{r}+1}{k}\right)/O(n^{q-1})}{O(n^{i_0-m})\Gamma\left(m+\frac{A+n+1}{k}\right)} \\ &= \frac{1}{O(n^{i_0-m})O(n^{q-1})}\int_0^1 t^{m+\frac{A+n-\tilde{r}+1}{k}-1}(1-t)^{\frac{\tilde{r}}{k}-1}dt \le \frac{1}{O(n^{i_0-m})O(n^{q})}. \end{aligned}$$

By the above inequalities, we obtain the desired inequality (2). □

Acknowledgements The author is most grateful to the referee for his careful reading and kind comments and suggestions to improve this paper. The author is partially supported by the Grant-in-Aid for Scientific Research No. 15K04898 of Japan Society for the Promotion of Science.

References

1. Balser, W.: From Divergent Power Series to Analytic Functions. Springer Lecture Notes, vol. 1582. Springer, Berlin (1994)
2. Braaksma, B.L.J.: Multisummability and stokes multipliers of linear meromorphic differential equations. J. Differ. Equ. **92**(1), 45–75 (1991)
3. Ichinobe, K.: Summability of formal solution of Cauchy problem for some PDE with variable coefficients, recent development of micro-local analysis for the theory of asymptotic analysis. RIMS Kôkyûroku Bessatsu **B40**, 081–094 (2013)
4. Ichinobe, K.: On a k-summability of formal solutions for a class of partial differential operators with time dependent coefficients. J. Differ. Equ. **257**(8), 3048–3070 (2014)
5. Ichinobe, K.: On k-summability of formal solutions for certain higher order partial differential operators with polynomial coefficients. Analytic, Algebraic and Geometric Aspects of Differential Equations. Trends in Mathematics, pp. 351–368. Birkhäuser/Springer, Charm (2017)
6. Ichinobe, K.: k-summability of formal solutions for certain partial differential equations and the method of successive approximation. RIMS Kôkyûroku, No. 2020 (to appear)
7. Ichinobe, K., Miyake, M.: On k-summability of formal solutions for certain partial differential operators with polynomial coefficients. Opusc. Math. **35**(5), 625–653 (2015)
8. Kitagawa, K., Sadamatsu, T.: A remark on a necessary condition of the Cauchy–Kowalevski theorem. Publ. Res. Inst. Math. Sci. **11**(2), 523–534 (1975/76)
9. Lutz, D., Miyake, M., Schäfke, R.: On the Borel summability of divergent solutions of the heat equation. Nagoya Math. J. **154**, 1–29 (1999)
10. Michalik, S.: On the summability of formal solutions to some linear partial differential equations. Manuscript, https://www.ma.utexas.edu/mp_arc/c/09/09-96.pdf (2009)
11. Michalik, S.: Analytic solutions of moment partial differential equations with constant coefficients. Funkcial. Ekvac. **56**, 19–50 (2013)
12. Miyake, M.: A remark on Cauchy–Kowalevski's theorem. Publ. Res. Inst. Math. Sci. **10**(1), 243–255 (1974/75)
13. Miyake, M.: Borel summability of divergent solutions of the Cauchy problem to non-Kowalevskian equations. Partial Differential Equations and their Applications (Wuhan 1999). World Scientific Publishing, Singapore (1999)

Singular Solutions to a System of Equations related to Ricci-Flat Kähler Metrics

Jose Ernie C. Lope and Mark Philip F. Ona

Abstract In the 1990s, Gérard and Tahara studied a class of singular partial differential equations and proved that such equations admit both holomorphic and singular solutions. Using an asymptotic approach, they showed that the obtained singular solution is unique in some space. In this paper, we will establish the existence of singular solutions to a system of partial differential equations considered by Bielawski. We then employ the asymptotic approach of Gérard and Tahara to prove its uniqueness in some space.

Keywords Singular solutions · System of PDEs · Singular nonlinear equation · Existence and uniqueness theorem · Majorant relations

MSC Primary 35A01 · Secondary 35A02, 35F50

1 Introduction

The existence and uniqueness of solutions of singular partial differential equations have been studied by a number of researchers over the last several decades. These studies stemmed from investigations done in the latter half of the 19th century by Fuchs, who worked on linear ordinary differential equations that have regular singularities, and by Briot and Bouquet, who studied first order nonlinear ordinary differential equations. In the 1970s, Baouendi and Goulaouic established the unique existence of solutions to a type of singular linear partial differential equations which are "natural" generalizations to partial differential equations of the ordinary differential equations studied by Fuchs. In the 1990s, Gérard and Tahara conducted

J. E. C. Lope (✉)
Institute of Mathematics, University of the Philippines Diliman, Quezon City, Philippines
e-mail: ernie@math.upd.edu.ph

M. P. F. Ona
Institute of Mathematics, University of the Philippines, Diliman Quezon City, Philippines
e-mail: mpona@math.upd.edu.ph

G. Filipuk et al. (eds.), *Formal and Analytic Solutions of Diff. Equations*,
Springer Proceedings in Mathematics & Statistics 256,
https://doi.org/10.1007/978-3-319-99148-1_3

extensive investigations on a generalization to partial differential equations of the Briot–Bouquet equation.

We consider a system of partial differential equations that consists of a first order singular partial differential equation which is similar to the equation studied by Gérard and Tahara [2] and several second order linear partial differential equations. Let $(t, x) \in \mathbb{C}_t \times \mathbb{C}^n_x$. We consider

$$\begin{aligned} t\partial_t u(t,x) &= F(t,x,u,\partial_x u,\{w_i\}_{i=1,\dots,N}) \\ \partial_t w_i(t,x) &= L_i(t,x;\partial_x)u + H_i(t,x), \quad i = 1,\dots,N. \end{aligned} \tag{1}$$

Here, $F(t, x, u, v, W)$ is holomorphic about the origin and satisfies $F(0, x, 0, 0, 0) \equiv 0$, and $\partial_{v_i} F(0, x, 0, 0, 0) \equiv 0$ for all $i = 1, \dots, n$. Moreover, L_i is a second order linear differential operator whose coefficients are holomorphic in t and x and H_i is a holomorphic function in t and x. This system is a slight generalization of the system considered by Bielawski [1] in his study of Ricci-flat Kähler metrics.

In this paper, we investigate the existence and uniqueness of solutions to (1) that admit a singularity at the origin. In particular, we construct a formal power series solution and prove that it converges by using a family of majorant functions that was employed earlier by Lax [4], Lope and Tahara [7] and Pongérard [9]. We will then show that this solution is unique in some space.

2 Preliminaries

Denote by $\mathbb{N}$ the set of all nonnegative integers and define $\mathbb{N}^* = \mathbb{N} \setminus \{0\}$. For a multivariable $x = (x_1, \dots, x_n)$ and a multi-index $\gamma = (\gamma_1, \dots, \gamma_n) \in \mathbb{N}^n$, we define $x^\gamma = x_1^{\gamma_1} x_2^{\gamma_2} \cdots x_n^{\gamma_n}$, $\partial_x u = (\partial_{x_1} u, \dots, \partial_{x_n} u)$, $|x| = x_1 + x_2 + \cdots + x_n$ and $|\gamma| = \gamma_1 + \gamma_2 + \cdots + \gamma_n$. For a single variable z, $|z|$ is just the modulus of z.

Let $a(z) = \sum a_\alpha z^\alpha$ and $A(z) = \sum A_\alpha z^\alpha$ with $A_\alpha \geq 0$. We say that $a(z)$ is majorized by $A(z)$, denoted by $a(z) \ll A(z)$, if for all α, $|a_\alpha| \leq A_\alpha$. In conjunction with the usual majorant, we introduce another majorant relation that will be used to prove the convergence of the formal solution. Define $P_m[y]$ to be the space of all polynomial functions of degree at most m in the indeterminate y. For any $f(y) = \sum_{i=0}^m f_i y^i \in P_m[y]$, and for $\lambda > 0$, we define a norm $\|\cdot\|_\lambda$ on $P_m[y]$ by

$$\|f(y)\|_\lambda := \sum_{i=0}^m |f_i| \lambda^i.$$

Let $u(y, Z) = \sum u_\alpha(y) Z^\alpha$ and $v(y, Z) = \sum v_\alpha(y) Z^\alpha$, for which u_α and v_α are in $P_m[y]$, for some m. We say that v majorizes u with respect to the norm $\|\cdot\|_\lambda$, written as $u \ll_\lambda v$, if and only if $\|u_\alpha\|_\lambda \leq \|v_\alpha\|_\lambda$ for all α.

Proposition 1 *Let* $\lambda > 0$ *and let* $u(y, Z) = \sum u_\alpha(y) Z^\alpha$ *and* $v(y, Z) = \sum v_\alpha(y) Z^\alpha$, *for which* u_α *and* v_α *are in* $P_{m_\alpha}[y]$, *for some* m_α. *Then the following are true:*

1. *If* $u \ll v$, *then* $u \ll_\lambda v$.
2. *Suppose* $u, v \gg 0$. *Then* $u(y, Z) \ll_\lambda v(y, Z)$ *if and only if* $u(\lambda, Z) \ll v(\lambda, Z)$.

Proof Let $u_\alpha = \sum_{i=0}^m u_{\alpha,i} y^i$ and $v_\alpha = \sum_{i=0}^m v_{\alpha,i} y^i$. If $u \ll v$, then $|u_{\alpha,i}| \le v_{\alpha,i}$. Thus $\|u_\alpha(y)\|_\lambda \le \|v_\alpha(y)\|_\lambda$. Moreover, if $u, v \gg 0$, then $\|u_\alpha(y)\|_\lambda = u_\alpha(\lambda)$ and $\|v_\alpha(y)\|_\lambda = v_\alpha(\lambda)$. The desired result easily follows. □

Let $\mathbb{C}\{x\}$ be the ring of all convergent power series with complex coefficients about the origin in $\mathbb{C}^n$ and define $D_r = \{x : |x_i| < r\}$. As in [2], let $\mathfrak{C}$ be the universal covering space for $\mathbb{C} \setminus \{0\}$, S_θ be the sector $\{t \in \mathfrak{C} : |\arg t| < \theta\}$, and $S(\epsilon(s)) = \{t \in \mathfrak{C} : 0 < |t| < \varepsilon(\arg t)\}$ for some positive-valued function $\varepsilon(s)$.

We define $\tilde{\mathscr{O}}_+$ to be the set of all $u(t, x)$ holomorphic in $S(\varepsilon(s)) \times D_\delta$ for some $\varepsilon(s)$ and $\delta > 0$, and satisfying

$$\max_{x \in K} |u(t, x)| = O(|t|^a) \quad (\text{as } t \to 0 \text{ in } S_\theta),$$

for some $a > 0$, for all $\theta > 0$ and for any compact subset K of D_δ. Note that the space $\tilde{\mathscr{O}}_+$ contains the germ of holomorphic functions that vanish with t and also series of the form $\sum_{0<i\le j\le 2i} t^i (\log t)^j$, where $\log t$ is the natural logarithmic function.

3 Statement of the Problem

Under the assumptions on F and L_i, Eq. (1) may be expanded as

$$\begin{aligned} (t\partial_t - \rho(x))u &= a(x)t + \sum_{i=1}^N b_i(x) w_i + G(x)(t, u, \partial_x u, \{w_i\}_{i=1,\dots,N}) \\ \partial_t w_i &= \sum_{|\gamma|\le 2} L_{i,\gamma}(t, x)\partial_x^\gamma u + H_i(t, x), \quad i = 1, \dots, N, \end{aligned} \tag{2}$$

where we set $\rho(x) = \partial_u F(0, x, 0, 0, 0)$ and $G(x)(t, u, v, W)$ is the collection of all the nonlinear terms with respect to t, u, v and W.

It is already shown in Lope and Ona [6] that (2) has a unique holomorphic solution $(u^*, w_1^*, \dots, w_N^*)$ provided that $\rho(0) \notin \mathbb{N}^*$. The following theorem is the main result of this paper.

Theorem 1 *Suppose* $\rho(0) \notin \mathbb{N}^*$ *and let* S_+ *be the set of all* $\tilde{\mathscr{O}}_+$ *solutions of (2).*

1. *If* $Re\rho(0) \le 0$, *then* $S_+ = \{(u^*, w_1^*, \dots, w_N^*)\}$.
2. *If* $Re\rho(0) > 0$, *then every solution* $(u, w_1, \dots, w_N)$ *in* S_+ *is uniquely determined by some holomorphic function* $\phi(x)$ *and is of the form*

$$u = U(\phi) = \sum_{\substack{2(m_1+m_2)\geq m_3+2 \\ m_1+m_2\geq 1}} \phi_{m_1,m_2,m_3}(x)t^{m_1+\rho(x)m_2}(\log t)^{m_3}, \tag{3}$$

$$w_i = W_i(\phi) = \sum_{\substack{2(m_1+m_2)\geq m_3+2 \\ 2m_1+m_2\geq 2}} \phi^i_{m_1,m_2,m_3}(x)t^{m_1+\rho(x)m_2}(\log t)^{m_3}, \tag{4}$$

for $i = 1, \ldots, N$, *with* $\phi_{0,1,0} = \phi(x)$.

The existence of an $\tilde{\mathscr{O}}_+$-solution will be shown in Sect. 4. A uniqueness result will be given in Sect. 5.

4 Existence of an $\tilde{\mathscr{O}}_+$ Solution

We introduce new variables $t_1 = t$, $t_2 = t^{\rho(x)}$ and $y = \log t$, and rewrite (2) as

$$\mathscr{T}^*(t_1\partial_{t_1}, t_2\partial_{t_2}, x)u = a(x)t_1 + \sum_{i=1}^{N} b_i(x)w_i + G(x)(t_1, u, \mathscr{X}(u), \{w_i\}_{i=1,\ldots,N}) \tag{5}$$

$$\mathscr{T}(t_1\partial_{t_1}, t_2\partial_{t_2}, x)w_i = \sum_{|\gamma|\leq 2} t_1 L_{i,\gamma}(t_1, x)\mathscr{X}^\gamma u + t_1 H_i(t_1, x), \quad i = 1, \ldots, N, \tag{6}$$

where $\mathscr{T}(m_1, m_2, x) := m_1 + \rho(x)m_2 + \partial_y$, $\mathscr{T}^* := \mathscr{T} - \rho(x)$, $\mathscr{X} = (\mathscr{X}_1, \ldots, \mathscr{X}_n)$ and $\mathscr{X}_i = \partial_{x_i} + \partial_{x_i}\rho(x) \cdot yt_2\partial_{t_2}$.

Let $P_m(x)[y]$ be the set of all polynomials of degree at most m in y with coefficients that are holomorphic with respect to x.

Proposition 2 *If $\rho(0) \notin \mathbb{N}^*$ and Re $\rho(0) > 0$, then the system (5)–(6) has a formal solution $(U, W_1, \ldots, W_N)$ of the form $U = \sum_{m_1+m_2\geq 1} U_{m_1,m_2}(x, y)t_1^{m_1}t_2^{m_2}$ and $W_i = \sum_{2m_1+m_2\geq 2} W^i_{m_1,m_2}(x, y)t_1^{m_1}t_2^{m_2}$, where $U_{0,1}(x, y)$ is independent of y and is chosen arbitrarily. All other coefficients are uniquely determined by $U_{0,1}(x, y)$ and are in $P_{2m_1+2m_2-2}(x)[y]$. Moreover, if $U_{0,1}(x, y)$ is holomorphic, then U and W_i are holomorphic as well.*

Proof If Re $\rho(0) > 0$, we can find an $R_0 > 0$ such that Re $\rho(x) > 0$, for $x \in D_{R_0}$. The existence of the formal solution follows from the property that for every $m \in \mathbb{N}$, the operators $\mathscr{T}(m_1, m_2, x)$ and $\mathscr{T}^*(m_1, m_2, x)$ are invertible over the space $P_m(x)[y]$ for any $2m_1 + m_2 \geq 2$ and $x \in D_{R_0}$. Using these, a recursive formula similar to the one in Sect. 2 of [2] can be obtained for the coefficients $U_{m_1,m_2}(x, y)$ and $W^i_{m_1,m_2}(x, y)$, that is, there will exist holomorphic functions h_{m_1,m_2} and $H^i_{m_1,m_2}$, which are uniquely determined by the previous coefficients and their derivatives, such that $U_{m_1,m_2} = h_{m_1,m_2}$ and $W^i_{m_1,,m_2} = H^i_{m_1,m_2}$. □

We now show that the formal solution converges. Choose a large enough $A > 0$ such that

$$A|m_1 + \rho(x)(m_2 - 1)| \geq m_1 + m_2 + 1.$$

Note that $m_1 + m_2 + 1 \leq A|m_1 + \rho(x)m_2|$ also holds for the same choice of A. We have the following lemma, whose proof can be found in [2]:

Lemma 1 *Let $m \in \mathbb{N}$ and $f \in P_m[y]$. Choose $\lambda > 2Am(m_1 + m_2 + 1)^{-1}$. Then for any m_1, m_2 satisfying $2m_1 + m_2 \geq 2$, the quantities $2A\|\mathscr{T}(m_1, m_2, x)f\|_\lambda$ and $2A\|\mathscr{T}^*(m_1, m_2, x)f\|_\lambda$ are both larger than $(m_1 + m_2 + 1)\|f\|_\lambda$.*

Let $\bar{\rho}(x) = \rho(x) - \rho(0)$. Then there exist holomorphic functions $\bar{\rho}_i(x)$ such that $\bar{\rho}(x) = \sum_{1\leq i\leq n} x_i\bar{\rho}_i(x)$. Choose a sufficiently large $M > 0$ such that $a(x)$, $b_i(x)$, $L_{i,\gamma}(t_1, x)$, $H_i(t_1, x)$, $G(x)(t_1, u, v, Z)$, $U_{0,1}(x)$, $\bar{\rho}_i(x)$ and $\partial_{x_i}\rho(x)$ are all bounded by M for $|t_1| \leq r_0$, $|x_i| \leq R_0$, $|u| \leq R_1$, $|v_i| \leq R_1$ and $|Z_i| \leq R_1$. Consider now the following system of majorant relations:

$$(t_1\partial_{t_1} + t_2\partial_{t_2} + 1)\mathscr{U} \gg_\lambda \frac{2AM(t_1 + t_2)}{1 - \frac{|x|}{R_0}} + \frac{2AM}{1 - \frac{|x|}{R_0}}\Bigg[|x|(t_2\partial_{t_2} + 1)\mathscr{U} + \sum_{1\leq i\leq N}\mathscr{W}_i$$
$$+ \sum_{p+q+|\alpha|+|\beta|\geq 2}\frac{t_1^p\mathscr{U}^q}{r_0^p R_1^{q+|\alpha|+|\beta|}}\prod_{1\leq j\leq n}\left(\partial_{x_j}\mathscr{U} + \frac{yM}{1 - \frac{|x|}{R_0}}t_2\partial_{t_2}\mathscr{U}\right)^{\alpha_j}\prod_{1\leq i\leq N}\mathscr{W}_i^{\beta_i}\Bigg], \tag{7}$$

$$(t_1\partial_{t_1} + t_2\partial_{t_2} + 1)\mathscr{W}_i \gg_\lambda \frac{2AM|x|}{1 - \frac{|x|}{R_0}}t_2\partial_{t_2}\mathscr{W}_i + \frac{2AMt_1}{1 - \frac{|x|}{R_0} - \frac{t_1}{r_0}}\times$$
$$\times\left[\sum_{0\leq|\gamma|\leq 2}\prod_{1\leq j\leq n}\left(\partial_{x_j} + \frac{yM}{1 - \frac{|x|}{R_0}}t_2\partial_{t_2}\right)^{\gamma_j}\mathscr{U} + 1\right]. \tag{8}$$

Proposition 3 *Let $\lambda \geq 4A$. Suppose that $\mathscr{U} = \sum \mathscr{U}_{m_1,m_2}(x, y)t_1^{m_1}t_2^{m_2}$ and $\mathscr{W}_i = \sum \mathscr{W}^i_{m_1,m_2}(x, y)t_1^{m_1}t_2^{m_2}$, with $\mathscr{U}_{m_1,m_2}$, $\mathscr{W}^i_{m_1,m_2} \in P_{2m_1+2m_2}(x)[y]$ satisfy (7) and (8). Then $U \ll_\lambda \mathscr{U}$ and $W_i \ll_\lambda \mathscr{W}_i$ for $i = 1, \ldots, N$.*

Proof Note that the right-hand side of relations (7) and (8) have the same form as the ones in (5) and (6), hence it is also seen that there exist functions $\mathfrak{h}_{m_1,m_2}$ and $\mathscr{H}^i_{m_1,m_2}$ (corresponding to $h_{m_1,m_2}$and H_{m_1,m_2}) such that $\mathscr{U}_{m_1,m_2} \gg_\lambda \mathfrak{h}_{m_1,m_2}$ and $\mathscr{W}^i_{m_1,m_2} \gg_\lambda \mathscr{H}^i_{m_1,m_2}$.

We then proceed by induction. Clearly, $\mathscr{U}_{i,,j} \gg_\lambda U_{i,,j}$ and $\mathscr{W}^i_{i,,j} \gg_\lambda W^i_{i,,j}$ for $i + j = 1$. Suppose $\mathscr{U}_{i,,j} \gg_\lambda U_{i,,j}$ and $\mathscr{W}^i_{i,,j} \gg_\lambda W^i_{i,,j}$ for $i \leq m_1$, $j \leq m_2$ and $i + j < m_1 + m_2$. It is easy to show that $\mathfrak{h}_{m_1,m_2} \gg_\lambda h_{m_1,m_2}$ and $\mathscr{H}_{m_1,m_2} \gg_\lambda H^i_{m_1,m_2}$, in view of Proposition 1(1).

By expanding the coefficients with respect to x, that is, by writing each $f(x, y)$ as $f(x, y) = \sum_{|m_3|\geq 0} f_{m_3}(y)x^{m_3}$, and by using Lemma 1, then for any $|m_3| \geq 0$,

$$(m_1 + m_2 + 1)\mathscr{U}_{m_1,m_2,m_3} \gg_\lambda 2A[\mathscr{T}(m_1, m_2, 0) - \rho(0)]U_{m_1,m_2,m_3}$$
$$\gg_\lambda (m_1 + m_2 + 1)U_{m_1,m_2,m_3},$$

as desired. The same goes for showing that $\mathscr{W}^i_{m_1,m_2,m_3} \gg_\lambda W^i_{m_1,m_2,m_3}$. □

We now wish to find a solution to the majorant relations (7) and (8).

Let us recall the majorant functions used in [6]. For $i \in \mathbb{N}$ and $s \in \mathbb{N}^*$, define

$$\varphi_i(z) := \frac{1}{4S}\sum_{n\geq 0}\frac{z^n}{(n+1)^{2+i}} \quad \text{and} \quad \Phi_i^s(t,z) := \sum_{p\geq 0} t^p \frac{D^{sp}\varphi_i(z)}{(sp)!},$$

where $S = \frac{1}{6}\pi^2$ and D is the derivative operator. For consistency with existing literature and for aesthetic reasons, we opted to adopt the notation for the derivative used by Pongérard in [9]. Note that for $i \in \mathbb{N}$ and $s \in \mathbb{N}^*$, $\varphi_i(z)$ converges when $|z| < 1$ and $\Phi_i^s(t, z/R)$ converges when $|t|^{1/s} + |z| < R$.

In addition, for all $i \in \mathbb{N}$, we have $\varphi_i(z)\varphi_i(z) \ll 2^i\varphi_i(z)$ and $2^{-3-i}\varphi_i(z) \ll \varphi'_{i+1}(z) \ll \varphi_i(z)$. Moreover, for a sufficiently small $R > 0$ and for all $p, q \in \mathbb{N}$, $(p+q)!D^p\varphi_i(z/R) \ll p!D^{p+q}\varphi_i(z/R)$ holds. The other properties of φ_i and Φ_i^s are found in [6] and are omitted here.

Proposition 4 *Let $r \in (0, r_0)$ and set $\tau = (t_1 + t_2)y^2$. We can find constants L_1, L_2, L_3, $c > 0$ and $R < \frac{1}{4}R_0$ such that the functions*

$$\mathscr{U}(t_1,t_2,x,y) = L_1\tau\Phi_2^2\Big(\frac{\tau}{cr},\frac{|x|}{R}\Big),$$

$$\mathscr{W}_i(t_1,t_2,x,y) = L_2(cr)^2\sum_{p\geq 1}(p+1)\Big(\frac{\tau}{cr}\Big)^{p+1}\frac{D^{2p}\varphi_2(|x|/R)}{(2p)!} + L_3\tau\Phi_2^2\Big(\frac{\tau}{cr},\frac{|x|}{R}\Big),$$

for $i = 1, \ldots, N$, satisfy (7) and (8).

Proof For brevity, we temporarily omit the arguments of Φ_i^s and φ_i. We first choose a sufficiently large $K > 0$ such that $(1 - \frac{|x|}{R_0})^{-1}D^j\varphi_i \ll KD^j\varphi_i$ and $(1 - \frac{|x|}{R_0} - \frac{\tau}{r_0})^{-1}\Phi_i^\eta \ll K\Phi_i^\eta$ for any $i \leq \eta$, for all $j \geq 0$ and for all $R < \frac{1}{4}R_0$.

Since $(t_1\partial_{t_1} + t_2\partial_{t_2})\tau = (\tau\partial_\tau)\tau$, the left-hand side of (8) majorizes

$$\begin{aligned}(t_1\partial_{t_1} + t_2\partial_{t_2})\mathscr{W}_i \gg_\lambda & L_2(cr)^2\sum_{p\geq 0}(p+2)^2\Big(\frac{\tau}{cr}\Big)^{p+2}\frac{D^{2p+2}\varphi_2}{(2p+2)!} + L_3\tau\Phi_2^2 \\ \gg_\lambda & \frac{L_2\tau^2\Phi_0^2}{2^9} + L_3\tau\Phi_2^2.\end{aligned}$$

As for the corresponding right-hand side, note that

$$\partial_{x_j}\mathscr{U} \ll_\lambda \frac{L_1\tau}{R}\sum_{p\geq 0}\Big(\frac{\tau}{cr}\Big)^p\frac{D^{2p}\varphi_1}{(2p)!} = \frac{L_1\tau}{R}\Phi_1^2$$

and

$$t_2\partial_{t_2}\mathscr{U} \ll_\lambda (t_1\partial_{t_1} + t_2\partial_{t_2})\mathscr{U} = L_1 cr \sum_{p\geq 0}(p+1)\Big(\frac{\tau}{cr}\Big)^{p+1}\frac{D^{2p}\varphi_2}{(2p)!}$$
$$\ll_\lambda L_1 cr \sum_{p\geq 0}(p+1)\Big(\frac{\tau}{cr}\Big)^{p+1}\frac{D^{2p+1}\varphi_2}{(2p+1)!} \ll_\lambda L_1\tau\Phi_1^2.$$

Thus by Proposition 1(2), for any $|\gamma| \leq 2$,

$$\Big(\partial_{x_j} + \frac{M}{1-\frac{|x|}{R_0}} y t_2\partial_{t_2}\Big)^{\gamma_j}\mathscr{U} \ll_\lambda \Big(\frac{1}{R} + MK\lambda\Big)^{\gamma_j} L_1\tau\Phi_0^2. \tag{9}$$

Therefore, since $\mathscr{U} \ll_\lambda L_1\tau\Phi_0^2$, $t_1 \ll_\lambda \tau$ and $R^{-1} + MK\lambda \geq 1$ for small values of R, it follows that there exists a $C_1 > 0$ such that

$$\frac{2AMt_1}{1-\frac{|x|}{R_0}-\frac{t_1}{r_0}}\Bigg[\sum_{|\gamma|\leq 2}\prod_{1\leq j\leq n}\Big(\partial_{x_j} + \frac{yM}{1-\frac{|x|}{R_0}} t_2\partial_{t_2}\Big)^{\gamma_j}\mathscr{U} + 1\Bigg]$$
$$\ll_\lambda C_1\tau\Big[\Big(\frac{1}{R} + MK\lambda\Big)^2 L_1\tau\Phi_0^2 + 4S\Phi_2^2\Big].$$

Finally, using the fact that $zD^\alpha\varphi_i(z) \ll 2^{2+i}D^\alpha\varphi_i(z)$ for any $\alpha \in \mathbb{N}$, we get

$$\frac{2AM|x|}{1-\frac{|x|}{R_0}} t_2\partial_{t_2}\mathscr{W}_i \ll_\lambda 32AKMRt_2\partial_{t_2}\mathscr{W}_i.$$

We now turn our attention to (7). The left-hand side satisfies

$$(t_1\partial_{t_1} + t_2\partial_{t_2} + 1)\mathscr{U} \gg_\lambda \frac{L_1 cr}{32}\sum_{p\geq 0}\Big(\frac{\tau}{cr}\Big)^{p+2}\frac{D^{2p+1}\varphi_1}{(2p+1)!} + L_1\tau\Phi_2^2$$
$$\gg_\lambda \frac{L_1\tau^2}{32cr}\Phi_1^2 + L_1\tau\Phi_2^2.$$

In addition, the left-hand side also majorizes

$$L_1 cr\sum_{p\geq 0}(p+2)\Big(\frac{\tau}{cr}\Big)^{p+2}\frac{D^{2p+2}\varphi_2}{(2p+2)!} + L_1\tau\Phi_2^2.$$

To deal with the right-hand side, we note the following estimates for $\mathscr{W}_i$:

$$\frac{2AM\mathscr{W}_i}{1-\frac{|x|}{R_0}} \ll_\lambda 2AKM\Big(L_2(cr)^2\sum_{p\geq 0}(p+2)\Big(\frac{\tau}{cr}\Big)^{p+2}\frac{D^{2p+2}\varphi_2}{(2p+2)!}+L_3\tau\Phi_2^2\Big),$$

$$\mathscr{W}_i \ll_\lambda L_2 cr\tau\sum_{p\geq 0}\Big(\frac{\tau}{cr}\Big)^p\frac{D^{2p}\varphi_1}{(2p)!}+L_3\tau\Phi_1^2 \ll_\lambda (KL_2+L_3)\tau\Phi_1^2. \qquad (10)$$

Here, we used the facts that $cr\tau \ll_\lambda crr_0(1-\frac{|x|}{R_0}-\frac{\tau}{r_0})^{-1}$ and $crr_0 < 1$. Using (9) and (10), by choosing the constant c small enough, and noting that $1 \ll_\lambda 4S\Phi_2^2$, the nonlinear term of the right-hand side of (7) can be majorized by

$$\frac{2AM}{1-\frac{|x|}{R_0}}\Bigg[\sum_{p\geq 1}\Big(\frac{t_1}{r_0}\Big)^p+\sum_{\substack{p+q+|\alpha|+|\beta|\geq 2\\ q+|\alpha|+|\beta|\geq 1}}\Big(\frac{t_1}{r_0}\Big)^p\Big(\frac{L_1\tau\Phi_2^2}{R_1}\Big)^q\times$$

$$\times\Big(\frac{(R^{-1}+MK\lambda)L_1\tau\Phi_1^2}{R_1}\Big)^\alpha\Big(\frac{(KL_2+L_3)\tau\Phi_1^2}{R_1}\Big)^\beta\Bigg]$$

$$\ll_\lambda \frac{2AM}{1-\frac{|x|}{R_0}}\Big(\frac{\tau}{r_0}\frac{4S\Phi_2^2}{1-\frac{\tau}{r_0}}+\frac{Q_R^2\tau^2\Phi_1^2}{1-Q_R\tau}\Big) \ll_\lambda 2AMK\Big(\frac{4S\tau}{r_0}\Phi_2^2+C_2Q_R^2\tau^2\Phi_1^2\Big)$$

where $Q_R=\frac{1}{r_0}+\frac{2L_1}{R_1}+\frac{2nL_1}{R_1}(R^{-1}+MK\lambda)+\frac{2N}{R_1}(KL_2+L_3)$ and C_2 is a constant obtained in the application of a property of Φ_1^2.

To ensure that (7) and (8) are satisfied, we simply compare the coefficients of Φ_1^2 and Φ_2^2 and force the inequality to happen. This can be done by suitably choosing constants L_3, L_1, R and L_2 in this order and fixing them, and finally by choosing a small enough c. □

Since $\mathscr{U}$ and $\mathscr{W}_i$ are majorized by $\Phi_2^2(\frac{\tau}{cr},\frac{|x|}{R})$ and $\Phi_1^2(\frac{\tau}{cr},\frac{|x|}{R})$, respectively, they converge on the set

$$\big\{(t_1,t_2,x,y): |t_1|\leq r_0,\ |t_2|\leq r_0,\ |t_1y^2|\leq cr(\tfrac{R}{3})^2,\ |t_2y^2|\leq cr(\tfrac{R}{3})^2,\ |x_i|\leq\tfrac{R}{3n}\big\}.$$

If $\lambda > 1$, then U will be bounded by

$$\sum_{m_1+m_2\geq 1,|m_3|\geq 0}\|U_{m_1,m_2,m_3}(y)\|_\lambda\eta(y)^{2(m_1+m_2)}|t_1|^{m_1}|t_2|^{m_2}\prod_{1\leq i\leq n}|x_i|^{m_{3,i}}$$

$$\leq \sum_{m_1+m_2\geq 1,|m_3|\geq 0}\|\mathscr{U}_{m_1,m_2,m_3}(y)\|_\lambda|\eta(y)^2t_1|^{m_1}|\eta(y)^2t_2|^{m_2}\prod_{1\leq i\leq n}|x_i|^{m_{3,i}},$$

where $\eta(y)=\max\{1,|y|\}$. Since $\mathscr{U}_{m_1,m_2,m_3}(y)=\mathscr{U}_{m_1,m_2,m_3}(1)y^{2(m_1+m_2)}$, the function U will converge on the set $\{(t_1,t_2,x,y): |t_1|\leq\varepsilon, |t_2|\leq\varepsilon, |t_1y^2|\leq\varepsilon, |t_2y^2|\leq\varepsilon, |x_i|\leq\frac{R}{3n}\}$, $\varepsilon < cr(\frac{R}{3\lambda})^2$. Likewise, each W_i will converge on the same set.

5 Uniqueness of the $\tilde{\mathscr{O}}_+$ Solution

Consider the integro-differential form of (2) which is given by:

$$(t\partial_t - \rho(x))u = \tilde{a}(x)t + \sum_{i=1}^{N} b_i(x) \int_0^t L_i(s, x; \partial_x)u ds$$
$$+ \tilde{G}(x)\Big(t, u, \partial_x u, \Big\{ \int_0^t L_i(s, x; \partial_x)u ds \Big\}_{1\le i\le N}\Big), \quad (11)$$

for some functions $\tilde{a}(x)$ and $\tilde{G}(x)(t, u, v, Z)$. It was already shown in [6] that (11) has a holomorphic solution u^*. Moreover, it has been shown in the previous section that whenever $\operatorname{Re}\rho(0) > 0$, an $\tilde{\mathscr{O}}_+$ solution exists. The following proposition establishes the uniqueness of the $\tilde{\mathscr{O}}_+$ solutions of (11).

Proposition 5 *Suppose $\rho(0) \notin \mathbb{N}^*$. Let $\mathscr{S}_+$ be the set of all $\tilde{\mathscr{O}}_+$ solutions of (11).*

1. *If Re $\rho(0) \le 0$, then $\mathscr{S}_+ = \{u^*\}$.*
2. *If Re $\rho(0) > 0$, then every solution $u \in \mathscr{S}_+$ is of the form $u = U(\phi)$, where*

$$U(\phi) = \sum_{\substack{2(m_1+m_2)\ge k+2\\ m_1+m_2\ge 1}} \phi_{m_1,m_2,k}(x)t^{m_1+m_2\rho(x)}(\log t)^k,$$

with $\phi_{0,1,0} = \phi(x)$ for some holomorphic $\phi(x)$ that is uniquely determined by u.

We present some necessary lemmas to prove the proposition above. We first prove a uniqueness result similar to the result of Lope [5]. Consider the following linear integro-differential equation:

$$\Big[t\partial_t - a(x) - a_0(t, x) - \sum_{i=1}^{n} a_i(t, x)\partial_{x_i}\Big]u = f(t, x) + \sum_{i=1}^{N} b_i(t, x) \int_0^t L_i(s, x; \partial_x)u ds, \quad (12)$$

where L_i is the linear differential operator described in Sect. 3. We assume that all the coefficients are continuous on $[0, T_0]$ and holomorphic in a neighborhood of D_{R_0}. In addition, we assume that for $i = 0, 1, \ldots, n$,

$$\max_{x\in K} |a_i(t, x)| = O(t^a) \ (t \to 0),$$

for some $a > 0$ and for any compact subset K of D_{R_0}.

Lemma 2 *If* Re $a(0) < 0$, *then for every $R \in (0, R_0)$, there exists $T \in (0, T_0)$ such that if (12) has a solution that is continuous on $[0, T]$ and holomorphic in D_R, then it is unique.*

Proof For simplicity, assume that T_0 and R_0 are less than 1. Choose $a \in (0, 1]$, $\delta \in (0, T_0)$ and $M > 0$ such that b_i and $L_{i\gamma}$ are bounded by M on $[0, T_0] \times D_{R_0}$, and that for $i = 0, 1, \ldots, n$ and for any $t \in [0, \delta]$, we have $\max_{x \in D_{R_0}} |a_i(t, x)| \leq M t^a$. In addition, assume that $\operatorname{Re} a(x) < L < 0$ for all $x \in D_{R_0}$.

Suppose $u(t, x)$ and $v(t, x)$ are solutions of (12) that are continuous on $[0, T_0]$ and holomorphic in D_{R_0}. Define $w = u - v$. Then there exists $W > 0$ such that $\sup_{(t,x) \in [0,T_0] \times D_{R_0}} |w(t, x)| \leq W$. In addition, w also satisfies the integro-differential equation

$$\big[t\partial_t - a(x)\big]w = a_0(t, x)w + \sum_{i=1}^{n} a_i(t, x)\partial_{x_i} w + \sum_{i=1}^{N} b_i(t, x) \int_0^t L_i(s, x; \partial_x) w ds.$$

Therefore, since $\operatorname{Re} a(0) < 0$, w will have the form

$$\begin{aligned} w(t, x) = &\int_0^t \left(\frac{\sigma}{t}\right)^{-a(x)} a_0(\sigma, x) w(\sigma, x) \frac{d\sigma}{\sigma} \\ &+ \sum_{i=1}^{n} \int_0^t \left(\frac{\sigma}{t}\right)^{-a(x)} a_i(\sigma, x) \partial_{x_i} w(\sigma, x) \frac{d\sigma}{\sigma} \\ &+ \sum_{i=1}^{N} \int_0^t \left(\frac{\sigma}{t}\right)^{-a(x)} b_i(\sigma, x) \int_0^\sigma L_i(s, x; \partial_x) w(s, x) ds \frac{d\sigma}{\sigma}. \end{aligned}$$

We claim that there exists a $C > 0$ such that for all $t \in [0, \delta]$ and for all $R \in (0, R_0)$,

$$\max_{x \in D_R} |w(t, x)| \leq W \Big(\frac{C t^a}{(R_0 - R)^2} \Big)^k$$

holds for all $k \in \mathbb{N}$. We shall prove this inductively. The inequality clearly holds when $k = 0$. Suppose it is true for $k = \ell$. Taking the maximum of both sides and using Nagumo's Lemma, we have

$$\begin{aligned} \max_{x \in D_R} |w(t, x)| \leq\ & t^L \int_0^t \sigma^{-L + a(\ell+1) - 1} \frac{C^\ell W M}{(R_0 - R)^{2\ell}} d\sigma \\ &+ t^L \sum_{i=1}^{n} \int_0^t \sigma^{-L + a(\ell+1) - 1} \frac{(2\ell + 1) C^\ell W M e}{(R_0 - R)^{2\ell+1}} d\sigma \\ &+ t^L \sum_{i=1}^{N} \sum_{|\gamma| \leq 2} \int_0^t M \sigma^{-L-1} \int_0^\sigma \frac{(2\ell + |\gamma|)^{|\gamma|} C^\ell W M e^{|\gamma|}}{(R_0 - R)^{2\ell + |\gamma|}} s^{a\ell} ds d\sigma \\ \leq\ & \frac{M}{-L + a(\ell+1)} \frac{W C^\ell t^{a(\ell+1)}}{(R_0 - R)^{2\ell}} + \frac{n(2\ell + 1) M e}{-L + a(\ell + 1)} \frac{W C^\ell t^{a(\ell+1)}}{(R_0 - R)^{2\ell+1}} \\ &+ \frac{(2\ell + 2)^2 (1 + n + n^2) N (Me)^2}{(a\ell + 1)(-L + a\ell + 1)} \frac{W C^\ell t^{a\ell+1}}{(R_0 - R)^{2\ell+2}}. \end{aligned}$$

We can choose C large enough such that it will be independent of ℓ. Thus,

$$\max_{x\in D_R}|w(t,x)| \le W\Big(\frac{Ct^a}{(R_0-R)^2}\Big)^{\ell+1}.$$

In conclusion, by choosing $0 < T \le \delta$ such that $CT^a(R_0-R)^{-2} < 1$, the claim then implies that $w(t,x) \equiv 0$ for $t \in [0,T]$ and $x \in D_R$. □

For the next lemmas, consider the following integro-differential equation:

$$(t\partial_t - \rho(x))u = \mathscr{A}_0(t,x)u + \sum_{i=1}^{n}\mathscr{A}_i(t,x)\partial_{x_i}u + \sum_{i=1}^{N}\mathscr{B}_i(t,x)\int_0^t L_i(s,x;\partial_x)u\,ds + \mathscr{G}(t,x)\Big(u,\partial_x u,\Big\{\int_0^t L_i(s,x;\partial_x)u\,ds\Big\}_{1\le i\le N}\Big), \tag{13}$$

where $\mathscr{G}(x,t)(u,v,Z)$ is the collection of all nonlinear terms in u, v and Z. We suppose that the coefficients $\mathscr{B}_i(t,x)$ are continuous for $|t| \le T_0$ and holomorphic in a neighborhood of D_{R_0} and that the coefficients $\mathscr{A}_i$ are in $O(t^s,\tilde{\mathscr{O}}_+)$ for some $s \in (0,1]$. Here, $O(t^s,\tilde{\mathscr{O}}_+)$ is the set of all $u(t,x)$ such that $t^{-s}u(t,x) \in \tilde{\mathscr{O}}_+$.

Let $R[u]$ be equal to the right-hand side of (13). Observe that if $u(t,x) \in O(t^a,\tilde{\mathscr{O}}_+)$ for $a > 0$, then $R[u] \in O(t^b,\tilde{\mathscr{O}}_+)$ for any $b < \min\{2a, a+s\}$. The following lemma is due to Gérard and Tahara [3].

Lemma 3 *Suppose $u(t,x) \in \tilde{\mathscr{O}}_+$ is a solution to (13). Then the following are true:*

1. *If $u \in O(t^a,\tilde{\mathscr{O}}_+)$ for some $a > 0$, then $u \in O(t^b,\tilde{\mathscr{O}}_+)$ for any $b < Re\,\rho(0)$.*
2. *If $u \in O(t^a,\tilde{\mathscr{O}}_+)$ for some $0 < a < Re\,\rho(0) < \min\{2a, a+s\}$, then u has the form*

$$u(t,x) = \phi(x)t^{\rho(x)} + O(t^b,\tilde{\mathscr{O}}_+),$$

for any $b > Re\,\rho(0)$ and for some $\phi(x) \in \mathbb{C}\{x\}$.

This lemma is an easy consequence of another result of Gérard and Tahara [3].

Lemma 4 *Let $u \in \tilde{\mathscr{O}}_+$ be a solution to (13). If $u \in O(t^a,\tilde{\mathscr{O}}_+)$ for some $a > Re\,\rho(0)$, then $u \equiv 0$ in $\tilde{\mathscr{O}}_+$.*

Proof Let $u \in O(t^a,\tilde{\mathscr{O}}_+)$ and set

$$\mathscr{A}_0^*(t,x) := \int_0^1 \partial_u\mathscr{G}(t,x)\Big(u\theta,\partial_x u\theta,\Big\{\int_0^t L_i(s,x;\partial_x)u\theta\,ds\Big\}_{1\le i\le N}\Big)d\theta,$$

$$\mathscr{A}_i^*(t,x) := \int_0^1 \partial_{v_i}\mathscr{G}(t,x)\Big(u\theta,\partial_x u\theta,\Big\{\int_0^t L_i(s,x;\partial_x)u\theta\,ds\Big\}_{1\le i\le N}\Big)d\theta,$$

$$\mathscr{B}_i^*(t,x) := \int_0^1 \partial_{Z_i}\mathscr{G}(t,x)\Big(u\theta,\partial_x u\theta,\Big\{\int_0^t L_i(s,x;\partial_x)u\theta\,ds\Big\}_{1\le i\le N}\Big)d\theta.$$

Observe that the functions $\mathscr{A}_i^*$ and $\mathscr{B}_i^*$ are in $O(t^a, \tilde{\mathscr{O}}_+)$. Define $w(t,x) = t^{-a}u(t,x)$, which is in $\tilde{\mathscr{O}}_+$ by definition. If we choose a small enough $0 < p < \min\{a, s\}$, then $A_0(t,x) = \mathscr{A}_0(t,x) + \mathscr{A}_0^*(t,x)$ and $A_i(t,x) = \mathscr{A}_i(t,x) + \mathscr{A}_i^*(t,x)$ are in $O(t^p, \tilde{\mathscr{O}}_+)$. Let $B_i(t,x) = \mathscr{B}_i(t,x) + \mathscr{B}_i^*(t,x)$. We can then rewrite (13) in terms of w as follows:

$$(t\partial_t + a - \rho(x))w = A_0(t,x)w + \sum_{i=0}^{n} A_i(t,x)\partial_{x_i} w + \sum_{i=1}^{N} B_i(t,x) \int_0^t L_i(s,x;\partial_x)w ds. \quad (14)$$

Since Re $(\rho(0) - a) < 0$, by Lemma 2, (14) will have the unique solution $w \equiv 0$. Therefore, $u \equiv 0$. □

We now prove Proposition 5. Let $u \in \mathscr{S}_+$ and $U = u - u^*$, where u^* is the holomorphic solution described in the beginning of this section. This means that U is in $O(t^a, \tilde{\mathscr{O}}_+)$ for some $a > 0$, and is a solution to the equation:

$$\begin{aligned}(t\partial_t - \rho(x))U &= \sum_{i=1}^{N} b_i(x) \int_0^t L_i(s,x;\partial_x)U ds \\ &\quad + \tilde{G}(x)\Big(t, U + u^*, \partial_x(U + u^*), \Big\{\int_0^t L_i(s,x;\partial_x)(U + u^*)ds\Big\}_{1 \le i \le N}\Big) \\ &\quad - \tilde{G}(x)\Big(t, u^*, \partial_x u^*, \Big\{\int_0^t L_i(s,x;\partial_x)u^* ds\Big\}_{1 \le i \le N}\Big).\end{aligned}$$

Note that this is a particular case of (13).

Suppose first that Re $\rho(0) \le 0$. Since $U \in O(t^a, \tilde{\mathscr{O}}_+)$ for some $a > 0 \ge$ Re $\rho(0)$, it follows from Lemma 4 that $U \equiv 0$.

On the other hand, if Re $\rho(0) > 0$, then choose a' such that $0 < a' <$ Re $\rho(0) < \min\{2a', a' + s\}$. Lemma 3 (1) assures us that $U \in O(t^{a'}, \tilde{\mathscr{O}}_+)$. By Lemma 3 (2), we have

$$U(t,x) = \phi(x)t^{\rho(x)} + O(t^b, \tilde{\mathscr{O}}_+)$$

for some Re $\rho(0) < b < \min\{2a', a' + s\}$ and for some holomorphic function $\phi(x)$. Thus, u is of the form $u = u^* + \phi(x)t^{\rho(x)} + O(t^b, \tilde{\mathscr{O}}_+)$ for some $\phi(x)$.

Now consider $v = u - U(\phi)$, where $U(\phi)$ is the solution we obtained in the Proposition 2. Then v satisfies the equation

$$\begin{aligned}(t\partial_t - \rho(x))v &= \sum_{i=1}^{N} b_i(x) \int_0^t L_i(s,x;\partial_x)v ds \\ &\quad + \tilde{G}(x)\Big(x, v + U(\phi), \partial_x(v + U(\phi)), \Big\{\int_0^t L_i(s,x;\partial_x)(v + U(\phi))ds\Big\}_{1 \le i \le N}\Big) \\ &\quad - \tilde{G}(x)\Big(x, U(\phi), \partial_x U(\phi), \Big\{\int_0^t L_i(s,x;\partial_x)U(\phi)ds\Big\}_{1 \le i \le N}\Big),\end{aligned}$$

which is again a particular form of (13). Note that $v \in O(t^b, \tilde{\mathscr{O}}_+)$, and by Lemma 4, $v \equiv 0$. Therefore, u is always of the form

$$u(t,x) = u^* + \sum_{2(i+j)\geq k+2,\, j\geq 1} \phi_{i,j,k}(x) t^{i+j\rho(x)} (\log t)^k.$$

It remains to show that each $\phi(x)$ is completely determined by u. To do so, we suppose $u = U(\phi(x)) = U(\psi(x))$. Thus,

$$(\phi(x) - \psi(x)) t^{\rho(x)} + O(t^b, \tilde{\mathscr{O}}_+) = 0.$$

Since $\operatorname{Re} \rho(0) < b$, it follows that $\phi(x) - \psi(x) \equiv 0$.

In summary, we have completely described the $\tilde{\mathscr{O}}_+$ solutions of (11) whenever $\rho(0) \notin \mathbb{N}^*$.

Acknowledgements The authors acknowledge the Office of the Chancellor of the University of the Philippines Diliman, through the Office of the Vice Chancellor for Research and Development, for funding support through the Outright Research Grant.

References

1. Bielawski, R.: Ricci-flat Kähler metrics on canonical bundles. Math. Proc. Camb. Philos. Soc. **132**(3), 471–479 (2002)
2. Gérard, R., Tahara, H.: Holomorphic and singular solutions of nonlinear singular first order partial differential equations. Publ. RIMS. Kyoto University **26**(6), 979–1000 (1990)
3. Gérard, R., Tahara, H.: Solutions holomorphes et singulières d'équations aux dérivées partielles singulières non linéaires. Publ. RIMS. Kyoto University **29**, 121–151 (1993)
4. Lax, P.: Nonlinear hyperbolic equations. Commun. Pure. Appl. Math. **6**, 231–258 (1953)
5. Lope, J.E.: Existence and uniqueness theorems for a class of linear Fuchsian partial differential equations. J. Math. Sci. Univ. Tokyo **6**(3), 527–538 (1999)
6. Lope, J.E., Ona, M.P.: Local solvability of a system of equations related to Ricci-flat Kähler metrics. Funkcial. Ekvac. **59**(1), 141–155 (2016)
7. Lope, J.E., Tahara, H.: On the analytic continuation of solutions to nonlinear partial differential equations. J. Math. Pures Appl. **81**(9), 811–826 (2002)
8. Tahara, H.: Singular hyperbolic systems, V. Asymptotic expansions for Fuchsian hyperbolic partial differential equations. J. Math. Soc. Jpn. **36**(3), 449–473 (1984)
9. Pongérard, P.: Sur une classe d'équations de Fuchs non linéaires. J. Math. Sci. Univ. Tokyo **7**(3), 423–448 (2000)

Hyperasymptotic Solutions for Certain Partial Differential Equations

Sławomir Michalik and Maria Suwińska

Abstract We present the hyperasymptotic expansions for a certain group of solutions of the heat equation. We extend this result to a more general case of linear PDEs with constant coefficients. The generalization is based on the method of Borel summability, which allows us to find integral representations of solutions for such PDEs.

Keywords Hyperasymptotic expansions · Heat equation · Linear PDEs with constant coefficients · Summability

MSC Primary 35C20 · Secondary 35G10, 35K05

1 Introduction

Errors generated in the process of estimating functions by a finite number of terms of their asymptotic expansions usually are of the form $\exp(-q/t)$ with $t \to 0$ and usually such a result is satisfactory. However, it is possible to obtain a refined information by means of finding the hyperasymptotic expansion of a given function, which amounts to expanding remainders of asymptotic expansions repeatedly.

More precisely, let us find the asymptotic expansion of a given function F. We receive

$$F(t) = A_0 + A_1 + \dots \quad \text{for} \quad t \to 0, \tag{1}$$

with $A_i = a_i t^i$. Once we truncate (1) after a certain amount of terms, we receive an approximation of F and

S. Michalik (✉) · M. Suwińska
Faculty of Mathematics and Natural Sciences, College of Science,
Cardinal Stefan Wyszyński University, Wóycickiego 1/3, 01-938 Warszawa, Poland
e-mail: s.michalik@uksw.edu.pl

M. Suwińska
e-mail: m.suwinska@op.pl

G. Filipuk et al. (eds.), *Formal and Analytic Solutions of Diff. Equations*,
Springer Proceedings in Mathematics & Statistics 256,
https://doi.org/10.1007/978-3-319-99148-1_4

$$F(t) = A_0 + A_1 + \ldots + A_{N_0-1} + R_{N_0}(t).$$

The optimal value of $N_0 = N_0(t)$ can be found by means of minimization of the remainder $R_{N_0}(t)$. After that we consider $R_{N_0}(t)$ as a function of two variables t and N_0 and expand it in a new asymptotic series

$$R_{N_0}(t) = B_0 + B_1 + \ldots,$$

which can be truncated optimally after N_1 terms. Thus we receive an estimation of F of the form

$$F(t) = A_0 + A_1 \ldots + A_{N_0-1} + B_0 + B_1 + \ldots + B_{N_1-1} + R_{N_1}(t),$$

and the remainder $R_{N_1}(t)$ appears to be exponentially small compared to $R_{N_0}(t)$.

After repeating the process n times we receive the *n-th level hyperasymptotic expansion* of F as $t \to 0$:

$$F(t) = A_0 + \ldots + A_{N_0-1} + B_0 + \ldots + B_{N_1-1} + C_0 + \ldots C_{N_2-1} + \ldots + R_{N_n}(t).$$

The concept of hyperasymptotic expansions emerged in 1990 as a topic of an article by M. V. Berry and C. J. Howls [2] and it was conceived as a way to estimate the solutions of Schrödinger-type equations. Methods of obtaining hyperasymptotic expansions were then developed mostly by A. B. Olde Daalhuis, who found an expansion for the confluent hypergeometric function [7, 8], linear ODEs with the singularity of rank one [9] and various nonlinear ODEs [10, 11].

Using the results from [7, 8], we will find a hyperasymptotic expansion for a certain group of solutions of the heat equation. To this end we will first obtain the optimal number of terms, after which the asymptotic expansion of the solution should be truncated. This will enable us to estimate the remainder using the Laplace method (see [12]). The reasoning then will be adapted to the case of n-level hyperasymptotic expansion.

Our main goal is to generalize those results to the case of linear PDEs with constant coefficients. To this end, first we reduce the general linear PDEs in two variables with constant coefficients to simple pseudodifferential equations using the methods of [4, 5]. Next, we apply the theory of summability, which allows to construct integral representations of solutions of such equations. Finally, in a similar way to the heat equation, we construct hyperasymptotic expansions for such integral representations of solutions.

Throughout the paper the following notation will be used.

A sector S in a direction $d \in \mathbb{R}$ with an opening $\alpha > 0$ in the universal covering space $\tilde{\mathbb{C}}$ of $\mathbb{C} \setminus \{0\}$ is defined by $S_d(\alpha) := \{z \in \tilde{\mathbb{C}} \colon z = re^{i\varphi},\ r > 0,\ \varphi \in (d - \alpha/2, d + \alpha/2)\}$. If the opening α is not essential, the sector $S_d(\alpha)$ is denoted briefly by S_d.

We denote by D_r a complex disc in $\mathbb{C}$ with radius $r > 0$ and the center in 0, i.e. $D_r := \{z \in \mathbb{C} : |z| < r\}$. In case that the radius r is not essential, the set D_r will be designated briefly by D.

If a function f is holomorphic on a domain $G \subset \mathbb{C}^n$, then it will be denoted by $f \in \mathcal{O}(G)$. Analogously, the space of holomorphic functions on a domain $G \subset \mathbb{C}^n$ with respect to the variable $z^{1/\gamma} := z_1^{1/\gamma_1}, \dots, z_n^{1/\gamma_n}$, where $1/\gamma := (1/\gamma_1, \dots, 1/\gamma_n)$ and $(\gamma_1, \dots, \gamma_n) \in \mathbb{N}^n$, is denoted by $\mathcal{O}_{1/\gamma}(G)$.

By ∂G we mean the boundary of the set G.

2 Hyperasymptotic Expansions for the Heat Equation

Let us consider the Cauchy problem for the heat equation

$$\begin{cases} u_t(t, z) - u_{zz}(t, z) = 0, \\ \qquad\qquad u(0, z) = \varphi(z). \end{cases} \tag{2}$$

We assume that the function φ has finitely many isolated singular points (single-valued and branching points) on $\mathbb{C}$. Without loss of generality we may assume that the set of singular points of φ is given by

$$\mathscr{A} := \{a_{ij} \in \mathbb{C}\colon \arg(a_{i1}) = \cdots = \arg(a_{iL_i}) = \lambda_i,\ |a_{i1}| < |a_{i2}| < \ldots < |a_{iL_i}|,$$
$$j = 1, \dots, L_i,\ i = 1, \dots, K\},$$

where $K \in \mathbb{N}$, $L_1, \dots, L_K \in \mathbb{N}$ and $\lambda_1, \dots, \lambda_K \in \mathbb{R}$ satisfy $\lambda_1 < \cdots < \lambda_K$.

Under these conditions we can define the set H as a sum of a finite number of half-lines (see Fig. 1) such that $H := \bigcup_{i=1}^{K}\{a_{i1}t : t \geq 1\}$. So we may assume that $\varphi \in \mathcal{O}(\mathbb{C} \setminus H)$ and $\mathscr{A}$ is the set of all singular points of φ. We denote it briefly by $\varphi \in \mathcal{O}_{\mathscr{A}}(\mathbb{C} \setminus H)$.

Moreover, let us assume that for any $\xi > 0$ there exist positive constants B and C such that $|\varphi(z)| \leq Ce^{B|z|^2}$ for all $z \in \mathbb{C} \setminus H_\xi$, where $H_\xi := \{z \in \mathbb{C}\colon \text{dist}\,(z, H) < \xi\}$. We write it $\varphi \in \mathcal{O}^2_{\mathscr{A}}(\mathbb{C} \setminus H)$ for short.

The solution of (2) is given by (see [6, Theorem 4])

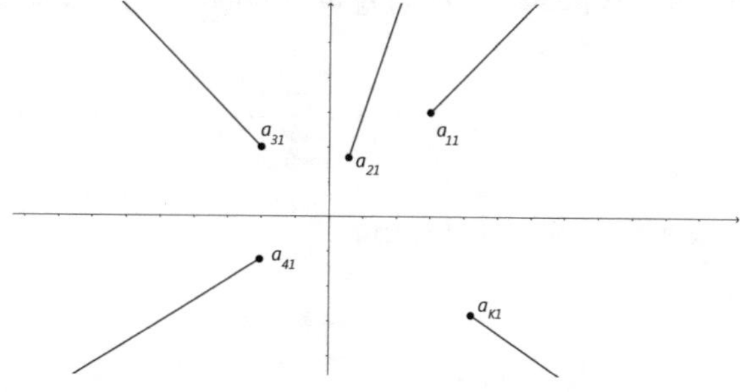

Fig. 1 The set H

$$u(t,z)=\frac{1}{2\sqrt{\pi t}}\int_{e^{i\frac{\theta}{2}}\mathbb{R}}e^{-\frac{s^2}{4t}}\varphi(z+s)\,ds, \tag{3}$$

under condition that θ is not the Stokes line for u (see [6, Definition 7], i.e. $\theta \neq 2\lambda_i \mod 2\pi$ for $i=1,\dots,K$.

To separate from the Stokes lines we fix a small positive number δ and we assume that

$$|(\theta-2\lambda_i) \mod 2\pi| \geq \delta \quad \text{for all} \quad i=1,\dots,K. \tag{4}$$

In other words we assume that $\theta \in [0,2\pi)\setminus\bigcup_{i=1}^{K}(2\lambda_i-\delta,\,2\lambda_i+\delta) \mod 2\pi$.

Our goal is to find a hyperasymptotic expansion of (3) for $t\to 0$ and $\arg t=\theta$ with z belonging to a small neighborhood of 0. To this end we fix a sufficiently small constant $\tilde{\varepsilon}$ such that $\varphi(z)\in\mathscr{O}(D_{\tilde{\varepsilon}})$.

2.1 0-Level Hyperasymptotic Expansion

To find the hyperasymptotic expansion of the solution of (2) we will use the method described in [7] (see also [8]) in the case of the confluent hypergeometric functions. In order to do so let us modify the right-hand side of (3) by

$$u(t,z)=\frac{1}{4\sqrt{\pi t}}\int_0^{e^{i\theta}\infty}e^{-\frac{s}{4t}}s^{-\frac{1}{2}}\left[\varphi\left(z+s^{\frac{1}{2}}\right)+\varphi\left(z-s^{\frac{1}{2}}\right)\right]ds.$$

Replacing s and t by $|s|e^{i\theta}$ and $|t|e^{i\theta}$, respectively, we obtain

$$u(t,z)=\frac{1}{2\sqrt{\pi|t|}}\int_0^{\infty}\frac{e^{-\frac{s}{4|t|}}}{2\sqrt{s}}\left[\varphi\left(z+e^{i\frac{\theta}{2}}\sqrt{s}\right)+\varphi\left(z-e^{i\frac{\theta}{2}}\sqrt{s}\right)\right]ds. \tag{5}$$

To find the asymptotic expansion of (5), we will expand the function

$$f_0(s,z):=\frac{1}{2}\left[\varphi\left(z+e^{i\frac{\theta}{2}}\sqrt{s}\right)+\varphi\left(z-e^{i\frac{\theta}{2}}\sqrt{s}\right)\right],$$

around the point $s=0$ using the complex Taylor formula. We receive

$$f_0(s,z)=\sum_{k=0}^{N_0-1}\frac{\varphi^{(2k)}(z)}{(2k)!}e^{ik\theta}s^k+f_1(s,z)s^{N_0}, \tag{6}$$

where $f_1(s,z)$ is of the form

$$f_1(s,z) := \frac{1}{2\pi i}\int_{\Omega_0(0,s)} \frac{f_0(w,z)}{w^{N_0}(w-s)}\,dw, \tag{7}$$

and the contour $\Omega_0(0,s)$ is a boundary of the sum of two discs such that all singular points of $f_0(w,z)$ are located outside of those discs and points 0 and s are both inside. More precisely, let us take $r := \min_{1\le i\le K} |a_{i1}| - \tilde{\varepsilon}$. In this case we can put $\Omega_0(0,s)$ as

$$\Omega_0(0,s) := \partial\Big(\{w\in\mathbb{C}\colon |w|\le r^2-\varepsilon\}\cup\{w\in\mathbb{C}\colon |w-s|\le \frac{\varepsilon}{2}\}\Big),$$

for some $\varepsilon \in (0, r^2/2)$ and ε separate from 0. It is possible to take such a contour, because by (4) we may choose so small $\tilde{\varepsilon} > 0$ that for $z \in D_{\tilde{\varepsilon}}$ the singularities $w_{ij}(z) := (a_{ij}-z)^2 e^{-i\theta}$ of $f_0(w,z)$ will never be positive real numbers. So we are able to choose ε satisfying additionally

$$\varepsilon < \frac{1}{2}\inf_{z\in D_{\tilde{\varepsilon}}}\min_{\substack{i=1,\dots K\\ j=1,\dots,L_i}} \operatorname{dist}(w_{ij}(z),\mathbb{R}_+),$$

and then $\Omega_0(0,s)$ satisfies the desired conditions.

Using (6) and basic properties of the gamma function, we can obtain an expansion of (5) of the form

$$u(t,z) = \sum_{k=0}^{N_0-1} \frac{\varphi^{(2k)}(z)}{k!} t^k + R_{N_0}(t,z), \tag{8}$$

where

$$R_{N_0}(t,z) = \frac{1}{2\sqrt{\pi |t|}}\int_0^\infty e^{-\frac{s}{4|t|}} s^{N_0-\frac{1}{2}} f_1(s,z)\,ds. \tag{9}$$

Seeing as $|w| \ge r^2-\varepsilon$, $|w-s| \ge \frac{\varepsilon}{2}$ and assuming that all the conditions given for the Cauchy datum hold, we can find the optimal value of $N_0 = N_0(t)$. The first step to do so is finding an estimation of $f_1(s,z)$. Let us note that there exist positive constants $\tilde{A}$ and $\tilde{B}$ such that $\left|\varphi\left(z \pm e^{i\frac{\theta}{2}}\sqrt{w}\right)\right| \le \tilde{A}e^{\tilde{B}s}$ for any $w\in\Omega_0(0,s)$, $s>0$ and $z \in D_{\tilde{\varepsilon}}$. Hence for $A_0 := 2\tilde{A}r^2/\varepsilon$ we estimate

$$\begin{aligned}|f_1(s,z)| &\le \frac{1}{2\pi}\int_{\Omega_0(0,s)} \frac{2\tilde{A}e^{\tilde{B}s}}{\varepsilon(r^2-\varepsilon)^{N_0}}\,d|w| \le \frac{2\tilde{A}e^{\tilde{B}s}}{\varepsilon(r^2-\varepsilon)^{N_0}}(r^2-\varepsilon+\frac{\varepsilon}{2})\\ &\le \frac{2\tilde{A}r^2e^{\tilde{B}s}}{\varepsilon(r^2-\varepsilon)^{N_0}} = \frac{A_0e^{\tilde{B}s}}{(r^2-\varepsilon)^{N_0}}.\end{aligned}$$

As a consequence,

$$|R_{N_0}(t,z)| \le \frac{A_0}{2\sqrt{\pi|t|}}\int_0^\infty e^{-\frac{s}{4|t|}} s^{N_0-\frac{1}{2}} \left(r^2-\varepsilon\right)^{-N_0} e^{\tilde{B}s}\,ds. \tag{10}$$

It is easy to check that the integrand of (10) has a maximum at a certain point $s = \sigma_1$ which satisfies the condition $N_0 = \sigma_1 \left(\frac{1}{4|t|} - \tilde{B}\right) + \frac{1}{2}$, and so now we can find the point where the minimum with respect to σ_1 of the function given by the formula

$$\sigma_1 \mapsto e^{-\sigma_1\left(\frac{1}{4|t|}-\tilde{B}\right)} \sigma_1^{\sigma_1\left(\frac{1}{4|t|}-\tilde{B}\right)} \left(r^2-\varepsilon\right)^{-\sigma_1\left(\frac{1}{4|t|}-\tilde{B}\right)-\frac{1}{2}},$$

is attained. This function is minimal at $\sigma_1 = r^2 - \varepsilon$. Because of these facts we can choose the optimal $N_0 := \left\lfloor (r^2 - \varepsilon)\left(\frac{1}{4|t|} - \tilde{B}\right) + \frac{1}{2}\right\rfloor$, where by $\lfloor\cdot\rfloor$ we denote the integer part of a real number. Next, we take $\sigma_1 := \frac{N_0 - \frac{1}{2}}{\frac{1}{4|t|} - \tilde{B}}$. Thus $\sigma_1 \le r^2 - \varepsilon$.

Thanks to that we are able to use the Laplace method, described at length in [12], and to estimate the right-hand side of (10). So, we conclude that

$$\left|R_{N_0}(t, z)\right| \sim O\left(\frac{e^{-\sigma_1\left(\frac{1}{4|t|}-\tilde{B}\right)}}{\sqrt{1-4\tilde{B}|t|}}\right) \quad \text{for} \quad t \to 0, \quad \arg t = \theta.$$

2.2 N-Level Hyperasymptotic Expansion

When the n-level asymptotic expansion is known, it is easy to compute the $(n+1)$-level expansion using the method presented in Sect. 2.1.

Observe that the remainder obtained in the n-level hyperasymptotic expansion is of the form

$$R_{N_n}(t, z) = \frac{1}{2\sqrt{\pi|t|}} \int_0^\infty e^{-\frac{s}{4|t|}} s^{N_0-\frac{1}{2}} (s-\sigma_1)^{N_1} \cdot \ldots \cdot (s-\sigma_n)^{N_n} f_{n+1}(s, z)\, ds, \tag{11}$$

where

$$f_{n+1}(s, z) := \frac{1}{2\pi i} \int_{\Omega_n(\sigma_n, s)} \frac{f_n(w, z)}{(w-\sigma_n)^{N_n}(w-s)}\, dw,$$

and

$$\Omega_n(\sigma_n, s) := \partial\Big(\{w \in \mathbb{C}\colon |w-\sigma_n| \le d(\sigma_n, \theta) - \rho_n\varepsilon\} \cup \{w \in \mathbb{C}\colon |w-s| \le 2^{-n-1}\varepsilon\}\Big),$$

with $d(w, \theta) := \inf_{z \in D_{\tilde{\varepsilon}}} \inf_{\zeta \in H} |w - e^{-i\theta}(z-\zeta)^2|$ and $\rho_n := 2 - 2^{-n}$. The contour is chosen in this way so that, when we express f_{n+1} in terms of f_0, that is as a multiple integral of the form

$$f_{n+1}(s,z) = \frac{1}{(2\pi i)^{n+1}} \int_{\Omega_n(\sigma_n,s)} \int_{\Omega_{n-1}(\sigma_{n-1},x_n)} \cdots \int_{\Omega_0(0,x_1)} f_0(x_0,z) \cdot \\ \cdot \frac{1}{x_0^{N_0} \prod_{k=1}^{n} \left[(x_k-\sigma_k)^{N_k} (x_{k-1}-x_k) \right] (x_n - s)} dx_0 \dots dx_n.$$

We show that all the singular points of $x_0 \mapsto f_0(x_0, z)$ are outside of the area surrounded by $\Omega_n(\sigma_n, s)$, $\Omega_{n-1}(\sigma_{n-1}, x_n)$,..., $\Omega_0(0, x_1)$. To this end we take $x_k \in \Omega_k(\sigma_k, x_{k+1})$ for $k = 0, \dots, n$ with the notation $\sigma_0 := 0$ and $x_{n+1} := s$. It is sufficient to prove that $d(x_0, \theta) \ge \varepsilon$. There are two possibilities.

In the first case $|x_k - x_{k-1}| = 2^{-k-1}\varepsilon$ for $k = 1, \dots, n$. Then

$$|x_0 - s| \le \sum_{k=0}^{n} |x_{k+1} - x_k| \le \sum_{k=0}^{n} 2^{-k-1}\varepsilon = (1 - 2^{-n-1})\varepsilon.$$

Since $d(s, \theta) \ge 2\varepsilon$ we get $d(x_0, \theta) \ge d(s, \theta) - |x_0 - s| \ge \varepsilon$.

In the second case there exists $m \in \{1, \dots, n\}$ such that $|x_k - x_{k-1}| = 2^{-k-1}\varepsilon$ for $k = 1, \dots, m-1$ and $|x_m - \sigma_m| = d(\sigma_m, \theta) - \rho_m\varepsilon$. Hence $|x_0 - x_m| \le (1 - 2^{-m})\varepsilon$ and $d(x_m, \theta) \ge d(\sigma_m, \theta) - |x_m - \sigma_m| \ge \rho_m\varepsilon$, so we conclude that $d(x_0, \theta) \ge d(x_m, \theta) - |x_0 - x_m| \ge \varepsilon$.

Using the same algorithm as in the case of the 0-level expansion, we can estimate $R_{N_n}(t, z)$ as follows

$$|R_{N_n}(t,z)| \le \frac{A_n}{2\sqrt{\pi|t|}} \int_0^\infty e^{-\frac{s}{4|t|}+\tilde{B}s} \frac{s^{N_0-\frac{1}{2}}}{\left(r^2-\varepsilon\right)^{N_0}} \cdot \frac{|s-\sigma_1|^{N_1}}{(d(\sigma_1,\theta)-\rho_1\varepsilon)^{N_1}} \cdot \dots \cdot \\ \cdot \frac{|s-\sigma_n|^{N_n}}{(d(\sigma_n,\theta)-\rho_n\varepsilon)^{N_n}} ds, \tag{12}$$

for a certain constant A_n.

Next, we find points where the integrand on the right-hand side of (12) attains its local maxima. Let us observe that this function has $n+1$ maxima in points $s_1, \dots, s_{n+1}$ such that $s_1 < \sigma_1 < s_2 < \dots < \sigma_n < s_{n+1}$ and all s_j satisfy the condition:

$$\frac{1}{4|t|} - \tilde{B} = \frac{N_0 - \frac{1}{2}}{s_j} + \frac{N_1}{s_j - \sigma_1} + \dots + \frac{N_n}{s_j - \sigma_n}. \tag{13}$$

From (13) we conclude that s_j are decreasing functions of N_n for $1 \le j \le n$ and s_{n+1} increases with respect to N_n. Moreover, the value of the integrand in (12) in the points s_j, $1 \le j \le n$, decreases with respect to N_n. However, it behaves differently in the point s_{n+1}. It decreases with respect to N_n when

$$\frac{s_{n+1} - \sigma_n}{d(\sigma_n, \theta) - \rho_n\varepsilon} < 1,$$

that is for $s_{n+1} \in (\sigma_n, \sigma_n + d(\sigma_n, \theta) - \rho_n \varepsilon)$, and increases when $s_{n+1} \in (\sigma_n + d(\sigma_n, \theta) - \rho_n \varepsilon, +\infty)$. Hence there exist $N_n \in \mathbb{N}$ and $s_{n+1} =: \sigma_{n+1}$ satisfying (13) for which the integrand reaches its minimal value (see [8]).

Again, we can use the Laplace method to obtain the estimation of R_{N_n} (compare [7, 8])

$$|R_{N_n}(t, z)| \sim O\left(\frac{e^{-\eta_n\left(\frac{1}{4|t|} - \tilde{B}\right)}}{\sqrt{1 - 4\tilde{B}|t|}}\right) \quad \text{for} \quad t \to 0, \quad \arg t = \theta.$$

We have the sequence of positive numbers $\eta_0 = \sigma_1 \sim r^2 < \eta_1 < \eta_2 < \eta_3 < \dots$, but it is not clear, whether or not, $\{\eta_n\}_{n \in \mathbb{N}}$ is an unbounded sequence (see [7, 8]).

To find the $(n + 1)$-level hyperasymptotic expansion we expand the function $s \mapsto f_{n+1}(s, z)$ around the point σ_{n+1}. As a result we receive a series

$$f_{n+1}(s, z) = \sum_{j=0}^{N_{n+1}-1} b_{n+1,j}(z)(s - \sigma_{n+1})^j + (s - \sigma_{n+1})^{N_{n+1}} f_{n+2}(s, z), \tag{14}$$

which, after substituting it in (11), gives us the $(n + 1)$-level expansion of the form

$$R_{N_n}(t, z) = \frac{1}{\sqrt{\pi |t|}} \sum_{j=0}^{N_{n+1}-1} b_{n+1,j}(z) \int_0^\infty \frac{e^{-\frac{s}{4|t|}} s^{N_0}}{2\sqrt{s}} (s - \sigma_1)^{N_1} \cdot \ldots \cdot$$
$$\cdot (s - \sigma_n)^{N_n} (s - \sigma_{n+1})^j ds + R_{N_{n+1}}(t, z).$$

Moreover, since

$$s^{N_0}(s - \sigma_1)^{N_1} \cdots (s - \sigma_n)^{N_n} (s - \sigma_{n+1})^j = \sum_{l=0}^{N_0+\cdots+N_n+j} a_{n,j,l} s^l, \tag{15}$$

is a polynomial of degree $N_0 + \cdots + N_n + j$, and by the properties of the gamma function

$$\frac{1}{\sqrt{\pi |t|}} \int_0^\infty \frac{e^{-\frac{s}{4|t|}}}{2\sqrt{s}} s^l \, ds = \frac{(2l)!}{l!} |t|^l = \frac{(2l)!}{l!} e^{-i\theta l} t^l \quad \text{for} \quad l = 0, 1, \dots,$$

we conclude that

$$R_{N_n}(t, z) = \sum_{j=0}^{N_{n+1}-1} b_{n+1,j}(z) \sum_{l=0}^{N_0+\cdots+N_n+j} \frac{(2l)!}{l!} a_{n,j,l} e^{-i\theta l} t^l + R_{N_{n+1}}(t, z).$$

Hence the hyperasymptotic expansion of u takes the form

$$u(t,z)=\sum_{l=0}^{N_0+\cdots+N_n-1}\psi_l(z)t^l+R_{N_n}(t,z), \tag{16}$$

for some functions $\psi_l(z)$ depending on $b_{n+1,j}(z)$ and $a_{n,j,l}$.

2.3 Conclusion

We can formulate the following theorem regarding the hyperasymptotic expansion of (2)

Theorem 1 *For any $n\in\mathbb{N}$ the solution (3) of the heat equation has the hyperasymptotic expansion as $t\to 0$ in the direction $\theta\in[0,2\pi)\setminus\bigcup_{i=1}^{K}(2\lambda_i-\delta,\ 2\lambda_i+\delta)$ mod 2π of the form*

$$u(t,z)=\sum_{j=0}^{N_0-1}\frac{\varphi^{(2j)}(z)}{j!}t^j+\sum_{m=1}^{n}\sum_{j=0}^{N_m-1}\frac{b_{m,j}(z)}{\sqrt{\pi|t|}}\int_0^\infty\frac{e^{-\frac{s}{4|t|}}s^{N_0}}{2\sqrt{s}}(s-\sigma_1)^{N_1}$$
$$\cdots(s-\sigma_{m-1})^{N_{m-1}}(s-\sigma_m)^j\,ds+R_{N_n}(t,z)=\sum_{l=0}^{N_0+\cdots+N_n-1}\psi_l(z)t^l+R_{N_n}(t,z),$$

where the remainder $R_{N_n}(t,z)$ is of the form

$$R_{N_n}(t,\ z)=\frac{1}{2\sqrt{\pi|t|}}\int_0^\infty e^{-\frac{s}{4|t|}}s^{N_0-\frac{1}{2}}(s-\sigma_1)^{N_1}\cdots(s-\sigma_n)^{N_n}f_{n+1}(s,z)\,ds,$$

and for any $m\le n$ and $j<N_m$

$$f_m(s,z)=\frac{1}{(2\pi i)^m}\int_{\Omega_{m-1}(\sigma_{m-1},s)}\int_{\Omega_{m-2}(\sigma_{m-2},x_{m-1})}\cdots\int_{\Omega_1(\sigma_1,x_2)}\int_{\Omega_0(0,x_1)}$$
$$\frac{f_0(x_0,z)\,dx_0\ldots dx_{m-1}}{x_0^{N_0}\prod_{k=1}^{m-1}\left[(x_k-\sigma_k)^{N_k}(x_{k-1}-x_k)\right](x_{m-1}-s)},$$

and

$$b_{m,j}(z)=\frac{1}{j!}\frac{\partial^j}{\partial s^j}f_m(s,z)\left.\right|_{s=\sigma_m}.$$

Moreover,

$$|R_{N_n}(t,z)| \sim O\left(\frac{e^{-\eta_n\left(\frac{1}{4|t|}-\tilde{B}\right)}}{\sqrt{1-4\tilde{B}|t|}}\right) \quad as \quad t \to 0, \ \arg t = \theta, \ z \in D_{\tilde{\varepsilon}},$$

for some sequence of positive numbers $\eta_0 = \sigma_1 \sim r^2 < \eta_1 < \eta_2 < \eta_3 < \ldots$.

3 Generalization to Linear PDEs with Constant Coefficients

In this section we show how to find the hyperasymptotic expansion for solutions of general linear non-Cauchy–Kowalevskaya type PDEs with constant coefficients. The result is based on the theory of summability which allows us to construct the actual solution, which is analytic in some sectorial neighborhood of the origin, from the divergent formal power series solution. Moreover this actual solution has an integral representation in the similar form to (3).

3.1 Summability

First, we define k-summability in a similar way to [6]. For more information about the theory of summability we refer the reader to [1].

We say that a formal power series $\hat{u}(t,z) = \sum_{n=0}^{\infty} \frac{u_n(z)}{n!} t^n$ with $u_n(z) \in \mathcal{O}_{1/\kappa}(D)$ is a *Gevrey series* of order q if there exist $A, B, r > 0$ such that $|u_n(z)| \le AB^n (n!)^{q+1}$ for every $|z| < r$ and every $n \in \mathbb{N}$. We denote by $\mathcal{O}_{1/\kappa}(D)[[t]]_q$ the set of such formal power series.

Moreover, for $k > 0$ and $d \in \mathbb{R}$, we say that $\hat{u}(t,z) \in \mathcal{O}_{1/\kappa}(D)[[t]]_{\frac{1}{k}}$ is *k-summable in a direction d* if its k-Borel transform

$$v(s,z) := (\mathcal{B}_k \hat{u})(s,z) := \sum_{n=0}^{\infty} \frac{u_n(z)}{\Gamma(1+(1+1/k)n)} s^n,$$

where $\Gamma(\cdot)$ denotes the gamma function, is analytically continued with respect to s to an unbounded sector S_d in a direction d and this analytic continuation has exponential growth of order k as s tends to infinity (i.e. $|v(s,z)| \le Ae^{B|s|^k}$ as $s \to \infty$). We denote it briefly by $v(s,z) \in \mathcal{O}^k_{1,1/\kappa}((D \cup S_d) \times D)$. In this case the *$k$-sum of $\hat{u}(t,z)$ in the direction d* is given by

$$u^d(t,z) := (\mathcal{L}_{k,d} v)(t,z) := t^{-k/(1+k)} \int_{e^{id}\mathbb{R}_+} v(s,z)\, C_{(k+1)/k}((s/t)^{\frac{k}{1+k}})\, ds^{\frac{k}{1+k}},$$

where $C_\alpha(\tau)$ is the *Ecalle kernel* defined by

$$C_\alpha(\tau) := \sum_{n=0}^{\infty} \frac{(-\tau)^n}{n!\,\Gamma\left(1 - \frac{n+1}{\alpha}\right)}.$$

3.2 Reduction of Linear PDEs with Constant Coefficients to Simple Pseudodifferential Equations

We consider the Cauchy problem

$$\begin{cases} P(\partial_t, \partial_z)u = 0, \\ \partial_t^j u(0, z) = \varphi_j(z) \in \mathcal{O}_{\mathscr{A}}(\mathbb{C} \setminus H), \end{cases} \tag{17}$$

where $P(\lambda, \zeta) := P_0(\zeta)\lambda^m - \sum_{j=1}^m P_j(\zeta)\lambda^{m-j}$ is a general polynomial of two variables, which is of order m with respect to λ.

First, we show how to use the methods from [4, 5] for the reduction of (17) to simple pseudodifferential equations.

If $P_0(\zeta)$ is not a constant, then a formal solution of (17) is not uniquely determined. To avoid this inconvenience we choose some special solution which is already uniquely determined. To this end we factorize the polynomial $P(\lambda, \zeta)$ as follows

$$P(\lambda, \zeta) = P_0(\zeta)(\lambda - \lambda_1(\zeta))^{m_1} \cdots (\lambda - \lambda_l(\zeta))^{m_l} =: P_0(\zeta)\widetilde{P}(\lambda, \zeta), \tag{18}$$

where $\lambda_1(\zeta), \dots, \lambda_l(\zeta)$ are the roots of the characteristic equation $P(\lambda, \zeta) = 0$ with multiplicity $m_1, \dots, m_l$ $(m_1 + \cdots + m_l = m)$ respectively.

Since $\lambda_\alpha(\zeta)$ are algebraic functions, we may assume that there exist $\kappa \in \mathbb{N}$ and $r_0 < \infty$ such that $\lambda_\alpha(\zeta)$ are holomorphic functions of the variable $\xi = \zeta^{1/\kappa}$ (for $|\zeta| \geq r_0$ and $\alpha = 1, \dots, l$) and, moreover, there exist $\lambda_\alpha \in \mathbb{C} \setminus \{0\}$ and $q_\alpha = \mu_\alpha/\nu_\alpha$ (for some relatively prime numbers $\mu_\alpha \in \mathbb{Z}$ and $\nu_\alpha \in \mathbb{N}$) such that $\lambda_\alpha(\zeta) \sim \lambda_\alpha \zeta^{q_\alpha}$ for $\alpha = 1, \dots, l$ (i.e. $\lim_{\zeta\to\infty} \frac{\lambda_\alpha(\zeta)}{\zeta^{q_\alpha}} = \lambda_\alpha$, λ_α and q_α are called respectively a *leading term* and a *pole order* of $\lambda_\alpha(\zeta)$). Observe that $\nu_\alpha | \kappa$ for $\alpha = 1, \dots, l$.

Following [5, Definition 13] we define the *pseudodifferential operators* $\lambda_\alpha(\partial_z)$ as

$$\lambda_\alpha(\partial_z)\varphi(z) := \frac{1}{2\kappa\pi i} \oint_{|w|=\varepsilon}^{\kappa} \varphi(w) \int_{e^{i\theta} r_0}^{e^{i\theta}\infty} \lambda_\alpha(\zeta)\mathbf{E}_{1/\kappa}(\zeta^{1/\kappa} z^{1/\kappa}) e^{-\zeta w}\, d\zeta\, dw, \tag{19}$$

for every $\varphi \in \mathcal{O}_{1/\kappa}(D_r)$ and $|z| < \varepsilon < r$, where $\mathbf{E}_{1/\kappa}(t) := \sum_{n=0}^{\infty} \frac{t^n}{\Gamma(1+n/\kappa)}$ is the *Mittag–Leffler function* of order $1/\kappa$, $\theta \in (-\arg w - \frac{\pi}{2}, -\arg w + \frac{\pi}{2})$ and $\oint_{|w|=\varepsilon}^{\kappa}$ means that we integrate κ times along the positively oriented circle of radius ε. Here the integration in the inner integral is taken over the ray $\{e^{i\theta} r : r \geq r_0\}$.

Under the above assumption, by a *normalized formal solution* $\hat{u}$ of (17) we mean such solution of (17), which is also a solution of the pseudodifferential equation $\widetilde{P}(\partial_t, \partial_z)\hat{u} = 0$ (see [4, Definition 10]).

Since the principal part of the pseudodifferential operator $\widetilde{P}(\partial_t, \partial_z)$ with respect to ∂_t is given by ∂_t^m, the Cauchy problem (17) has a unique normalized formal power series solution $\hat{u} \in \mathcal{O}(D)[[t]]$.

Next, we reduce the Cauchy problem (17) of a general linear partial differential equation with constant coefficients to a family of the Cauchy problems of simple pseudodifferential equations. Namely we have

Proposition 1 ([5, Theorem 1]) *Let $\hat{u}$ be the normalized formal solution of (17). Then $\hat{u} = \sum_{\alpha=1}^{l} \sum_{\beta=1}^{m_\alpha} \hat{u}_{\alpha\beta}$ with $\hat{u}_{\alpha\beta}$ being a formal solution of a simple pseudodifferential equation*

$$\begin{cases} (\partial_t - \lambda_\alpha(\partial_z))^\beta \hat{u}_{\alpha\beta} = 0, \\ \partial_t^j \hat{u}_{\alpha\beta}(0, z) = 0 \ \ (j = 0, \dots, \beta - 2), \\ \partial_t^{\beta-1} \hat{u}_{\alpha\beta}(0, z) = \lambda_\alpha^{\beta-1}(\partial_z)\varphi_{\alpha\beta}(z), \end{cases} \tag{20}$$

where $\varphi_{\alpha\beta}(z) := \sum_{j=0}^{m-1} d_{\alpha\beta j}(\partial_z)\varphi_j(z) \in \mathcal{O}_{1/\kappa}(D)$ and $d_{\alpha\beta j}(\zeta)$ are some holomorphic functions of the variable $\xi = \zeta^{1/\kappa}$ and of polynomial growth.

Moreover, if q_α is a pole order of $\lambda_\alpha(\zeta)$ and $\overline{q}_\alpha = \max\{0, q_\alpha\}$, then a formal solution $\hat{u}_{\alpha\beta}$ is a Gevrey series of order $\overline{q}_\alpha - 1$ with respect to t.

For this reason we will study the following simple pseudodifferential equation

$$\begin{cases} (\partial_t - \lambda(\partial_z))^\beta u = 0, \\ \partial_t^j u(0, z) = 0 \ (j = 0, \dots, \beta - 2), \\ \partial_t^{\beta-1} u(0, z) = \lambda^{\beta-1}(\partial_z)\varphi(z) \in \mathcal{O}_{1/\kappa}(D), \end{cases} \tag{21}$$

where $\lambda(\zeta) \sim \lambda\zeta^q$ for some $q \in \mathbb{Q}$, $q > 1$. So we assume that $q = \mu/\nu$ for some relatively prime $\mu, \nu \in \mathbb{N}$, $\mu > \nu$.

3.3 Summable Solutions of Simple Pseudodifferential Equations

We have the following representation of summable solutions of (21).

Theorem 2 *Let $k := (q-1)^{-1}$ and $d \in \mathbb{R}$. Suppose that $\hat{u}(t, z)$ is a unique formal power series solution of the Cauchy problem (21) and*

$$\varphi(z) \in \mathcal{O}_{1/\kappa}^{qk}\big(D \cup \bigcup_{l=0}^{q\kappa-1} S_{(d+\arg\lambda+2l\pi)/q}\big). \tag{22}$$

Then $\hat{u}(t,z)$ is k-summable in the direction d and its k-sum is given by

$$u(t,z) = u^d(t,z) = \frac{1}{t^{1/q}} \int_{e^{\frac{id}{q}}\mathbb{R}_+} v(s^q, z) C_q(s/t^{1/q})\, ds, \tag{23}$$

where

$$v(t,z) := \hat{\mathscr{B}}_{1/k}\hat{u}(t,z) = \hat{\mathscr{B}}_{1/k}\left(\sum_{n=0}^{\infty}\frac{u_n(z)}{n!}t^n\right) = \sum_{n=0}^{\infty}\frac{u_n(z)}{\Gamma(1+qn)}t^n \in \mathscr{O}^q_{1,1/\kappa}((D\cup S_d)\times D), \tag{24}$$

has the integral representation

$$v(t,z) = \frac{t^{\beta-1}}{(\beta-1)!}\partial_t^{\beta-1}\frac{1}{2\kappa\pi i}\oint^{\kappa}_{|w|=\varepsilon}\varphi(w)\int_{e^{i\theta}r_0}^{e^{i\theta}\infty}\mathbf{E}_q(t\lambda(\zeta))\mathbf{E}_{1/\kappa}(\zeta^{1/\kappa}z^{1/\kappa})e^{-\zeta w}\,d\zeta\,dw, \tag{25}$$

with $\theta \in (-\arg w - \frac{\pi}{2}, -\arg w + \frac{\pi}{2})$. Moreover, if $\varphi \in \mathscr{O}_{\mathscr{A}}(\mathbb{C}\setminus H)$ and $z \in D_{\tilde{\varepsilon}}$ for some $\tilde{\varepsilon} > 0$ then the function $t \mapsto v(t,z)$ is holomorphic for $|t| < \frac{(r-\tilde{\varepsilon})^q}{|\lambda|}$, where $r := \min_{1\le i\le K}|a_{i1}|$.

Proof First, observe that by Proposition 1 we get $\hat{u}(t,z) \in \mathscr{O}_{1/\kappa}(D)[[t]]_{q-1}$. Moreover, by [5, Proposition 7] the function $v(t,z) = \hat{\mathscr{B}}_{1/k}\hat{u}(t,z) \in \mathscr{O}_{1,1/\kappa}(D^2)$ satisfies the moment partial differential equation

$$\begin{cases} (\partial_{t,\Gamma_q} - \lambda(\partial_z))^{\beta} v = 0, \\ \partial^j_{t,\Gamma_q} v(0,z) = 0 \ (j = 0,\ldots,\beta-2), \\ \partial^{\beta-1}_{t,\Gamma_q} v(0,z) = \lambda^{\beta-1}(\partial_z)\varphi(z) \in \mathscr{O}_{1/\kappa}(D), \end{cases} \tag{26}$$

where Γ_q is a moment function defined by $\Gamma_q(n) := \Gamma(1+nq)$ for $n \in \mathbb{N}_0$ and ∂_{t,Γ_q} is so called *Γ_q-moment differential operator* defined by (see [5, Definition 12])

$$\partial_{t,\Gamma_q}\Big(\sum_{n=0}^{\infty}\frac{a_n(z)}{\Gamma_q(n)}t^n\Big) := \sum_{n=0}^{\infty}\frac{a_{n+1}(z)}{\Gamma_q(n)}t^n.$$

Hence by [5, Lemma 3] with $m_1(n) = \Gamma_q(n)$ and $m_2(n) = \Gamma(1+n)$ we get the integral representation (25) of $v(t,z)$.

Since $\varphi(z)$ satisfies (22), by [5, Lemma 4] we conclude that $v(t,z) \in \mathscr{O}^q_{1,1/\kappa}((D\cup S_d)\times D)$. So, the function $u^d(t,z) := \mathscr{L}_{k,d}v(t,z)$ is well-defined and is given by (23).

Since the Mittag–Leffler function is the entire function satisfying the estimation $|\mathbf{E}_q(z)| \le Ce^{|z|^{1/q}}$ (see [1, Appendix B.4]), the integrand in the inner integral in (25) is estimated for $|z| < \widetilde{\varepsilon}$ by

$$|\mathbf{E}_q(t\lambda(\zeta))\mathbf{E}_{1/\kappa}(\zeta^{1/\kappa}z^{1/\kappa})e^{-\zeta w}| \le \tilde{C}e^{|\zeta|(|\lambda|^{1/q}|t|^{1/q}-|w|+\widetilde{\varepsilon})},$$

as $\zeta \to \infty, \arg\zeta = \theta = -\arg w$. By the hypothesis $\varphi(w)$ is holomorphic for $|w| < r$, so we may deform the path of integration in the outer integral in (25) from $|w| = \varepsilon$ to $|w| = \tilde{r}$ for any $\tilde{r} < r$. It means that the inner integral in (25) is convergent for any t satisfying $|t| < \frac{(r-\widetilde{\varepsilon})^q}{|\lambda|}$ and the function $t \mapsto v(t,z)$ is holomorphic for such t.

3.4 *Hyperasymptotic Expansion of Solution of Simple Pseudodifferential Equations*

Using the change of variables to (23), as in the case of the heat equation we obtain

$$u^{\theta}(t,z) = \frac{1}{qt^{1/q}} \int_0^{e^{i\theta}\infty} \frac{1}{s^{1-\frac{1}{q}}} v(s,z) C_q((s/t)^{1/q})\, ds,$$

so as $t \to 0$, $\arg t = \theta$ we conclude that

$$u^{\theta}(t,z) = \frac{1}{q|t|^{1/q}} \int_0^{\infty} \frac{1}{s^{1-\frac{1}{q}}} v(se^{i\theta},z) C_q((s/|t|)^{1/q})\, ds,$$

for any θ different from the Stokes lines, i.e. $\theta \neq q\lambda_i - \arg\lambda \mod 2\pi$ for $i = 1,\dots,K$.

Now we are ready to repeat the construction of the hyperasymptotic expansion for the heat equation under condition that $\varphi \in \mathcal{O}^{kq}_{\mathcal{A}}(\mathbb{C}\setminus H)$ (i.e. $\varphi \in \mathcal{O}_{\mathcal{A}}(\mathbb{C}\setminus H)$ and $\varphi(z)$ has the exponential growth of order kq as $z \to \infty$, $z \in \mathbb{C}\setminus H$). We also assume that the direction θ is separated from the Stokes lines, i.e. that

$$\theta \in [0,2\pi) \setminus \bigcup_{i=1}^{K} (q\lambda_i - \arg\lambda - \delta, q\lambda_i - \arg\lambda + \delta) \mod 2\pi \quad \text{for fixed } \delta > 0.$$

We put $f_0(s,z) := v(se^{i\theta},z)$, $r := \min_{1\le i\le K} |a_{i1}| - \widetilde{\varepsilon}$ and

$$\Omega_0(0,s) := \partial\Big(\{w \in \mathbb{C}\colon |w| \le \frac{r^q}{|\lambda|} - \varepsilon\} \cup \{w \in \mathbb{C}\colon |w-s| \le \frac{\varepsilon}{2}\}\Big),$$

for some $\varepsilon \in \Big(0, \frac{r^q}{2|\lambda|}\Big)$.

Observe that by Theorem 2 for any $z \in D_{\widetilde{\varepsilon}}$ the function $w \mapsto f_0(w,z)$ is holomorphic in the domain bounded by $\Omega_0(0,s)$. By [4, Lemma 2]

$$u(t,z) = \sum_{j=\beta-1}^{N_0-1} \binom{j}{\beta-1} \frac{\lambda^j(\partial_z)\varphi(z)}{j!} t^j + R_{N_0}(t,z).$$

Moreover, as in the case of the heat equation

$$R_{N_0}(t,z) = \frac{1}{q|t|^{1/q}} \int_0^{\infty} s^{N_0-1+\frac{1}{q}} C_q((s/|t|)^{1/q}) f_1(s,z)\,ds,$$

where $f_1(s,z)$ is defined as in (7).

By (24) there exist positive constants A' and B' such that

$$|f_0(w,z)| \le A' e^{B'|s|^k} \quad \text{for any} \quad w \in \Omega_0(0,s).$$

Hence

$$\begin{aligned} |f_1(s,z)| &\le \frac{1}{2\pi} \int_{\Omega_0(0,s)} \frac{2A' e^{B'|s|^k}}{\varepsilon(\frac{r^q}{|\lambda|}-\varepsilon)^{N_0}}\, d|w| \le \frac{2A' e^{B'|s|^k}}{\varepsilon(\frac{r^q}{|\lambda|}-\varepsilon)^{N_0}} (\frac{r^q}{|\lambda|} - \varepsilon + \frac{\varepsilon}{2}) \\ &\le \frac{2A' r^q e^{B'|s|^k}}{|\lambda|\varepsilon(\frac{r^q}{|\lambda|}-\varepsilon)^{N_0}} = \frac{A_0' e^{B'|s|^k}}{(\frac{r^q}{|\lambda|}-\varepsilon)^{N_0}}, \end{aligned}$$

where $A_0' := \frac{2A'r^q}{|\lambda|\varepsilon}$.

Moreover, by the properties of the Ecalle kernel (see [3, Lemma 6]) we may estimate

$$|C_q(\tau)| \le C e^{-(\tau^{k+1}/c_q)} \quad \text{with} \quad c_q = (k+1)^{k+1} k^{-k}.$$

So

$$|R_{N_0}(t,z)| \le \frac{A_0'}{q|t|^{1/q}} \int_0^{\infty} s^{N_0-1+\frac{1}{q}} e^{-s^k(\frac{1}{c_q|t|^k}-B')} \left(\frac{r^q}{|\lambda|} - \varepsilon\right)^{-N_0} ds. \qquad (27)$$

Similarly to the heat equation case we conclude that the integrand of (27) has a maximum at certain point $s = \sigma_1$ satisfying $N_0 = k\sigma_1^k(\frac{1}{c_q|t|^k} - B') + 1 - \frac{1}{q}$. Now, the minimum with respect to σ_1 is given at $\sigma_1 = \frac{r^q}{|\lambda|} - \varepsilon$. Hence we take $N_0 := \lfloor k(\frac{r^q}{|\lambda|} - \varepsilon)^k(\frac{1}{c_q|t|^k} - B') + 1 - \frac{1}{q} \rfloor$ and $\sigma_1 := \left(\frac{N_0+\frac{1}{q}-1}{k(\frac{1}{c_q|t|^k}-B')}\right)^{1/k}$. Observe that $\sigma_1 \le \frac{r^q}{|\lambda|} - \varepsilon$. So we are able to use the Laplace method and to conclude that

$$|R_{N_0}(t,z)| \sim O\left(\frac{e^{-\sigma_1^k(\frac{1}{c_q|t|^k}-B')}}{|t|^{1/q}\sqrt{\frac{1}{c_q|t|^k} - B'}}\right) \quad \text{for} \quad t \to 0,\ \arg t = \theta,\ z \in D_{\tilde{\varepsilon}}.$$

Next, we construct the n-level hyperasymptotic expansion as for the heat equation. The remainder obtained in the n-level hyperasymptotic expansion is of the form

$$R_{N_n}(t,z) = \frac{1}{q|t|^{1/q}} \int_0^\infty C_q\Big(\big(\frac{s}{|t|}\big)^{\frac{1}{q}}\Big) s^{N_0-1+\frac{1}{q}} (s-\sigma_1)^{N_1} \cdots (s-\sigma_n)^{N_n} f_{n+1}(s,z)\, ds, \tag{28}$$

where

$$f_{n+1}(s,z) := \frac{1}{2\pi i} \int_{\Omega_n(\sigma_n, s)} \frac{f_n(w,z)}{(w-\sigma_n)^{N_n}(w-s)}\, dw.$$

Here we take

$$\Omega_n(\sigma_n, s) := \partial\Big(\{w \in \mathbb{C} \colon |w-\sigma_n| \le d(\sigma_n,\theta) - \rho_n\varepsilon\} \cup \{w \in \mathbb{C} \colon |w-s| \le 2^{-n-1}\varepsilon\}\Big),$$

with $d(\sigma_n,\theta) := \inf_{z \in D_{\tilde{\varepsilon}}} \inf_{\zeta \in H} |\sigma_n - e^{-i\theta}\lambda(z-\zeta)^q|$ and $\rho_n := 2 - 2^{-n}$.

Using the same algorithm as in the case of the heat equation, we can estimate $R_{N_n}(t,z)$ as follows

$$|R_{N_n}(t,z)| \le \frac{A'_n}{q|t|^{1/q}} \int_0^\infty e^{-s^k(\frac{1}{c_q|t|^k} - B')} \frac{s^{N_0-1+\frac{1}{q}}}{\left(\frac{r^q}{|\lambda|} - \varepsilon\right)^{N_0}} \cdot \frac{|s-\sigma_1|^{N_1}}{(d(\sigma_1,\theta)-\rho_1\varepsilon)^{N_1}} \cdot \ldots \frac{|s-\sigma_n|^{N_n}}{(d(\sigma_n,\theta)-\rho_n\varepsilon)^{N_n}}\, ds, \tag{29}$$

for a certain constant A'_n.

Let us observe that the integrand on the right-hand side of (28) has $n+1$ maxima in points $s_1, \ldots, s_{n+1}$ such that $s_1 < \sigma_1 < s_2 < \ldots < \sigma_n < s_{n+1}$ and all s_j satisfy the condition:

$$k s_j^{k-1}\left(\frac{1}{c_q|t|^k} - B'\right) = \frac{N_0 - 1 + \frac{1}{q}}{s_j} + \frac{N_1}{s_j - \sigma_1} + \ldots + \frac{N_n}{s_j - \sigma_n}.$$

From this, as in the case of the heat equation, we conclude that s_j are decreasing functions of N_n for $1 \le j \le n$ and s_{n+1} increases to infinity as $N_n \to \infty$. Similarly, the value of the integrand in (29) in the points s_j, $1 \le j \le n$, decreases with respect to N_n. Moreover, this value in the point s_{n+1} decreases with respect to N_n for $s_{n+1} < \sigma_n + d(\sigma_n,\theta) - \rho_n\varepsilon$ and increases when $s_{n+1} > \sigma_n + d(\sigma_n,\theta) - \rho_n\varepsilon$.

Hence, as in the case of the heat equation there exists $N_n \in \mathbb{N}$ and s_{n+1} for which the integrand reaches its minimal value. We denote such s_{n+1} by σ_{n+1}.

Again, using the Laplace method we obtain the estimation of R_{N_n}

$$|R_{N_n}(t,z)| \sim O\left(\frac{e^{-\tilde{\eta}_n^k(\frac{1}{c_q|t|^k} - B')}}{|t|^{1/q}\sqrt{\frac{1}{c_q|t|^k} - B'}}\right) \quad \text{for} \quad t \to 0,\ \arg t = \theta,\ z \in D_{\tilde{\varepsilon}},$$

where, as previously $\tilde{\eta}_0 = \sigma_1 \sim \frac{r^q}{|\lambda|} < \tilde{\eta}_1 < \tilde{\eta}_2 < \tilde{\eta}_3 < \dots$ is some increasing sequence of positive numbers (see [7, 8]).

To find the $(n+1)$-level hyperasymptotic expansion we expand the function $s \mapsto f_{n+1}(s, z)$ around the point σ_{n+1} as in (14), which, after substituting it in (28), gives us the $(n+1)$-level expansion

$$R_{N_n}(t, z) = \frac{1}{q|t|^{1/q}} \sum_{j=0}^{N_{n+1}-1} b_{n+1,j}(z) \int_0^\infty C_q((\frac{s}{|t|})^{\frac{1}{q}}) s^{N_0-1+\frac{1}{q}} (s-\sigma_1)^{N_1} \\ \cdots (s-\sigma_n)^{N_n} (s-\sigma_{n+1})^j \, ds + R_{N_{n+1}}(t, z).$$

Since the Laplace transform $\mathscr{L}_{k,d}$ is inverse to k-Borel transform $\hat{\mathscr{B}}_k$, we conclude that $\mathscr{L}_{k,d}(t^l) = \frac{\Gamma(1+ql)}{l!} t^l$ for $l = 0, 1, \dots$. It means that

$$\frac{1}{q|t|^{1/q}} \int_0^\infty \frac{C_q\left(\left(\frac{s}{|t|}\right)^{\frac{1}{q}}\right)}{s^{1-\frac{1}{q}}} s^l \, ds = \frac{\Gamma(1+ql)}{l!} |t|^l = \frac{\Gamma(1+ql)}{l!} e^{-i\theta l} t^l,$$

and using (15) we get

$$R_{N_n}(t, z) = \sum_{j=0}^{N_{n+1}-1} b_{n+1,j}(z) \sum_{l=0}^{N_0+\cdots+N_n+j} a_{n,j,l} \frac{\Gamma(1+ql)}{l!} e^{-i\theta l} t^l + R_{N_{n+1}}(t, z).$$

Hence, as in the case of the heat equation, we conclude that the hyperasymptotic expansion of u takes also the form (16) for some functions $\psi_l(z)$.

Finally, similarly to the heat equation, we get as the conclusion

Theorem 3 (Hyperasymptotic expansion for the simple equation) *For every $n \in \mathbb{N}$ the solution of the Eq. (21) with $\varphi \in \mathscr{O}^{kq}_{\mathscr{A}}(\mathbb{C} \setminus H)$ has the hyperasymptotic expansion as t tends to zero in a direction $\theta \in [0, 2\pi) \setminus \bigcup_{i=1}^K (q\lambda_i - \arg\lambda - \delta, q\lambda_i - \arg\lambda + \delta) \mod 2\pi$, which has the form*

$$u^\theta(t, z) = \sum_{j=\beta-1}^{N_0-1} \binom{j}{\beta-1} \frac{\lambda^j(\partial_z)\varphi(z)}{j!} t^j + \sum_{m=1}^{n} \sum_{j=0}^{N_m-1} \frac{b_{m,j}(z)}{|t|^{1/q}} \int_0^\infty \frac{1}{q s^{1-\frac{1}{q}}} \\ \cdot C_q((s/|t|)^{1/q}) s^{N_0} (s-\sigma_1)^{N_1} \cdots (s-\sigma_{m-1})^{N_{m-1}} (s-\sigma_m)^j \, ds + R_{N_n}(t, z) \\ = \sum_{l=0}^{N_0+\cdots+N_n-1} \psi_l(z) t^l + R_{N_n}(t, z),$$

where

$$b_{m,j}(z) = \frac{1}{j!}\frac{\partial^j}{\partial s^j} f_m(s,z)|_{s=\sigma_m},$$

$$R_{N_n}(t,z) = \frac{1}{|t|^{1/q}}\int_0^\infty \frac{1}{qs^{1-\frac{1}{q}}} C_q((\frac{s}{|t|})^{\frac{1}{q}}) s^{N_0}(s-\sigma_1)^{N_1}\cdots(s-\sigma_n)^{N_n} f_{n+1}(s,z)\,ds,$$

$$f_m(s,z) = \frac{1}{(2\pi i)^m}\int_{\Omega_{m-1}(\sigma_{m-1},s)}\int_{\Omega_{m-2}(\sigma_{m-2},x_{m-1})}\cdots\int_{\Omega_1(\sigma_1,x_2)}\int_{\Omega_0(0,x_1)} \frac{v(x_0e^{i\theta},z)\,dx_0\ldots dx_{m-1}}{x_0^{N_0}\big[\prod_{k=1}^{m-1}(x_k-\sigma_k)^{N_k}(x_{k-1}-x_k)\big](x_{m-1}-s)},$$

and $v(s,z)$ *is defined by (25).*

Moreover $R_{N_n}(t,z) \sim O\Bigg(\frac{e^{-\tilde{\eta}_n^k(\frac{1}{c_q|t|^k}-B')}}{|t|^{1/q}\sqrt{\frac{1}{c_q|t|^k}-B'}}\Bigg)$ *as* $t\to 0$, $\arg t = \theta$, $z \in D_{\tilde{\varepsilon}}$ *for some sequence of positive numbers* $\tilde{\eta}_0 = \sigma_1 \sim \frac{r^q}{|\lambda|} < \tilde{\eta}_1 < \tilde{\eta}_2 < \tilde{\eta}_3 < \ldots$.

Acknowledgements The authors would like to thank the anonymous referee for valuable comments, suggestions, and especially for indication of the form of hyperasymptotic expansion of the solution $u(t,z)$ presented in (16).

References

1. Balser, W.: Formal Power Series and Linear Systems of Meromorphic Ordinary Differential Equations. Springer, New York (2000)
2. Berry, M.V., Howls, C.J.: Hyperasymptotics. Proc. R. Soc. Lond. Ser. A **430**, 653–668 (1990)
3. Martinet, J., Ramis, J.-P.: Elementary acceleration and multisummability. I. Ann. de l'I. H. P. Sect. A **54**, 331–401 (1991)
4. Michalik, S.: Analytic solutions of moment partial differential equations with constant coefficients. Funkcial. Ekvac. **56**, 19–50 (2013)
5. Michalik, S.: Summability of formal solutions of linear partial differential equations with divergent initial data. J. Math. Anal. Appl. **406**, 243–260 (2013)
6. Michalik, S., Podhajecka, B.: The Stokes phenomenon for certain partial differential equations with meromorphic initial data. Asymptot. Anal. **99**, 163–182 (2016)
7. Olde Daalhuis, A.B.: Hyperasymptotic expansions of confluent hypergeometric functions. IMA J. Appl. Math. **49**, 203–216 (1992)
8. Olde Daalhuis, A.B.: Hyperasymptotics and the Stokes' phenomenon. Proc. R. Soc. Edinb. Sect. A **123**, 731–743 (1993)
9. Olde Daalhuis, A.B.: Hyperasymptotic solutions of higher order linear differential equations with a singularity of rank one. Proc. R. Soc. Lond. Ser. A **454**, 1–29 (1998)
10. Olde Daalhuis, A.B.: Hyperasymptotics for nonlinear ODEs I. A Riccati equation. Proc. R. Soc. Lond. Ser. A **461**, 2503–2520 (2005)
11. Olde Daalhuis, A.B.: Hyperasymptotics for nonlinear ODEs II. The first Painlevé equation and a second-order Riccati equation. Proc. R. Soc. Lond. Ser. A **461**, 3005–3021 (2005)
12. Olver, F.W.J.: Asymptotics and Special Functions. Academic, New York (1974)

The Stokes Phenomenon for Certain PDEs in a Case When Initial Data Have a Finite Set of Singular Points

Bożena Tkacz

Abstract We study the Stokes phenomenon via hyperfunctions for the solutions of the 1-dimensional complex heat equation under the condition that the Cauchy data are holomorphic on $\mathbb{C}$ but a finitely many singular or branching points with the appropriate growth condition at the infinity. The main tool are the theory of summability and the theory of hyperfunctions, which allows us to describe jumps across Stokes lines.

Keywords Stokes phenomenon · Hyperfunctions

MSC Primary 35C10 · Secondary 35C20, 35K05, 35E15, 40G10

1 Introduction

This paper deals with the 1-dimensional complex heat equation $\partial_t u(t,z) = \partial_z^2 u(t,z)$, $u(0,z) = \varphi(z)$. The aim of this work is to describe jumps across the Stokes lines in terms of hyperfunctions in the case when the initial data $\varphi(z)$ have a finite set of singular points. First, we consider the function $\varphi(z)$ which has a single-valued singular point and we derive the jump in a form of convergent series (see Theorem 1). Then we discuss the case when the function $\varphi(z)$ has a multi-valued singular point and we give the integral representation of the jump (see Theorem 2). Thus we obtain a full characterization of the Stokes phenomenon for the considered equation. At the end, we extend our results to the generalization of the heat equation.

The important point to note here is that D.A. Lutz, M. Miyake and R. Schäfke in [7] considered the similar problem for the heat equation when the Cauchy data is a function $\varphi(z) = 1/z$ with singularity at 0. They proved that the heat kernel was given by a function as a jump of Borel sum (see [[7], Theorem 5.1]).

B. Tkacz (✉)
Faculty of Mathematics and Natural Sciences, College of Science,
Cardinal Stefan Wyszyński University, Wóycickiego 1/3, 01-938 Warszawa, Poland
e-mail: bpodhajecka@o2.pl

G. Filipuk et al. (eds.), *Formal and Analytic Solutions of Diff. Equations*,
Springer Proceedings in Mathematics & Statistics 256,
https://doi.org/10.1007/978-3-319-99148-1_5

It is worth pointing out that this work is a continuation of the paper [9] in which we study the heat equation with the Cauchy data given by a meromorphic function with a simple pole or finitely many poles.

2 Notation. Gevrey's Asymptotics and k-Summability

In the paper we use the following notation.

A set of the form

(1)

$$S = S_d(\alpha, R) = \{z \in \tilde{\mathbb{C}} \colon\ z = re^{i\phi},\ r \in (0, R),\ \phi \in (d - \alpha/2, d + \alpha/2)\}.$$

defines a sector S in a direction $d \in \mathbb{R}$ with an opening $\alpha > 0$ and a radius $R \in \mathbb{R}_+$ in the universal covering space $\tilde{\mathbb{C}}$ of $\mathbb{C} \setminus \{0\}$,

(2)

$$D_r = \{z \in \mathbb{C} : |z| < r\}.$$

defines a complex disc D_r in $\mathbb{C}$ with a radius $r > 0$.

In the case that

1. $R = +\infty$, then this sector is called unbounded and one can write $S = S_d(\alpha)$ for short,
2. the opening α is not essential, then the sector $S_d(\alpha)$ is denoted briefly by S_d,
3. the radius r is not essential, the set D_r will be designate by D.

To simplify the notation, we abbreviate a set $S_d(\alpha) \cup D$ (resp. $S_d \cup D$) to $\widehat{S}_d(\alpha)$ (resp. $\widehat{S}_d$).

If f is a holomorphic function on a domain $G \subset \mathbb{C}^n$, then it will be written as $f \in \mathcal{O}(G)$.

The set of all formal power series (i.e. a power series $\sum_{n=0}^{\infty} a_n t^n$ created for a sequence of complex numbers $(a_n)_{n=0}^{\infty}$) will be represented by the symbol $\mathbb{C}[[t]]$. Similarly, $\mathcal{O}(D_r)[[t]]$ stands for the set of all formal power series $\sum_{n=0}^{\infty} a_n(z) t^n$ with $a_n(z) \in \mathcal{O}(D_r)$ for all $n \in \mathbb{N}_0$.

Definition 1 Assume that $k > 0$ and $f \in \mathcal{O}(S)$. The function f is called of *exponential growth of order at most* k, if for every proper subsector $S^* \prec S$ (i.e. $\overline{S^*} \setminus \{0\} \subseteq S$) there exist constants $C_1, C_2 > 0$ such that $|f(x)| \leq C_1 e^{C_2|x|^k}$ for every $x \in S^*$.

If the function f is of exponential growth of order at most k, then one can write $f \in \mathcal{O}^k(S)$.

Definition 2 A power series $\sum_{n=0}^{\infty} a_n t^n \in \mathbb{C}[[t]]$ is called a *formal power series of Gevrey order* s ($s \in \mathbb{R}$), if there exist positive constants $A, B > 0$ such that $|a_n| \leq$

$AB^n(n!)^s$ for every $n \in \mathbb{N}_0$. The set of all such formal power series is denoted by $\mathbb{C}[[t]]_s$ (resp. $\mathcal{O}(D_r)[[t]]_s$).

Remark 1 (*see* [1]) If $k < 0$ then $u \in \mathbb{C}[[t]]_k \iff u$ is convergent and $u \in \mathcal{O}^{-\frac{1}{k}}(\mathbb{C})$.

Definition 3 Assume that $s \in \mathbb{R}$, S is a given sector in $\tilde{\mathbb{C}}$ and $f \in \mathcal{O}(S)$. A power series $\hat{f}(t) = \sum_{n=0}^{\infty} a_n t^n \in \mathbb{C}[[t]]_s$ is called *Gevrey's asymptotic expansion of order* s of the function f in S (in symbols $f(t) \sim_s \hat{f}(t)$ in S) if for every $S^* \prec S$ there exist positive constants $A, B > 0$ such that for every $N \in \mathbb{N}_0$ and every $t \in S^*$

$$|f(t) - \sum_{n=0}^{N} a_n t^n| \le AB^N (N!)^s |t|^{N+1}.$$

To introduce the notion of summability, by Balser's theory of general moment summability ([1, Sect. 6.5], in particular [1, Theorem 38]), we may take Ecalle's acceleration and deceleration operators instead of the standard Laplace and Borel transform.

Definition 4 (see [1, Sect. 11.1]) Let $d \in \mathbb{R}$, $\tilde{k} > \bar{k} > 0$ and $k := (1/\bar{k} - 1/\tilde{k})^{-1}$.

The *acceleration operator in a direction d with indices* $\tilde{k}$ *and* $\bar{k}$, denoted by $\mathcal{A}_{\tilde{k},\bar{k},d}$, is defined for every $g(t) \in \mathcal{O}^k(\widehat{S}_d)$ by

$$(\mathcal{A}_{\tilde{k},\bar{k},d}\, g)(t) := t^{-\tilde{k}} \int_{e^{id}\mathbb{R}_+} g(s) C_{\tilde{k}/\bar{k}}\big((s/t)^{\tilde{k}}\big)\, ds^{\tilde{k}},$$

where the *Ecalle kernel* C_α is defined by

$$C_\alpha(\tau) := \sum_{n=0}^{\infty} \frac{(-\tau)^n}{n!\,\Gamma\big(1 - \frac{n+1}{\alpha}\big)} \quad \text{for} \quad \alpha > 1 \tag{1}$$

and the integration is taken over the ray $e^{id}\mathbb{R}_+ := \{re^{id} : r \ge 0\}$.

The *formal deceleration operator with indices* $\tilde{k}$ *and* $\bar{k}$, denoted by $\hat{\mathcal{D}}_{\tilde{k},\bar{k}}$, is defined for every $\hat{f}(t) = \sum_{n=0}^{\infty} a_n t^n \in \mathbb{C}[[t]]$ by

$$(\hat{\mathcal{D}}_{\tilde{k},\bar{k}} \hat{f})(t) := \sum_{n=0}^{\infty} a_n t^n \frac{\Gamma(1+n/\tilde{k})}{\Gamma(1+n/\bar{k})}.$$

Definition 5 Let $k > 0$ and $d \in \mathbb{R}$. A formal power series $\hat{f}(t) = \sum_{n=0}^{\infty} a_n t^n \in \mathbb{C}[[t]]$ is called *k-summable in a direction d* if

$$g(t) = (\hat{\mathcal{D}}_{1,\frac{k}{k+1}} \hat{f})(t) = \sum_{n=0}^{\infty} a_n \frac{\Gamma(1+n)}{\Gamma(1+\frac{n(k+1)}{k})} t^n \in \mathcal{O}^k(\widehat{S}_d(\varepsilon)) \quad \text{for some} \quad \varepsilon > 0.$$

Moreover, the *k-sum of* $\hat{f}(t)$ *in the direction d* is given by

$$f^d(t) = \mathcal{S}_{k,d}\hat{f}(t) := (\mathcal{A}_{1,\frac{k}{k+1},\theta}\hat{\mathcal{D}}_{1,\frac{k}{k+1}}\hat{f})(t) \quad \text{with} \quad \theta \in (d - \varepsilon/2, d + \varepsilon/2). \quad (2)$$

Definition 6 If $\hat{f}$ is k-summable in all directions d but (after identification modulo 2π) finitely many directions $d_1, \dots, d_n$ then $\hat{f}$ is called *k-summable* and $d_1, \dots, d_n$ are called *singular directions of* $\hat{f}$.

3 The Stokes Phenomenon and Hyperfunctions

3.1 The Stokes Phenomenon for k-Summable Formal Power Series

Now let us recall the concept of the Stokes phenomenon [9, Definition 7].

Definition 7 Assume that $\hat{f} \in \mathbb{C}[[t]]_{1/k}$ (resp. $\hat{u} \in \mathcal{O}(D)[[t]]_{1/k}$) is k-summable with finitely many singular directions $d_1, d_2, \dots, d_n$. Then for every $l = 1, \dots, n$ a set $\mathcal{L}_{d_l} = \{t \in \tilde{\mathbb{C}} \colon \arg t = d_l\}$ is called a *Stokes line for* $\hat{f}$ *(resp.* $\hat{u}$*)*. Of course every such Stokes line $\mathcal{L}_{d_l}$ for $\hat{f}$ (resp. $\hat{u}$) determines so called *anti-Stokes lines* $\mathcal{L}_{d_l \pm \frac{\pi}{2k}}$ for $\hat{f}$ *(resp.* $\hat{u}$*)*.

Moreover, if d_l^+ (resp. d_l^-) denotes a direction close to d_l and greater (resp. less) than d_l, and let $f^{d_l^+} = \mathcal{S}_{k,d_l^+}\hat{f}$ (resp. $f^{d_l^-} = \mathcal{S}_{k,d_l^-}\hat{f}$) then the difference $f^{d_l^+} - f^{d_l^-}$ is called a *jump* for $\hat{f}$ across the Stokes line $\mathcal{L}_{d_l}$. Analogously we define the jump for $\hat{u}$.

Remark 2 Let $r(t) := f^{d_l^+}(t) - f^{d_l^-}(t)$ for all $t \in S = S_{d_l}(\frac{\pi}{k})$. Then $r(t) \sim_{1/k} 0$ on S.

3.2 Laplace Type Hyperfunctions

We will describe jumps across the Stokes lines in terms of hyperfunctions. The similar approach to the Stokes phenomenon one can find in [3, 8, 10]. For more information about the theory of hyperfunctions we refer the reader to [5].

We will consider the space

$$\mathcal{H}^k(\mathcal{L}_d) := \mathcal{O}^k(D \cup (S_d \setminus \mathcal{L}_d)) \Big/ \mathcal{O}^k(\widehat{S}_d)$$

of Laplace type hyperfunctions supported by $\mathcal{L}_d$ with exponential growth of order k. It means that every hyperfunction $G \in \mathcal{H}^k(\mathcal{L}_d)$ may be written as

$$G(s) = [g(s)]_d = \{g(s) + h(s) \colon h(s) \in \mathcal{O}^k(\widehat{S}_d)\}$$

for some defining function $g(s) \in \mathcal{O}^k(D \cup (S_d \setminus \mathcal{L}_d))$.

By the Köthe type Theorem [6] one can treat the hyperfunction $G = [g(s)]_d$ as the analytic functional defined by

$$G(s)[\varphi(s)] := \int_{\gamma_d} g(s)\varphi(s)\, ds \quad \text{for sufficiently small} \quad \varphi \in \mathcal{O}^{-k}(\widehat{S_d}) \tag{3}$$

with γ_d being a path consisting of the half-lines from $e^{id^-}\infty$ to 0 and from 0 to $e^{id^+}\infty$, i.e. $\gamma_d = -\gamma_{d^-} + \gamma_{d^+}$ with $\gamma_{d^\pm} = \mathcal{L}_{d^\pm}$.

3.3 The Description of Jumps Across the Stokes Lines in Terms of Hyperfunctions

Assume that $\hat{f}$ is k-summable and d is a singular direction. By (2) the jump for $\hat{f}$ across the Stokes line $\mathcal{L}_d$ is given by

$$f^{d^+}(t) - f^{d^-}(t) = (\mathcal{A}_{1,\frac{k}{k+1},d^+} - \mathcal{A}_{1,\frac{k}{k+1},d^-})\hat{\mathcal{D}}_{1,\frac{k}{k+1}}\hat{f}(t).$$

We will describe this jump in terms of hyperfunctions. To this end, observe that we can treat $g(t) := \hat{\mathcal{D}}_{1,\frac{k}{k+1}}\hat{f}(t) \in \mathcal{O}^k(D \cup (S_d \setminus \mathcal{L}_d))$ as a defining function of the hyperfunction $G(s) := [g(s)]_d \in \mathcal{H}^k(\mathcal{L}_d)$.

So, for sufficiently small $r > 0$ and $t \in S_d(\frac{\pi}{k}, r)$ this jump is given as the Ecalle acceleration operator $\mathcal{A}_{1,\frac{k}{k+1},d}$ acting on the hyperfunction $G(s)$. Precisely, we have

$$\begin{aligned} f^{d^+}(t) - f^{d^-}(t) &= (\mathcal{A}_{1,\frac{k}{k+1},d}G)(t) := G(s)\Big[t^{\frac{-k}{1+k}}C_{\frac{k+1}{k}}((s/t)^{\frac{k}{1+k}})\frac{k}{1+k}s^{-\frac{1}{1+k}}\Big] \\ &= G(s^{\frac{1+k}{k}})\Big[t^{-k/(1+k)}C_{\frac{k+1}{k}}(s/t^{\frac{k}{1+k}})\Big], \end{aligned}$$

where $G(s)[\varphi(s)]$ is defined by (3), and the last equality holds by the change of variables, because if $G(s) = [g(s)]_d$ then $G(s^p) = [g(s^p)]_{d/p}$ for every $p > 0$.

4 Characterization of the Stokes Phenomenon in a Case When the Initial Data Have a Finite Set of Singular Points

In this section we specify a form of the jumps across the Stokes lines based on the solution of the heat equation in a case when the initial data have a finite set of singular points. Due to the linearity of the equation, it is enough to consider the case that the singularity occurs only at one point – singular or branching point.

Recall the following proposition

Proposition 1 *([9, Theorem 4]) Suppose that $\hat{u}$ is a unique formal solution of the Cauchy problem of the heat equation*

$$\begin{cases} \partial_t u = \partial_z^2 u \\ u(0,z) = \varphi(z) \end{cases} \tag{4}$$

with

$$\varphi \in \mathcal{O}^2\left(D \cup S_{\frac{d}{2}}\left(\frac{\varepsilon}{2}\right) \cup S_{\frac{d}{2}+\pi}\left(\frac{\varepsilon}{2}\right)\right) \quad \textit{for some} \quad \varepsilon > 0. \tag{5}$$

Then $\hat{u}$ is 1-summable in the direction d and for every $\theta \in (d - \frac{\varepsilon}{2}, d + \frac{\varepsilon}{2})$ and for every $\tilde{\varepsilon} \in (0, \varepsilon)$ there exists $r > 0$ such that its 1-sum $u^\theta \in \mathcal{O}(S_\theta(\pi - \tilde{\varepsilon}, r) \times D)$ is represented by

$$u(t,z) = u^\theta(t,z) = \frac{1}{\sqrt{4\pi t}} \int_0^{e^{i\frac{\theta}{2}}\infty} \big(\varphi(z+s) + \varphi(z-s)\big)\, e^{\frac{-s^2}{4t}}\, ds \tag{6}$$

for $t \in S_\theta(\pi - \tilde{\varepsilon}, r)$ and $z \in D_r$.

Now consider the heat equation (4) with $\varphi(z) \in \mathcal{O}^2(\widetilde{\mathbb{C} \setminus \{z_0\}})$ for some $z_0 \in \mathbb{C} \setminus \{0\}$. First, observe that in this case $\mathcal{L}_\delta$ with $\delta := 2\theta := 2\arg z_0$ is a Stokes line for $\hat{u}$.

For every sufficiently small $\varepsilon > 0$ there exists $r > 0$ such that for every fixed $z \in D_r$ the jump is given by

$$u^{\delta^+}(t,z) - u^{\delta^-}(t,z) = F_z(s)\left[\frac{1}{\sqrt{4\pi t}} e^{-\frac{s^2}{4t}}\right],$$

where $t \in S_\delta(\pi - \varepsilon, r)$ and

$$\begin{aligned} F_z(s) &= \Big[\varphi(s+z) + \varphi(z-s)\Big]_{\theta_z} \\ &= \Big[\varphi(s+z)\Big]_{\theta_z} \in \mathcal{O}^2\big(D \cup (S_\theta(\alpha) \setminus \mathcal{L}_{\theta_z})\big) / \mathcal{O}^2\big(D \cup S_\theta(\alpha)\big) \end{aligned}$$

and $\theta_z = \arg(z_0 - z)$.

Remark 3 In the remainder of this section we assume that $t \in S_\delta(\pi - \varepsilon, r)$ and fixed $z \in D$ ($\varepsilon, r > 0$).

Now we consider the case when z_0 is a single-valued singular point of the function $\varphi(z) \in \mathcal{O}^2(\mathbb{C} \setminus \{z_0\})$.

Theorem 1 *Suppose that* $\varphi(z) = \sum_{n=1}^{\infty} \frac{a_n}{(z-z_0)^n} + \phi(z)$, *where* $a_1, a_2, \ldots \in \mathbb{C}$ *and* $\lim_{n\to\infty} \sqrt[n]{|a_n|} < 1$, $z_0 \in \mathbb{C} \setminus \{0\}$, $\phi(z) \in \mathcal{O}^2(\mathbb{C})$. *Then*

$$F_z(s) = \left[\sum_{n=1}^{\infty} \frac{a_n}{(z+s-z_0)^n}\right]_{\theta_z} = -2\pi i \sum_{n=1}^{\infty} \frac{a_n(-1)^{n-1}}{(n-1)!}\delta^{(n-1)}(z+s-z_0),$$

where δ *is the Dirac function and* $\delta^{(n-1)}$ *denotes its* $(n-1)$*-th derivative.*

Moreover, the jump is given by the convergent series

$$u^{\delta^+}(t,z) - u^{\delta^-}(t,z) = -i\sqrt{\frac{\pi}{t}} \sum_{n=1}^{\infty} \frac{a_n(-1)^{n-1}}{(n-1)!} \frac{\mathrm{d}^{n-1}}{\mathrm{d}s^{n-1}} e^{-\frac{s^2}{4t}}\bigg|_{s=z_0-z}.$$

Proof Observe that since $\delta(x) = \left[-\frac{1}{2\pi i s}\right]$ (see [5]), then $\delta(x-a) = \left[-\frac{1}{2\pi i(s-a)}\right]$ (where $a \in \mathbb{R}$) and differentiating it n-times one can easily obtain

$$\delta^{(n)}(x-a) = \left[-\frac{(-1)^n n!}{2\pi i(s-a)^{n+1}}\right] \Longrightarrow -\frac{2\pi i\, \delta^{(n-1)}(x-a)}{(-1)^{n-1}(n-1)!} = \left[\frac{1}{(s-a)^n}\right].$$

Notice that the same holds for $a = z_0 - z \in \mathbb{C}$.

Hence we derive

$$F_z(s) = \left[\sum_{n=1}^{\infty} \frac{a_n}{(s+z-z_0)^n}\right]_{\theta_z} = -2\pi i \sum_{n=1}^{\infty} \frac{a_n(-1)^{n-1}}{(n-1)!}\delta^{(n-1)}(s+z-z_0).$$

Thus

$$\begin{aligned} u^{\delta^+}(t,z) - u^{\delta^-}(t,z) &= F_z(s)\left[\frac{1}{\sqrt{4\pi t}} e^{-\frac{s^2}{4t}}\right] \\ &= -2\pi i \sum_{n=1}^{\infty} \frac{a_n(-1)^{n-1}}{(n-1)!}\delta^{(n-1)}(s+z-z_0)\left[\frac{1}{\sqrt{4\pi t}} e^{-\frac{s^2}{4t}}\right] \\ &= -i\sqrt{\frac{\pi}{t}} \sum_{n=1}^{\infty} \frac{a_n(-1)^{n-1}}{(n-1)!} \frac{\mathrm{d}^{n-1}}{\mathrm{d}s^{n-1}} e^{-\frac{s^2}{4t}}\bigg|_{s=z_0-z}. \end{aligned}$$

It remains to prove the convergence of the series above.

Notice that by Remark 1 since $s \mapsto e^{-\frac{s^2}{4t}} \in \mathcal{O}^2(\mathbb{C})$, then there exist $\tilde{A}, \tilde{B} > 0$ such that for every $t \in S(\theta, \pi - \tilde{\varepsilon}, r)$ and $z \in D_r$ (for every sufficiently small $\varepsilon > 0$ and $\tilde{\varepsilon} \in (0, \varepsilon)$) we have that

$$\left|\delta^{(n-1)}(z+s-z_0)\left[e^{-\frac{s^2}{4t}}\right]\right| = \left|\frac{\mathrm{d}^{n-1}}{\mathrm{d}s^{n-1}} e^{-\frac{s^2}{4t}}\Big|_{s=z_0-z}\right| \le \tilde{A}\tilde{B}^{(n-1)}((n-1)!)^{\frac{1}{2}},$$

so

$$\begin{aligned}\left|u^{\delta^+}(t,z)-u^{\delta^-}(t,z)\right| &= \left|-i\sqrt{\frac{\pi}{t}}\sum_{n=1}^{\infty}\frac{a_n(-1)^{(n-1)}}{(n-1)!}\delta^{(n-1)}(z+s-z_0)\left[e^{-\frac{s^2}{4t}}\right]\right| \\ &\le \sqrt{\frac{\pi}{t}}\sum_{n=1}^{\infty}\frac{|a_n|}{(n-1)!}\left|\delta^{(n-1)}(z+s-z_0)\left[e^{-\frac{s^2}{4t}}\right]\right| \\ &\le \sqrt{\frac{\pi}{t}}\sum_{n=1}^{\infty}\frac{|a_n|}{(n-1)!}\tilde{A}\tilde{B}^{(n-1)}((n-1)!)^{\frac{1}{2}} = \sqrt{\frac{\pi}{t}}\tilde{A}\sum_{n=1}^{\infty}\frac{|a_n|\tilde{B}^{(n-1)}}{((n-1)!)^{\frac{1}{2}}} < \infty,\end{aligned}$$

because $\lim_{n\to\infty}\sqrt[n]{|a_n|} < 1$. Thus this implies the convergence of $u^{\delta^+}(t,z) - u^{\delta^-}(t,z)$. □

In particular, from the above theorem we obtain the following examples.

Example 1 Assume now $\varphi(z) = \sum_{n=1}^{N}\frac{a_n}{(z-z_0)^n} + \phi(z)$, for some $z_0 \in \mathbb{C}\setminus\{0\}$, where $N \in \mathbb{N}\setminus\{0\}$, $a_1, a_2, \ldots a_N \in \mathbb{C}$ and $\phi(z) \in \mathcal{O}^2(\mathbb{C})$. Then

$$F_z(s) = \left[\sum_{n=1}^{N}\frac{a_n}{(z+s-z_0)^n}\right]_{\theta_z} = -2\pi i\sum_{n=1}^{N}\frac{a_n(-1)^{n-1}}{(n-1)!}\delta^{(n-1)}(z+s-z_0),$$

and the jump is given by

$$u^{\delta^+}(t,z) - u^{\delta^-}(t,z) = -i\sqrt{\frac{\pi}{t}}\sum_{n=1}^{N}\frac{a_n(-1)^{n-1}}{(n-1)!}\frac{\mathrm{d}^{n-1}}{\mathrm{d}s^{n-1}}e^{-\frac{s^2}{4t}}\bigg|_{s=z_0-z}.$$

Example 2 Let $\varphi(z) = e^{\frac{1}{z-z_0}}$ for some $z_0 \in \mathbb{C}\setminus\{0\}$. Then

$$F_z(s) = \left[e^{\frac{1}{z+s-z_0}}\right]_{\theta_z} = -2\pi i\sum_{k=0}^{\infty}\frac{(-1)^k}{k!(k+1)!}\delta^{(k)}(z+s-z_0),$$

and the jump is given by

$$u^{\delta^+}(t,z) - u^{\delta^-}(t,z) = -i\sqrt{\frac{\pi}{t}}\sum_{k=0}^{\infty}\frac{(-1)^k}{k!(k+1)!}\frac{\mathrm{d}^k}{\mathrm{d}s^k}e^{-\frac{s^2}{4t}}\bigg|_{s=z_0-z}.$$

Let us now consider the general case. For this purpose fix $z \in D$. For $s \in \mathcal{L}_{\theta_z}$ define (similarly as in [3, 10]) a function on $\mathcal{L}_{\theta_z}$ by

$$\operatorname{var} F_z(s) = \begin{cases} 0 & \text{, if } |s| < |z_0 - z| \\ \varphi\big(z_0 + (s+z-z_0)e^{2\pi i}\big) - \varphi(s+z) & \text{, if } |s| > |z_0 - z|, \end{cases}$$

and a Heaviside function in a direction θ_z by

$$H_{\theta_z}(xe^{i\theta_z}) = \begin{cases} 1 \text{ , for } x > 0 \\ 0 \text{ , for } x < 0, \end{cases}$$

thus $F_z(s) = \Big[\varphi(s+z)\Big]_{\theta_z} = -\operatorname{var} F_z(s) = -\operatorname{var} F_z(s) H_{\theta_z}(s+z-z_0)$.

So $u^{\delta^+}(t,z) - u^{\delta^-}(t,z) = -\operatorname{var} F_z(s)\Big[\frac{1}{\sqrt{4\pi t}}\, e^{-\frac{s^2}{4t}}\Big]$ where, in general, $-\operatorname{var} F_z(s)$ is an analytic functional on $\mathcal{L}_{\theta_z}$.

Notation. The set of all measurable functions $f : \mathcal{L}_{\theta_z} \to \mathbb{C}$ such that $\int_K |f| dx < \infty$ for all compact sets $K \subset \mathcal{L}_{\theta_z}$ will be denoted by $L^1_{\text{loc}}(\mathcal{L}_{\theta_z})$.

Theorem 2 *Under the above assumptions we have several cases to discuss*

(1) $\operatorname{var} F_z(s) \in L^1_{\text{loc}}(\mathcal{L}_{\theta_z})$ *and is an analytic function of exponential growth of order at most 2 for* $|s| > |z_0 - z|$.

Then for every sufficiently small $\varepsilon > 0$ *there exists* $r > 0$ *such that the jump is given by*

$$u^{\delta^+}(t,z) - u^{\delta^-}(t,z) = -\int_{z_0-z}^{e^{i\theta_z}\infty} \operatorname{var} F_z(s)\, \frac{1}{\sqrt{4\pi t}}\, e^{-\frac{s^2}{4t}}\, ds,$$

for $(t,z) \in S_\delta(\pi - \varepsilon, r) \times D$.

(2) $\operatorname{var} F_z(s)$ *is a distribution on* $\mathcal{L}_{\theta_z}$ *and is an analytic function of exponential growth of order at most 2 for* $|s| > |z_0 - z|$.

Then there exist $m \in \mathbb{N}$ *and* $\operatorname{var} \tilde{F}_z(s)$ *satisfying the assumptions of the case (1) such that*

$$\frac{\mathrm{d}^m}{\mathrm{d}s^m} \operatorname{var} \tilde{F}_z(s) = \operatorname{var} F_z(s).$$

Moreover, for every sufficiently small $\varepsilon > 0$ *there exists* $r > 0$ *such that the jump is given by*

$$\begin{aligned} u^{\delta^+}(t,z) - u^{\delta^-}(t,z) &= -\operatorname{var} F_z(s)\Big[\frac{1}{\sqrt{4\pi t}} e^{-\frac{s^2}{4t}}\Big] \\ &= -\frac{\mathrm{d}^m}{\mathrm{d}s^m} \operatorname{var} \tilde{F}(s)\Big[\frac{1}{\sqrt{4\pi t}} e^{-\frac{s^2}{4t}}\Big] = -\operatorname{var} \tilde{F}(s)\Big[(-1)^m \frac{\mathrm{d}^m}{\mathrm{d}s^m}\Big(\frac{1}{\sqrt{4\pi t}} e^{-\frac{s^2}{4t}}\Big)\Big] \\ &= -\int_{z_0-z}^{e^{i\theta_z}\infty} \operatorname{var} \tilde{F}_z(s)\, (-1)^m \frac{\mathrm{d}^m}{\mathrm{d}s^m}\Big(\frac{1}{\sqrt{4\pi t}}\, e^{-\frac{s^2}{4t}}\Big)\, ds, \end{aligned}$$

for $(t,z) \in S_\delta(\pi - \varepsilon, r) \times D$.

(3) $\operatorname{var} F_z(s)$ *is an analytic functional on* $\mathcal{L}_{\theta_z}$.
Then $\operatorname{var} F_z(s) = \sum_{n=0}^{\infty} \operatorname{var} F_{z,n}(s)$, *where* $\operatorname{var} F_{z,n}(s)$ *satisfy the assumptions of the case (2). So for every* $n \in \mathbb{N}$ *there exists* $k_n \in \mathbb{N}$ *and* $\operatorname{var} \tilde{F}_{z,n}(s)$ *satisfying the assumptions of the case (1) such that*

$$\operatorname{var} F_{z,n}(s) = \frac{\mathrm{d}^{k_n}}{\mathrm{d}s^{k_n}} \operatorname{var} \tilde{F}_{z,n}(s).$$

Moreover, for every sufficiently small $\varepsilon > 0$ *there exists* $r > 0$ *such that the jump is given by*

$$\begin{aligned} u^{\delta^+}(t,z) - u^{\delta^-}(t,z) &= -\operatorname{var} F_z(s)\left[\frac{1}{\sqrt{4\pi t}} e^{-\frac{s^2}{4t}}\right] = -\sum_{n=0}^{\infty} \operatorname{var} F_{z,n}(s)\left[\frac{1}{\sqrt{4\pi t}} e^{-\frac{s^2}{4t}}\right] \\ &= -\sum_{n=0}^{\infty} \int_{z_0-z}^{e^{i\theta_z}\infty} \operatorname{var} \tilde{F}_{z,n}(s)\,(-1)^{k_n} \frac{\mathrm{d}^{k_n}}{\mathrm{d}s^{k_n}}\left(\frac{1}{\sqrt{4\pi t}}\, e^{-\frac{s^2}{4t}}\right) ds, \end{aligned}$$

for $(t,z) \in S_\delta(\pi - \varepsilon, r) \times D$.

Proof Ad.(1) First observe that for every $z \in D$ the function $s \mapsto \operatorname{var} F_z(s)$ is analytic on $\mathcal{L}_{\theta_z} \setminus \{z_0 - z\}$, locally integrable and has an exponential growth of order at most 2 as $s \to \infty$, $s \in \mathcal{L}_{\theta_z}$. Hence for every sufficiently small $\varepsilon > 0$ there exists $r > 0$ such that the integral $u^{\delta^+}(t,z) - u^{\delta^-}(t,z)$ is well defined for $(t,z) \in S_\delta(\pi - \varepsilon, r) \times D$.

For $z_0 - z = x_0 e^{i\theta_z}$ and $s = x e^{i\theta_z}$, where $x_0, x > 0$, we obtain

$$\begin{aligned} u^{\delta^+}(t,z) - u^{\delta^-}(t,z) &= F_z(s)\left[\frac{1}{\sqrt{4\pi t}} e^{-\frac{s^2}{4t}}\right] \\ &= \frac{1}{\sqrt{4\pi t}} \lim_{\varepsilon \to 0^+} \left\{ \int_0^\infty \varphi\big((x+i\varepsilon)e^{i\theta_z} + z_0 - x_0 e^{i\theta_z}\big) e^{-\frac{1}{4t}\left((x+i\varepsilon)e^{i\theta_z}\right)^2} e^{i\theta_z} dx \right. \\ &\quad \left. - \int_0^\infty \varphi\big((x-i\varepsilon)e^{i\theta_z} + z_0 - x_0 e^{i\theta_z}\big) e^{-\frac{1}{4t}\left((x-i\varepsilon)e^{i\theta_z}\right)^2} e^{i\theta_z} dx \right\} \\ &= \frac{1}{\sqrt{4\pi t}} \int_0^\infty e^{-\frac{1}{4t}(xe^{i\theta_z})^2} e^{i\theta_z} \lim_{\varepsilon \to 0^+} \left\{ \varphi\big((x+i\varepsilon-x_0)e^{i\theta_z} + z_0\big) \right. \\ &\quad \left. - \varphi\big((x-i\varepsilon-x_0)e^{i\theta_z} + z_0\big) \right\} dx = (*) \end{aligned}$$

Observe that

- for $x - x_0 > 0$, we have

$$\begin{aligned} &\lim_{\varepsilon \to 0^+} \left\{ \varphi\big((x+i\varepsilon-x_0)e^{i\theta_z} + z_0\big) - \varphi\big((x-i\varepsilon-x_0)e^{i\theta_z} + z_0\big) \right\} \\ &\quad = \varphi\big((x-x_0)e^{i\theta_z} + z_0\big) - \varphi\big((x-x_0)e^{i\theta_z} e^{2\pi i} + z_0\big) \end{aligned}$$

- for $x - x_0 < 0$, we have

$$\lim_{\varepsilon \longrightarrow 0^+} \left\{ \varphi\big((x + i\varepsilon - x_0)e^{i\theta_z} + z_0\big) - \varphi\big((x - i\varepsilon - x_0)e^{i\theta_z} + z_0\big) \right\} = 0.$$

Hence

$$(*) = \frac{1}{\sqrt{4\pi t}} \int_{x_0}^{\infty} \left\{ \varphi\big((x - x_0)e^{i\theta_z} + z_0\big) - \varphi\big((x - x_0)e^{i\theta_z}e^{2\pi i} + z_0\big) \right\} e^{-\frac{(xe^{i\theta_z})^2}{4t}} e^{i\theta_z} dx$$

$$= \frac{1}{\sqrt{4\pi t}} \int_{z_0 - z}^{e^{i\theta_z}\infty} \Big(\varphi(s + z) - \varphi\big((s + z - z_0)e^{2\pi i} + z_0\big) \Big) e^{-\frac{s^2}{4t}} ds$$

$$= - \int_{z_0 - z}^{e^{i\theta_z}\infty} \operatorname{var} F_z(s) \, \frac{1}{\sqrt{4\pi t}} \, e^{-\frac{s^2}{4t}} \, ds.$$

Ad.(2) Observe that since $\operatorname{var} F_z(s)$ is continuous on $\mathcal{L}_{\theta_z} \setminus \{z_0 - z\}$, by the locally structure theorem for distributions (see Proposition 7.1 [2]), there exist $m \in \mathbb{N}$ and $\operatorname{var} \tilde{F}_z(s) \in L^1_{\text{loc}}(\mathcal{L}_{\theta_z})$ such that

$$\frac{\mathrm{d}^m}{\mathrm{d}s^m} \operatorname{var} \tilde{F}_z(s) = \operatorname{var} F_z(s).$$

Furthermore, $\operatorname{var} F_z(s)$ has exponential growth of order at most 2 as $s \to \infty$, $s \in \mathcal{L}_{\theta_z}$, then also $\operatorname{var} \tilde{F}_z(s)$ has an exponential growth of order at most 2 as $s \to \infty$, $s \in \mathcal{L}_{\theta_z}$. The rest of the proof is analogous to the proof of the case (1).

Ad.(3) Notice that since $\operatorname{var} F_{z,n}(s)$ obey the assumptions of the case (2) and based on results in [4] we can write $\operatorname{var} F_z(s) = \sum_{n=0}^{\infty} \operatorname{var} F_{z,n}(s)$, where $\operatorname{var} F_{z,n}(s)$ satisfy the assumptions of the case (2). Then for every $n \in \mathbb{N}$ there exists $k_n \in \mathbb{N}$ and $\operatorname{var} \tilde{F}_{z,n}(s)$ satisfying the assumptions of the case (1) such that $\operatorname{var} F_{z,n}(s) = \frac{\mathrm{d}^{k_n}}{\mathrm{d}s^{k_n}} \operatorname{var} \tilde{F}_{z,n}(s)$. The rest of the proof is also similar to the proof of the case (1). □

Now we give two examples of the function $\varphi(z)$ satisfying the case (1) of Theorem 2.

Example 3 Assume that $\varphi(z) = \ln(z - z_0)$ for some $z_0 \in \mathbb{C} \setminus \{0\}$. Then

$$\operatorname{var} F_z(s) = 2\pi i H_{\theta_z}(s + z - z_0),$$

and the jump is given by

$$u^{\delta^+}(t, z) - u^{\delta^-}(t, z) = -i\sqrt{\frac{\pi}{t}} \int_{z_0 - z}^{e^{i\theta_z}\infty} e^{-\frac{s^2}{4t}} ds.$$

Indeed, for $|s| > |z_0 - z|$ we derive

$$\begin{aligned}\operatorname{var} F_z(s) &= \varphi\big(z_0 + (s+z-z_0)e^{2\pi i}\big) - \varphi(s+z) = \\ &= \ln\Big(z_0 + (s+z-z_0)e^{2\pi i} - z_0 \Big) - \ln\Big((s+z) - z_0 \Big) = \\ &= \ln\big((s+z-z_0)e^{2\pi i}\big) - \ln(s+z-z_0) = 2\pi i.\end{aligned}$$

Example 4 Let $\varphi(z) = (z - z_0)^\lambda$ for some $z_0 \in \mathbb{C} \setminus \{0\}$, $\lambda \notin \mathbb{Z}$ and $\lambda > -1$. Then

$$\operatorname{var} F_z(s) = 2i H_{\theta_z}(z + s - z_0)(-s - z + z_0)^\lambda \sin(\lambda\pi),$$

and the jump is given by

$$u^{\delta^+}(t, z) - u^{\delta^-}(t, z) = -\frac{i}{\sqrt{\pi t}} \int_{z_0 - z}^{e^{i\theta_z}\infty} e^{-\frac{s^2}{4t}} (-s - z + z_0)^\lambda \sin(\lambda\pi) ds.$$

More precisely, for $|s| > |z_0 - z|$

$$\begin{aligned}\operatorname{var} F_z(s) &= \varphi\big(z_0 + (z+s-z_0)e^{2\pi i}\big) - \varphi(z+s) \\ &= \Big(\big(z_0 + (z+s-z_0)e^{2\pi i}\big) - z_0\Big)^\lambda - \Big((z+s) - z_0\Big)^\lambda \\ &= \big((z+s-z_0)e^{2\pi i}\big)^\lambda - (z+s-z_0)^\lambda = (z+s-z_0)^\lambda (e^{2\pi i\lambda} - 1) \\ &= 2i(-1)^\lambda (z+s-z_0)^\lambda \sin(\pi\lambda),\end{aligned}$$

because

$$\begin{aligned}\sin(\pi\lambda) = \frac{e^{i\pi\lambda} - e^{-i\pi\lambda}}{2i} &= \frac{e^{2i\pi\lambda} - 1}{2i e^{i\pi\lambda}} \\ &\Longrightarrow e^{2i\pi\lambda} - 1 = 2i e^{i\pi\lambda} \sin(\pi\lambda) = 2i(-1)^\lambda \sin(\pi\lambda).\end{aligned}$$

Now we present an example of the function $\varphi(z)$ satisfying the case (2) of Theorem 2.

Example 5 Let again $\varphi(z) = (z - z_0)^\lambda$ for some $z_0 \in \mathbb{C} \setminus \{0\}$, $\lambda \notin \mathbb{Z}$ and $\lambda < -1$. Then for $m = \lfloor -\lambda \rfloor$ we can define $\operatorname{var} \tilde{F}_z(s) \in L^1_{\text{loc}}(\mathcal{L}_{\theta_z})$ by

$$\operatorname{var} \tilde{F}_z(s) = \frac{2i(-1)^{\lambda+m} \sin(\pi(\lambda+m))}{(\lambda+1)(\lambda+2)\dots(\lambda+m)} (s + z - z_0)^{\lambda+m} H_{\theta_z}(s + z - z_0),$$

thus

$$\begin{aligned}\operatorname{var} F_z(s) &= \frac{\mathrm{d}^m}{\mathrm{d}s^m} \operatorname{var} \tilde{F}_z(s) \\ &= \frac{\mathrm{d}^m}{\mathrm{d}s^m} \left\{ \frac{2i(-1)^{\lambda+m} \sin(\pi(\lambda+m))}{(\lambda+1)(\lambda+2)\dots(\lambda+m)} (s+z-z_0)^{\lambda+m} H_{\theta_z}(s+z-z_0) \right\},\end{aligned}$$

and the jump is given by

$$u^{\delta^+}(t,z) - u^{\delta^-}(t,z) = \frac{-i}{\sqrt{\pi t}} \int_{z_0-z}^{e^{i\theta_z}\infty} \frac{(-1)^{\lambda}(s+z-z_0)^{\lambda+m}\sin((\lambda+m)\pi)}{(\lambda+1)(\lambda+2)\dots(\lambda+m)} \frac{\mathrm{d}^m}{\mathrm{d}s^m}\left(e^{-\frac{s^2}{4t}}\right)ds.$$

Finally, we give an example of the function $\varphi(z)$ that satisfies the case (3) of Theorem 2.

Example 6 Assume now that $\varphi(z) = e^{\frac{1}{(z-z_0)^{\lambda}}}$ where $z_0 \in \mathbb{C}\setminus\{0\}$, $\lambda \notin \mathbb{Q}$ and $\lambda > 0$. Then for $k_n = \lfloor \lambda n \rfloor$ we can define functions $\operatorname{var}\tilde{F}_{z,n}(s) \in L^1_{\mathrm{loc}}(\mathcal{L}_{\theta_z})$ by

$$\operatorname{var}\tilde{F}_{z,n}(s) = \frac{2i(-s-z+z_0)^{-\lambda n+k_n}\sin((-\lambda n+k_n)\pi)}{n!(-\lambda n+1)(-\lambda n+2)\dots(-\lambda n+k_n)} H_{\theta_z}(s+z-z_0),$$

and

$$\begin{aligned}\operatorname{var}F_z(s) &= \sum_{n=0}^{\infty}\operatorname{var}F_{z,n}(s) = \sum_{n=0}^{\infty}\frac{\mathrm{d}^{k_n}}{\mathrm{d}s^{k_n}}\operatorname{var}\tilde{F}_{z,n}(s)\\ &= \left(\sum_{n=0}^{\infty}\frac{\mathrm{d}^{k_n}}{\mathrm{d}s^{k_n}}\frac{2i(-s-z+z_0)^{-\lambda n+k_n}\sin((-\lambda n+k_n)\pi)}{n!(-\lambda n+1)(-\lambda n+2)\dots(-\lambda n+k_n)}\right) H_{\theta_z}(s+z-z_0).\end{aligned}$$

Then the jump is given by

$$u^{\delta^+}(t,z) - u^{\delta^-}(t,z) = -\int_{z_0-z}^{e^{i\theta_z}\infty}\sum_{n=0}^{\infty}\frac{i(-s-z+z_0)^{-\lambda n+k_n}\sin((-\lambda n+k_n)\pi)}{\sqrt{\pi t}n!(-\lambda n+1)(-\lambda n+2)\dots(-\lambda n+k_n)}\left[(-1)^{k_n}\frac{\mathrm{d}^{k_n}}{\mathrm{d}s^{k_n}}\left(e^{-\frac{s^2}{4t}}\right)\right]ds.$$

Observe that by Remark 1 since $s \mapsto e^{-\frac{s^2}{4t}} \in \mathcal{O}^2(\mathbb{C})$, then there exist $A, B > 0$ such that for every $t \in S(\theta, \pi-\tilde{\varepsilon}, r)$ and $z \in D_r$ (for every sufficiently small $\varepsilon > 0$ and $\tilde{\varepsilon} \in (0,\varepsilon)$) we have that
$\left|(-1)^{k_n}\frac{\mathrm{d}^{k_n}}{\mathrm{d}s^{k_n}}\left(e^{-\frac{s^2}{4t}}\right)\right| \le AB^{\lambda n}(n!)^{\frac{\lambda}{2}}$ and

$$\left|\frac{1}{n!(-\lambda n+1)(-\lambda n+2)\dots(-\lambda n+k_n)}\right| \le \frac{1}{(n!)^{\frac{\lambda}{2}+1}} < \infty$$

hence analogously to the proof of Theorem 1 we obtain the convergence of the above series.

At the end of this section, we similarly derive jumps for the following generalization of the heat equation

$$\begin{cases} \partial_t^p u(t,z) = \partial_z^q u(t,z), \ p,q \in \mathbb{N}, \ 1 \le p < q \\ u(0,z) = \varphi(z) \in \mathcal{O}(D) \\ \partial_t^j u(0,z) = 0 \text{ for } j = 1,2,\ldots,p-1, \end{cases} \tag{7}$$

with $\varphi(z) \in \mathcal{O}^{\frac{q}{q-p}}\Big(D \cup \bigcup_{l=0}^{q-1} S_{\frac{dp}{q}+\frac{2\pi l}{q}}(\frac{\varepsilon p}{q})\Big)$ for some $\varepsilon > 0$.

Then a unique formal solution $\hat{u}(t,z)$ of this Cauchy problem is $\frac{p}{q-p}$-summable in the direction d and for every $\psi \in (d - \frac{\varepsilon}{2}, d + \frac{\varepsilon}{2})$ and for every $\tilde{\varepsilon} \in (0,\varepsilon)$ there exists $r > 0$ such that its $\frac{p}{q-p}$-sum $u \in \mathcal{O}(S_d(\frac{\pi(q-p)}{p} - \tilde{\varepsilon}, r) \times D)$ is given by (see [9, Theorem 6])

$$u(t,z) = u^{\psi}(t,z) = \frac{1}{q\sqrt[q]{t^p}} \int_0^{e^{\frac{i\psi p}{q}}\infty} \big(\varphi(z+s) + \cdots + \varphi(z + e^{\frac{2(q-1)\pi i}{q}} s)\big) C_{\frac{q}{p}}(\frac{s}{\sqrt[q]{t^p}}) ds. \tag{8}$$

As in the case of the heat equation (4), we assume that $\varphi(z) \in \mathcal{O}^{\frac{q}{q-p}}(\widetilde{\mathbb{C} \setminus \{z_0\}})$.

Then $\mathcal{L}_\delta$ with $\delta := q\theta/p := q \arg z_0/p$ is a separate Stokes line for $\hat{u}$, such that $\delta_z = q \arg(z_0 - z)/p$ for every sufficiently small z.

For every sufficiently small $\varepsilon > 0$ there exists $r > 0$ such that for every fixed $z \in D_r$ the jump is given by

$$\begin{aligned} u^{\delta^+}(t,z) - u^{\delta^-}(t,z) &= F_z(s)\Big[\frac{1}{q\sqrt[q]{t^p}} C_{\frac{q}{p}}(s/\sqrt[q]{t^p})\Big] \\ &= \Big[\varphi(z+s) + \cdots + \varphi(z + e^{\frac{2(q-1)\pi i}{q}} s)\Big]_{\theta_z} \Big[\frac{1}{q\sqrt[q]{t^p}} C_{\frac{q}{p}}(s/\sqrt[q]{t^p})\Big] \\ &= \Big[\varphi(z+s)\Big]_{\theta_z} \Big[\frac{1}{q\sqrt[q]{t^p}} C_{\frac{q}{p}}(s/\sqrt[q]{t^p})\Big], \end{aligned}$$

(the last equality arising from the fact that in this case all singular points appear in the function $\varphi(z+s)$), where the hyperfunction $F_z(s) = \Big[\varphi(z+s) + \cdots + \varphi(z + e^{\frac{2(q-1)\pi i}{q}} s)\Big]_{\theta_z}$ belongs to the space $\mathcal{O}^{\frac{q}{q-p}}\big(D \cup (S_\theta(\alpha) \setminus \mathcal{L}_{\theta_z})\big) / \mathcal{O}^{\frac{q}{q-p}}\big(D \cup S_\theta(\alpha)\big)$ with $\theta_z = \arg(z_0 - z)$.

Thus, we obtain analogous results as for the heat equation (4), only that in Theroems 1 and 2 we replace $\frac{1}{\sqrt{4\pi t}} e^{-\frac{s^2}{4t}}$ by $\frac{1}{q\sqrt[q]{t^p}} C_{\frac{q}{p}}(s/\sqrt[q]{t^p})$.

Acknowledgements The author would like to thank the anonymous referee for valuable comments and suggestions.

References

1. Balser, W.: Formal Power Series and Linear Systems of Meromorphic Ordinary Differential Equations. Springer, New York (2000)
2. El Kinani, A., Oudadess, M.: Distributions Theory and Applications. World Scientific, Singapore (2010)
3. Immink, G.: Multisummability and the Stokes phenomenon. J. Dyn. Control Syst. **1**, 483–534 (1995)
4. Kaneko, A.: On the structure of hyperfunctions with compact supports. Proc. Jpn. Acad. **II**, 956–959 (1971)
5. Kaneko, A.: Mathematics and its applications. In: Introduction to Hyperfunctions, vol. 3. Kluwer, Dordrecht (1988)
6. Köthe, G.: Dualität in der Funktionentheorie. J. Reine Angew. Math. **191**, 30–49 (1953)
7. Lutz, D., Miyake, M., Schäfke, R.: On the Borel summability of divergent solutions of the heat equation. Nagoya Math. J. **154**, 1–29 (1999)
8. Malek, S.: On the Stokes phenomenon for holomorphic solutions of integro-differential equations with irregular singularity. J. Dyn. Control Syst. **14**, 371–408 (2008)
9. Michalik, S., Podhajecka, B.: The Stokes phenomenon for certain partial differential equations with meromorphic initial data. Asymptot. Anal. **99**, 163–182 (2016)
10. Sternin, B.Y., Shatalov, V.E.: Borel-Laplace Transform and Asymptotic Theory. CRC Press, Boca Raton (1995)

Soliton Resolution for the Focusing Integrable Discrete Nonlinear Schrödinger Equation

Hideshi Yamane

Abstract We study the long-time asymptotics for the focusing integrable discrete nonlinear Schrödinger equation. The soliton resolution conjecture holds true for this equation: under generic assumptions, the solution is a sum of 1-solitons up to a small error term. The phase shifts of solitons are described in detail by using the terminology of inverse scattering. In $|n| < 2t$, they are determined by the eigenvalues and the reflection coefficient corresponding to the initial potential, while the reflection coefficient becomes irrelevant in $|n| \geq 2t$. If solitons are absent, the asymptotic behavior as $t \to \infty$ is damped oscillation of the Zakharov–Manakov type in $|n| < 2t$, but the solution decays more slowly along $|n| = 2t$, and in $|n| > 2t$ the solution decays faster than any negative power of n.

Keywords Discrete nonlinear Schrödinger equation · Soliton · Inverse scattering

MSC Primary 35Q55 · Secondary 35Q15

1 Introduction

In this article we announce our recent results about the long-time behavior of the solutions to the focusing integrable discrete nonlinear Schrödinger equation (IDNLS) introduced in [1]:

$$i\frac{d}{dt}R_n + (R_{n+1} - 2R_n + R_{n-1}) + |R_n|^2(R_{n+1} + R_{n-1}) = 0 \quad (n \in \mathbf{Z}). \tag{1}$$

This work was partially supported by JSPS KAKENHI Grant Number 26400127.

H. Yamane (✉)
Department of Mathematical Sciences, Kwansei Gakuin University,
Gakuen 2-1 Sanda, Hyogo 669-1337, Japan
e-mail: yamane@kwansei.ac.jp

G. Filipuk et al. (eds.), *Formal and Analytic Solutions of Diff. Equations*,
Springer Proceedings in Mathematics & Statistics 256,
https://doi.org/10.1007/978-3-319-99148-1_6

It is an integrable discretization of the focusing nonlinear Schrödinger equation (NLS)

$$iu_t + u_{xx} + 2u|u|^2 = 0. \tag{2}$$

These two equations can be solved by the inverse scattering transform.

Recall that (2) has pairs of eigenvalues and bright soliton solutions and that 1-solitons are written in terms of an oscillatory exp factor multiplied by a traveling sech factor. Equation (1) has quartets of eigenvalues of the form $(\pm z_j, \pm \bar{z}_j^{-1})$ and admits soliton solutions ([1, 2]). The 1-soliton corresponding to the quartet of eigenvalues including $z_1 = \exp(\alpha_1 + i\beta_1)$ with $\alpha_1 > 0$ is

$$\begin{aligned} R_n(t) &= \mathrm{BS}(n, t; z_1, C_1(0)) \\ &= \exp\Big(-i[2\beta_1(n+1) - 2w_1 t + \arg C_1(0)]\Big) \\ &\qquad \times \sinh(2\alpha_1)\operatorname{sech}[2\alpha_1(n+1) - 2v_1 t - \theta_1], \\ v_1 &= -\sinh(2\alpha_1)\sin(2\beta_1), \\ w_1 &= \cosh(2\alpha_1)\cos(2\beta_1) - 1, \\ \theta_1 &= \log|C_1(0)| - \log\sinh(2\alpha_1), \end{aligned} \tag{3}$$

where $C_1(0)$ is the norming constant. Notice that the soliton is the product of an oscillatory exp factor and a traveling sech factor. The velocity of the latter is $\mathrm{tw}(z_1) = \alpha_1^{-1} v_1$. If there are other quartets of eigenvalues or if the reflection coefficient is not identically zero, $C_1(0)$ is replaced by the product of itself and some other quantity. It causes phase shifts in the oscillatory and the traveling factors.

The soliton resolution conjecture [16] about nonlinear dispersive equations is a statement that any reasonable solution becomes a sum of solitons up to a small error term when t is large. Although this conjecture is known to be true about some non-integrable equations as well, integrable ones are the prototypes and the most important cases. The study of these equations is particularly interesting and fruitful, especially because detailed description of the long-time behavior can be obtained by using notions of inverse scattering. The phase shifts of solitons are written in terms of the eigenvalues and the reflection coefficient. In the works [3, 8, 11, 12] about the continuous NLS (2), the reflection coefficient contributes to the phase shifts in the entire half-plane $t > 0$, while there is no such contribution in the 'spacelike' domain $|n| \geq 2t$ in the discrete case.

Roughly speaking, our main result is as follows: if the quartets of eigenvalues are $(\pm z_j, \pm \bar{z}_j^{-1})$ with $\mathrm{tw}(z_j) < \mathrm{tw}(z_{j'})$ $(j < j')$, then we have, formally,

$$
\begin{aligned}
R_n(t) \sim & \sum_{j\in G_1} \mathrm{BS}\left(n, t; z_j, \delta_{n/t}(0)\delta_{n/t}(z_j)^2 p_j T(z_j)^{-2} C_j(0)\right) \\
& + \sum_{j\in G_2} \mathrm{BS}\left(n, t; z_j, p_j T(z_j)^{-2} C_j(0)\right), \\
p_j = & \prod_{k>j} z_k^2 \bar{z}_k^{-2}, \\
T(z_j) = & \prod_{k>j} \frac{z_k^2(z_j^2 - \bar{z}_k^{-2})}{z_j^2 - z_k^2}
\end{aligned}
$$

under generic assumptions. Here $G_1 = \{j;\ |\mathrm{tw}(z_j)| < 2\}$ and $G_2 = \{j;\ |\mathrm{tw}(z_j)| \geq 2\}$. The function $\delta_{n/t}(z) = \delta(z)$ is defined in (12) below in terms of the reflection coefficient and reduces to the constant 1 in the reflectionless case.

We give a brief review, far from exhaustive, of some known results about the long-time asymptotics of integrable equations based on the method of nonlinear steepest descent.[1] First, this method was employed in [6] to study the MKdV equation. In [4], the defocusing NLS was studied and soliton resolution for the focusing NLS (2) was proved, in different situations, in [3, 8, 11, 12] as is mentioned above. The KdV equation was studied in [9]. The Toda lattice was the topic of [10, 13]. The present author derived some results about the defocusing IDNLS in [17, 18].

2 Inverse Scattering Transform

To fix notation, we review the inverse scattering transform for (1) following [1] and [2, Chap. 3]. The Lax pair of (1) is

$$
X_{n+1} = \begin{bmatrix} z & -\bar{R}_n \\ R_n & z^{-1} \end{bmatrix} X_n, \tag{4}
$$

$$
\frac{d}{dt} X_n = \begin{bmatrix} -iR_{n-1}\bar{R}_n - \frac{i}{2}(z - z^{-1})^2 & i(z\bar{R}_n - z^{-1}\bar{R}_{n-1}) \\ i(z^{-1}R_n - zR_{n-1}) & iR_n\bar{R}_{n-1} + \frac{i}{2}(z - z^{-1})^2 \end{bmatrix} X_n. \tag{5}
$$

Eigenfunctions of (4) can be constructed for any fixed t in $|z| \geq 1$ and $|z| \leq 1$. More precisely, it is possible to define the eigenfunctions $\phi_n(z, t), \psi_n(z, t) \in \mathscr{O}(|z| > 1) \cap \mathscr{C}^0(|z| \geq 1)$ and $\psi_n^*(z, t), \phi_n^*(z, t) \in \mathscr{O}(|z| < 1) \cap \mathscr{C}^0(|z| \leq 1)$ such that

$$
\begin{aligned}
\phi_n(z, t) \sim z^n \begin{bmatrix} 1 \\ 0 \end{bmatrix}, \quad \phi_n^*(z, t) \sim z^{-n} \begin{bmatrix} 0 \\ 1 \end{bmatrix} \quad & \text{as } n \to -\infty, \\
\psi_n(z, t) \sim z^{-n} \begin{bmatrix} 0 \\ 1 \end{bmatrix}, \quad \psi_n^*(z, t) \sim z^n \begin{bmatrix} 1 \\ 0 \end{bmatrix} \quad & \text{as } n \to \infty.
\end{aligned}
$$

[1] There were earlier results based on other methods like [14, 15]. See the introduction of [6] for other works.

On the unit circle $C: |z| = 1$, there are unique functions $a(z),\ a^*(z), b(z,t), b^*(z,t)$ such that

$$\phi_n(z,t) = b(z,t)\psi_n(z,t) + a(z)\psi_n^*(z,t),$$
$$\phi_n^*(z,t) = a^*(z)\psi_n(z,t) + b^*(z,t)\psi_n^*(z,t)$$

holds. One can prove that a and a^* are independent of t. We have

$$a(z) \in \mathscr{O}(|z| > 1) \cap \mathscr{C}^0(|z| \geq 1),\ a^*(z) \in \mathscr{O}(|z| < 1) \cap \mathscr{C}^0(|z| \leq 1),$$
$$a^*(z) = \bar{a}(1/\bar{z})\ (0 < |z| \leq 1),$$
$$b(z), b^*(z) \in \mathscr{C}^0(|z| = 1),\ b^*(z) = -\bar{b}(1/\bar{z})\ (|z| = 1).$$

If $R_n(t)$ decays rapidly as $|n| \to \infty$, then a, a^*, b and b^* are smooth on $|z| = 1$.

We assume that $a(z)$ and $a^*(z)$ never vanish on the unit circle. Their zeros in $|z| > 1$ and $|z| < 1$ are called *eigenvalues*. They are independent of t and appear in quartets of the form $(\pm z_j, \pm \bar{z}_j^{-1})$ $(1 \leq j \leq J)$. We assume that the eigenvalues are all simple zeros. If $a(z_j) = 0$, we have $\phi_n(z_j) = b_j\psi_n(z_j)$ for some constant b_j. The *norming constant* C_j associated with z_j is defined by

$$C_j = C_j(t) = \frac{b_j}{\frac{d}{dz}a(z_j)}.$$

The time evolution of the norming constant is given by

$$C_j(t) = C_j(0)\exp(2i\omega_j t),\ \omega_j = (z_j - z_j^{-1})^2/2.$$

The *reflection coefficient* $r(z,t)$ is defined by

$$r(z,t) = \frac{b(z,t)}{a(z)},\ |z| = 1.$$

Its time evolution is given by

$$r(z,t) = r(z)\exp\left(it(z - z^{-1})^2\right) = r(z)\exp\left(it(z - \bar{z})^2\right),$$

where $r(z) = r(z,0)$. The potential $R_n(t)$ is said to be *reflectionless* if $r(z,t)$ vanishes for any z. For any fixed t, $R_n(t)$ is reflectionless if and only if $R_n(0)$ is reflectionless. In this case, $R_n(t)$ is a 1- or multi-soliton.

Let us introduce the phase function

$$\varphi = \varphi(z) = \varphi(z;n,t) = \frac{1}{2}it(z - z^{-1})^2 - n\log z.$$

By using the scattering data (the eigenvalues, the norming constants and the reflection coefficient) corresponding to the initial potential $R_n(0)$, we formulate the following Riemann–Hilbert problem (RHP):

$$m_+(z) = m_-(z)v(z) \text{ on } C\colon |z| = 1, \tag{6}$$

$$v(z) = v(z,t) = \begin{bmatrix} 1+|r(z)|^2 & e^{-2\varphi(z)}\bar{r}(z) \\ e^{2\varphi(z)}r(z) & 1 \end{bmatrix}, \tag{7}$$

$$m(z) \to I \text{ as } z \to \infty, \tag{8}$$

$$\text{Res}(m(z); \pm z_j) = \lim_{z\to\pm z_j} m(z) \begin{bmatrix} 0 & 0 \\ C_j(0)\exp[2\varphi(z_j)] & 0 \end{bmatrix} \ (1 \le j \le J), \tag{9}$$

$$\text{Res}(m(z); \pm \bar{z}_j^{-1}) = \lim_{z\to\pm \bar{z}_j^{-1}} m(z) \begin{bmatrix} 0 & \bar{z}_j^{-2}\overline{C_j(0)}\exp\left[2\overline{\varphi(z_j)}\right] \\ 0 & 0 \end{bmatrix} \ (1 \le j \le J) \tag{10}$$

Here m_+ and m_- are the boundary values from the *outside* and *inside* of C respectively (C is oriented clockwise following [2].)

The potential $R_n = R_n(t)$ can be reconstructed from the solution $m(z)$ of the RHP above. We have [2, (3.2.91c)]

$$R_n(t) = -\left.\frac{d}{dz}m(z)_{21}\right|_{z=0}. \tag{11}$$

Proposition 1 *If* $r(z) \equiv 0$ *and there is only one quartet of eigenvalues* $(\pm z_1, \pm \bar{z}_1^{-1})$ *with* $z_1 = \exp(\alpha_1 + i\beta_1)$, $\alpha_1 > 0$, *then the RHP (6)–(10) has a unique solution. The potential* $R_n(t)$ *obtained from it through (11) is the bright 1-soliton solution (3).*

Remark 1 Information about the velocities of solitons is contained in (9) and (10) since

$$\text{Re}\,\varphi(z_j) = \alpha_j t\left[\text{tw}(z_j) - n/t\right],\ \text{tw}(z_j) = -\alpha_j^{-1}\sinh(2\alpha_j)\sin(2\beta_j),$$

where $z_j = \exp(\alpha_j + i\beta_j)$, $\alpha_j > 0$.

3 Main Results

In this section we state our main results. See [19] for details.

Throughout this section, we make the following three generic assumptions:

- $a(z)$ never vanishes on the unit circle. It implies that $a^*(z)$ never vanishes there either.
- The eigenvalues are all simple.
- $\text{tw}(z_j)$'s are mutually distinct. We may assume that $\text{tw}(z_j) < \text{tw}(z_{j+1})$ for any j without loss of generality.

In the region $|n| < 2t$, the function $\varphi(z)$ has four saddle points (stationary points of the first order) on the unit circle $|z| = 1$. They are

$$S_1 = e^{-\pi i/4}A,\ S_2 = e^{-\pi i/4}\bar{A},\ S_3 = -S_1,\ S_4 = -S_2,$$
$$A = 2^{-1}\big(\sqrt{2+n/t} - i\sqrt{2-n/t}\,\big).$$

We set

$$\delta(z) = \exp\left(\frac{-1}{2\pi i}\left[\int_{S_1}^{S_2} + \int_{S_3}^{S_4}\right](\tau - z)^{-1}\log(1+|r(\tau)|^2)\,d\tau\right), \qquad (12)$$

where the contours are the arcs $\subset \{|z| = 1\}$. Notice that $\delta(z)$ is identically equal to 1 in the reflectionless case. Our main results are as follows.

Theorem 1 *Assume $\{R_n(0)\} \in \bigcap_{p=0}^{\infty} \ell^{1,p}$. Then in the 'timelike' region $|n| < 2t$, the asymptotic behavior of the solution to (1) is as follows:*
(soliton case) *In the region $-d \le \mathrm{tw}(z_j) - n/t \le d,\ j \in \{1, \ldots, J\}$, with sufficiently small d, we have*

$$R_n(t) = \mathrm{BS}\left(n, t; z_j, \delta(0)\delta(z_j)^{-2}p_j T(z_j)^{-2}C_j(0)\right) + O(t^{-1/2}),$$
$$p_j = \prod_{k>j} z_k^2\bar{z}_k^{-2}, \quad T(z_j) = \prod_{k>j}\frac{z_k^2(z_j^2 - \bar{z}_k^{-2})}{z_j^2 - z_k^2}.$$

(solitonless case) *If $\{\mathrm{tw}(z_j);\ j = 1, \ldots, J\} \cap [n/t - d, n/t + d] = \emptyset$, then there exist $C_k = C_k(n/t) \in \mathbf{C}$ and $p_k = p_k(n/t)$, $q_k = q_k(n/t) \in \mathbf{R}$ $(k = 1, 2)$ depending only on the ratio n/t such that*

$$R_n(t) = \sum_{k=1}^{2} C_k t^{-1/2} e^{-i(p_k t + q_k \log t)} + O(t^{-1}\log t) \quad as \quad t \to \infty. \qquad (13)$$

When we study other regions, we may assume $n > 0$ without loss of generality, since the Eq. (1) is invariant under $n \mapsto -n$.

Theorem 2 *Assume that $\mathrm{tw}(z_j) = 2$ for some j. Then in the region $2t - Mt^{1/3} < n < 2t + Mt^{1/3}$ $(M > 0)$, we have*

$$R_n(t) = \mathrm{BS}\left(n, t; z_j, p_j T(z_j)^{-2}C_j(0)\right) + O(t^{-1/3}) \quad as\ t \to \infty.$$

If $\mathrm{tw}(z_j) \ne 2$ for any j (the solitonless case), the behavior is damped oscillation as follows:

$$R_n(t) = \frac{e^{2p' - \pi i/4}\alpha'}{(3t')^{1/3}}u\left(\frac{4q'}{3^{1/3}}; \hat{r}, -\hat{r}, 0\right) + O(t'^{-2/3}).$$

Here $u(s; \mathrm{p}, \mathrm{q}, \mathrm{r})$ *is the solution of the Painlevé II equation* $u''(s) - su(s) - 2u^3(s) = 0$ *parametrized as in [6]. We have chosen* t_0 *so that* $\arg r(e^{-\pi i/4})\overline{T(e^{-\pi i/4})}^{-2} - 2t_0$ *is an integer multiple of* π *and have set* $t' = t - t_0$, $p' = d + i(-4t' + \pi n)/4$, $\alpha' = [12t'/(6t' - n)]^{1/3}$, $q' = -2^{-4/3}3^{1/3}(6t' - n)^{-1/3}(2t' - n)$ *and* $\hat{r} = r(e^{-\pi i/4})\overline{T(e^{-\pi i/4})}^{-2}$.

Remark 2 Notice that q' tends to 0 as t and n tends to ∞ along the ray $n = 2t, t > 0$. Therefore the Painlevé factor is almost constant.

Theorem 3 *In the region* $2 < \mathrm{tw}(z_s) - d \leq n/t \leq \mathrm{tw}(z_s) + d$ *with sufficiently small d,*

$$R_n(t) = \mathrm{BS}\left(n, t; z_s, p_s T(z_s)^{-2} C_s(0)\right) + O(n^{-k}) \quad as \ |n| \to \infty$$

for any positive integer k.

If $\mathrm{tw}(z_j) \notin [n/t - d, n/t + d]$ *for any* j, *we have* $R_n(t) = O(n^{-k})$ *as* $|n| \to \infty$ *for any positive integer k.*

4 Open Problem

We would like to prove the surjectivity of the scattering transform. It means that any (reasonable) $\{(\pm z_j, \pm \bar{z}_j^{-1}), C_j\}_j \cup \{r(z)\}$ is the scattering data of some potential. A partial answer may be given by Darboux transformations, which enable us to add and remove eigenvalues. A complete answer would be obtained by using Fredholm arguments.

References

1. Ablowitz, M.J., Ladik, J.F.: Nonlinear differential-difference equations and Fourier analysis. J. Math. Phys. **17**, 1011–1018 (1976)
2. Ablowitz, M.J., Prinari, B., Trubatch, A.D.: Discrete and Continuous Nonlinear Schrödinger Systems. Cambridge University Press, Cambridge (2004)
3. Borghese, M., Jenkins, R., McLaughlin, K.D.T.-R.: Long Time Asymptotic Behavior of the Focusing Nonlinear Schrödinger Equation. Ann. Inst. H. Poincaré Anal. Non Linéaire (in press)
4. Deift, P.A., Its, A.R., Zhou, X.: Long-time asymptotics for integrable nonlinear wave equations. In: Fokas, A.S., Zakharov, V.E. (eds.) Important Developments in Soliton Theory, 1980–1990, pp. 181–204. Springer, Berlin (1993)
5. Deift, P., Kamvissis, S., Kriecherbauer, T., Zhou, X.: The Toda rarefaction problem. Commun. Pure Appl. Math. **49**(1), 35–83 (1996)
6. Deift, P.A., Zhou, X.: A steepest descent method for oscillatory Riemann–Hilbert problems. Asymptotics for the MKdV equation. Ann. Math. (2) **137**(2), 295–368 (1993)
7. Deift, P.A., Zhou, X.: Long-time asymptotics for solutions of the NLS equation with initial data in a weighted Sobolev space. Commun. Pure Appl. Math. **56**(8), 1029–1077 (2003)

8. Fokas, A.S., Its, A.R.: The linearization of the initial-boundary value problem of the nonlinear Schrödinger equation. SIAM J. Math. Anal. **27**(3), 738–764 (1996)
9. Grunert, K., Teschl, G.: Long-time asymptotics for the Korteweg-de Vries equation via nonlinear steepest descent. Math. Phys. Anal. Geom. **12**(3), 287–324 (2009)
10. Kamvissis, S.: On the long time behavior of the doubly infinite Toda lattice under initial data decaying at infinity. Commun. Math. Phys. **153**(3), 479–519 (1993)
11. Kamvissis, S.: Focusing NLS with infinitely many solitons. J. Math. Phys. **36**(8), 4175–4180 (1995)
12. Kamvissis, S.: Long time behavior for the focusing nonlinear Schroedinger equation with real spectral singularities. Commun. Math. Phys. **180**(2), 325–341 (1996)
13. Krüger, H., Teschl, G.: Long-time asymptotics of the Toda lattice in the soliton region. Math. Z. **262**(3), 585–602 (2009)
14. Novokshënov, V. Yu.: Asymptotic behavior as $t \to \infty$ of the solution of the Cauchy problem for a nonlinear differential-difference Schrödinger equation. Differentsialnye Uravneniya, **21**(11), 1915–1926 (1985). (in Russian); Differential Equations, **21**(11), 1288–1298 (1985)
15. Tanaka, S.: Korteweg-de Vries equation: asymptotic behavior of solutions. Publ. RIMS, Kyoto Univ. **10**, 367–379 (1975)
16. Tao, T.: Why are solitons stable? Bull. Amer. Math. Soc. (N. S.) **46**(1), 1–33 (2009)
17. Yamane, H.: Long-time asymptotics for the defocusing integrable discrete nonlinear Schrödinger equation. J. Math. Soc. Jpn. **66**, 765–803 (2014)
18. Yamane, H.: Long-time asymptotics for the defocusing integrable discrete nonlinear Schrödinger equation II. SIGMA **11**, 020, 17 (2015)
19. Yamane, H.: Long-time Asymptotics for the Integrable Discrete Nonlinear Schrödinger Equation: the Focusing Case. arXiv:1512.01760 [math-ph], to appear in Funk. Ekvac

Complicated and Exotic Expansions of Solutions to the Painlevé Equations

Alexander D. Bruno

Abstract We consider the complicated and exotic asymptotic expansions of solutions to a polynomial ordinary differential equation (ODE). They are such series on integral powers of the independent variable, which coefficients are the Laurent series on decreasing powers of the logarithm of the independent variable and on its pure imaginary power correspondingly. We propose an algorithm for writing ODEs for these coefficients. The first coefficient is a solution of a truncated equation. For some initial equations, it is an usual or Laurent polynomial. Question: will the following coefficients be such polynomials? Here the question is considered for the third (P_3), fifth (P_5) and sixth (P_6) Painlevé equations. These 3 Painlevé equations have 8 families of complicated expansions and 4 families of exotic expansions. I have calculated several first polynomial coefficients of expansions for all these 12 families. Second coefficients in 7 of 8 families of complicated expansions are polynomials, as well in 2 families of exotic expansions, but one family of complicated and two families of exotic expansions demand some conditions for polynomiality of the second coefficient. Here we give a detailed presentation with proofs of all results.

Keywords Expansions of solutions to ODE · Complicated expansions · Exotic expansions · Polynomiality of coefficients · Painlevé equations

MSC Primary 33E17 · Secondary 34E05, 41E58

1 Introduction

In 2004 I proposed a method for calculation of asymptotic expansions of solutions to a polynomial ordinary differential equation (ODE) [1]. It allowed to compute power expansions and power-logarithmic expansions (or Dulac series) of solutions, where coefficients of powers of the independent variable x are either constants or

---

A. D. Bruno (✉)
Keldysh Institute of Applied Mathematics of RAS, Miusskaya Sq. 4, Moscow 125047, Russia
e-mail: abruno@keldysh.ru

G. Filipuk et al. (eds.), *Formal and Analytic Solutions of Diff. Equations*, Springer Proceedings in Mathematics & Statistics 256,
https://doi.org/10.1007/978-3-319-99148-1_7

polynomials of logarithm of x. I will remind the method lately. Later it is appeared that such equations have solutions with other expansions: they can have coefficients of powers of x as Laurent series either in increasing powers of $\log x$ or in increasing and decreasing imaginary powers of x. They are correspondingly complicated (psi-series) [2] or exotic [3] expansions. Methods from [1] are not suitable for their calculation. Now I have found a method to writing down ODE for each coefficient of such series (Sect. 2). The equations are linear and contain high and low variations from some parts of the initial equation. The first coefficient is a solution of the truncated equation, and usually it is a Laurent series in $\log x$ or in $x^{i\gamma}$. But it is a polynomial or a Laurent polynomial for some equations.

Question: *Will be the following coefficients of the same structure?*

I consider this question for three Painlevé equations P_3, P_5 and P_6, because among 6 Painlevé equations P_1–P_6 there are 3 equations P_3, P_5, P_6 having complicated and exotic expansions of solutions [4–6]. First coefficients for equations P_3, P_5 and P_6 are polynomials in $\log x$ in complicated expansions and Laurent polynomials in $x^{i\gamma}$ in exotic expansions [4, 6]. Each of the Painlevé equations P_3, P_5 and P_6 has 4 complex parameters a, b, c, d. Two of them are included into the truncated equation. These three Painlevé equations have 8 families of complicated expansions and 4 families of exotic expansions. I have calculated several first polynomial coefficients for all these 12 families, sometimes under some simplifications. Second coefficients in 7 of 8 families of complicated expansions are polynomials, as well in 2 families of exotic expansions, but one family of complicated and two families of exotic expansions demand some conditions for polynomiality of the second coefficient. The third coefficient is a polynomial ether always, either under some additional restrictions on parameters, or never. Results for equation P_3, P_5, P_6 are given in Sects. 3, 4 and 5, 6 correspondingly.

2 Writing ODEs for Coefficients

2.1 Algebraic Case

Let we have the polynomial

$$f(x, y) \tag{1}$$

and the series

$$y = \sum_{k=0}^{\infty} \varphi_k x^k , \tag{2}$$

where coefficients φ_k are functions of some quantities. Let we put the series (2) into the polynomial (1) and will select all addends with fixed power exponent of x. For that, we break up the polynomial (1) into the sum $f(x, y) = \sum_{i=0}^{m} f_i(y) x^i$, and we write the series (2) in the form $y = \varphi_0 + \sum\limits_{k=1}^{\infty} \varphi_k x^k \overset{def}{=} \varphi_0 + \Delta$. Then $\Delta^j = \sum\limits_{k=j}^{\infty} c_{jk} x^k$, where coefficients c_{jk} are definite sums of products of j coefficients φ_l and corresponding multinomial coefficients [7]. At last, each item $f_i(\varphi_0 + \Delta)$ can be expanded into the Taylor series

$$f_i = \sum_{j=0}^{\infty} \frac{1}{j!} \left. \frac{d^j f_i}{dy^j} \right|_{y=\varphi_0} \Delta^j .$$

So the result of the substitution of series (2) into the polynomial (1) can be written as the sum

$$\sum_{i=0}^{m} x^i \left[f_i(\varphi_0) + \sum_{j=1}^{\infty} \frac{1}{j!} \frac{d^j f_i(\varphi_0)}{dy^j} \sum_{k=j}^{\infty} c_{jk} x^k \right]$$

of items of the form

$$x^i \frac{1}{j!} \frac{d^j f_i(\varphi_0)}{dy^j} c_{jk} x^k . \tag{3}$$

Here integral indexes $i, j, k \geq 0$ are such

$$k \geq j; \quad \text{if} \quad j = 0, \quad \text{then} \quad k = 0 . \tag{4}$$

Set of such points $(i, j, k) \in \mathbb{Z}^3$ will be denoted as $\mathbf{M}$. At last, all items (3) with fixed power exponent x^n are selected by the equation $i + k = n$. The set $\mathbf{M}$ can be considered as a part of the integer lattice $\mathbb{Z}^3$ in $\mathbb{R}^3$ with points (i, j, k), which satisfy (4).

If we look for expansion (2) as a solution of the equation $f(x, y) = 0$ and want to use the method of indeterminate coefficients, then we obtain the equation $f_0(\varphi_0) = 0$ for the coefficient φ_0, and equation

$$\frac{df_0(\varphi_0)}{dy} \varphi_n x^n + \sum_{(i,j,k) \in \mathbf{N}(n)} x^i \frac{1}{j!} \frac{d^j f_i(\varphi_0)}{dy^j} c_{jk} x^k + x^n f_n(\varphi_0) = 0 , \tag{5}$$

for the coefficient φ_n with $n > 0$, where $\mathbf{N}(n) = \mathbf{M} \cap \{j > 0, \ i + k = n \text{ and } j > 1, \ \text{if } i = 0\}$. That equation can be canceled by x^n and be written in the form

$$\frac{df_0(\varphi_0)}{dy} \varphi_n + \sum_{(i,j,k) \in \mathbf{N}(n)} \frac{1}{j!} \frac{d^j f_i(\varphi_0)}{dy^j} c_{jk} + f_n(\varphi_0) = 0 . \tag{6}$$

Theorem 1 ([8]) *If $df_0(\varphi_0)/dy \neq 0$, then coefficients φ_n can be found from Eq.* (6) *successfully with increasing n.*

2.2 Case of ODE

If $f(x, y)$ is a differential polynomial, i.e. it contains derivatives $d^l y/dx^l$, then the job of derivatives $\dfrac{d^j f_i}{dy^j}$ play variations $\dfrac{\delta^j f_i}{\delta y^j}$, which are derivatives of Frechet or Gateaux. Here the j-variation $\dfrac{\delta^j f}{\delta y^j} = \dfrac{d^j f}{dy^j}$, if the polynomial does not contain derivatives, and variation of a derivation is $\dfrac{\delta}{\delta y}\left(\dfrac{d^k y}{dx^k}\right) = \dfrac{d^k}{dx^k}$, and for products

$$\frac{\delta(f\cdot g)}{\delta y} = f\frac{\delta g}{\delta y} + \frac{\delta f}{\delta y}\cdot g\,, \quad \frac{\delta}{\delta y}\left(\frac{d^k y}{dx^k}\cdot\frac{d^l}{dx^l}\right) = \frac{d^{k+l}}{dx^{k+l}}\,.$$

Analog of the Taylor formula is correct for variations

$$f(y+\Delta) = \sum_{j=0}^{\infty}\frac{1}{j!}\frac{\delta^j f(y)}{\delta y^j}\,\Delta^j\,.$$

Let now we have the differential polynomial $f(x, y)$ and we look for solution of the equation $f(x, y) = 0$ in the form of expansion (2). Here the technique, described above for algebraic equation, can be used, but with the following refinements.

(1) According to [1], differential polynomial $f(x, y)$ is a sum of differential monomials $a(x, y)$, which are products of a usual monomial $\mathrm{const}\cdot x^r y^s$ and several derivatives $d^l y/dx^l$. Each monomial $a(x, y)$ corresponds to its vectorial power exponent $Q(a) = (q_1, q_2)$ under the following rules: $Q(\mathrm{const}) = 0$, $Q(x^r y^s) = (r, s)$, $Q(d^l y/dx^l) = (-l, 1)$, vectorial power exponent of a product of differential monomials is a vectorial sum of their vectorial power exponents $Q(ab) = Q(a) + Q(b)$. Set $\mathbf{S}(f)$ of all vectorial power exponents $Q(a)$ of all differential monomials $a(x, y)$ containing in $f(x, y)$ is called as *support* of f. Its convex hull $\Gamma(f)$ is a *Newton polygon* of f. Its boundary $\partial\Gamma$ consists of vertices $\Gamma_j^{(0)}$ and edges $\Gamma_j^{(1)}$. To each boundary element $\Gamma_j^{(d)}$ corresponds the *truncated equation* $\hat f_j^{(d)} = 0$, where $\hat f_j^{(d)}$ is a sum of all monomials with power exponents $Q \in \Gamma_j^{(d)}$. The first term of solution's expansion to the full equation is a solution to the corresponding truncated equation. Now the part $f_i(x, y)$ contains all such differential monomials $a(x, y)$, for which in $Q(a)$ the first coordinate $q_1 = i$. Besides, we assume that $f(x, y)$ has no monomials with $q_1 < 0$, and $f_0(y) \not\equiv 0$. Then all formula of the algebraic case with variations instead of derivations are correct.

(2) Variations are operators, which are not commute with differential polynomials. So the formulae (5) takes the form

$$\frac{\delta f_0}{\delta y}x^n\varphi_n + \sum_{(i,j,k)\in\mathbf{N}(n)} x^i\frac{1}{j!}\frac{\delta^j f_i}{\delta y^j}x^k c_{jk} + x^n f_n = 0\,, \tag{7}$$

but in it we cannot cancel by x^n and obtain an analog of formulae (6). In (7) all $\delta^j f_i/\delta y^j$ are taken for $y = \varphi_0$.

Theorem 2 ([8]) *In the expansion* (2) *coefficient* φ_n *satisfies Eq.* (7).

(3) Rules of commutation of variations with functions of different classes exist. If φ_k is a series in $\log x$, then $\xi = \log x$ and $x^s = e^{s\xi}$.

Lemma 1 ([4])

$$\frac{d^n}{d\xi^n}\left[e^{s\xi}\varphi(\xi)\right] = e^{s\xi}\sum_{k=0}^{n}\binom{n}{k}s^{n-k}\varphi^{(k)}(\xi)\,,$$

where $\binom{n}{k}$ *are binomial coefficients and* $\varphi^{(k)}$ *is the kth derivation of* $\varphi(\xi)$ *along* ξ.

Proof follows from the Leibniz's formula for derivation of a product.

Corollary 1

$$\frac{d}{d\xi}\left[x^s\varphi(\xi)\right] = x^s[s\varphi(\xi) + \dot{\varphi}(\xi)]\,,$$
$$\frac{d^2}{d\xi^2}\left[x^s\varphi(\xi)\right] = x^s[s^2\varphi(\xi) + 2s\dot{\varphi}(\xi) + \ddot{\varphi}(\xi)]\,.$$

If φ_k *is a series in* $x^{i\gamma}$*, then* $\xi = x^{i\gamma}$ *and* $x^s = \xi^{s/(i\gamma)}$.

Lemma 2 ([9])

$$\frac{d^n}{d\xi^n}\left[\xi^{s/(i\gamma)}\varphi(\xi)\right] =$$
$$= \xi^{s/(i\gamma)}\left[\sum_{k=0}^{n-1}\binom{n}{k}\frac{s}{i\gamma}\left(\frac{s}{i\gamma}-1\right)\ldots\left(\frac{s}{i\gamma}-n+k+1\right)\varphi^{(k)}(\xi)\frac{1}{\xi^{n-k}} + \varphi^{(n)}\right].$$

Corollary 2

$$\xi\frac{d}{d\xi}\left[x^n\varphi(\xi)\right] = x^n\left[\frac{n}{i\gamma}\varphi + \xi\dot{\varphi}\right],$$
$$\xi^2\frac{d^2}{d\xi^2}\left[x^n\varphi(\xi)\right] = x^n\left[\frac{n}{i\gamma}\left(\frac{n}{i\gamma}-1\right)\varphi + \frac{2n}{i\gamma}\xi\dot{\varphi} + \xi^2\ddot{\varphi}\right].$$

These Lemmas give rules of commutation of an operator with x^s. Applying them in Eq. (7), we can cancel the equation by x^n and obtain an equation without x, only with ξ. So the algorithm consists of the following steps.

Step 0 From the initial equation $f(x, y) = 0$, we select such truncated equation $\hat{f}_1^{(1)}(x, y) = 0$, which corresponds to edge $\Gamma_1^{(1)}$ of the polygon Γ of the differential sum $f(x, y)$ and has a complicated or exotic solution depending from $\log x$ or $x^{i\gamma}$, $\gamma \in \mathbb{R}$ correspondingly.

Step 1 We make a power transformation of the variables $y = x^l z$ to make the truncated equation correspond to the vertical edge.

Step 2 We divide the transformed equation $g(x, z) = 0$ into parts $g_i(x, y)x^i$, corresponding to different verticals of its support.

Step 3 In these parts $g_i(x, y)x^i$ we change the independent variable x by $\log x$ or by $x^{i\gamma}$.

Step 4 We write down equations for several first coefficients φ_k.

Step 5 Using the rules of commutation, we exclude powers of x from these equations and we obtain linear ODEs for coefficients with independent variable $\log x$ or $x^{i\gamma}$. Their solutions are power expansions and can be computed by known methods from [1].

3 The Third Painlevé Equation P_3

3.1 Truncated Equation and its Logarithmic Solutions

The third Painlevé equation P_3 is

$$y'' = \frac{y'^2}{y} - \frac{y'}{x} + \frac{ay^2 + b}{x} + cy^3 + \frac{d}{y}.$$

Let multiply it by its denominator xy and translate the left hand side into right side. Then we obtain the equation P_3, written as a differential polynomial

$$f(x, y) \overset{def}{=} -xyy'' + xy'^2 - yy' + ay^3 + by + cxy^4 + dx = 0, \tag{8}$$

where a, b, c, d are complex parameters. Its support and polygon for $a, b, c, d \neq 0$ are shown in Fig. 1. The edge $\Gamma_1^{(1)}$ corresponds to the truncated equation

$$\hat{f}_1^{(1)} \overset{def}{=} -xyy'' + xy'^2 - yy' + by + dx = 0. \tag{9}$$

After the power transformation $y = xz$ and canceling by x, the full Eq. (8) became

$$g \overset{def}{=} -x^2zz'' + x^2z'^2 - xzz' + bz + d + ax^2z^3 + cx^4z^4 = 0. \tag{10}$$

Here the truncated Eq. (9) takes the form

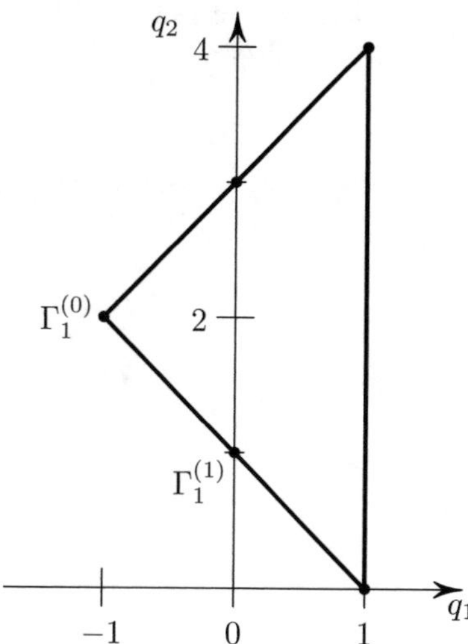

Fig. 1 Support and polygon of the Eq. (8) for $a, b, c, d \neq 0$

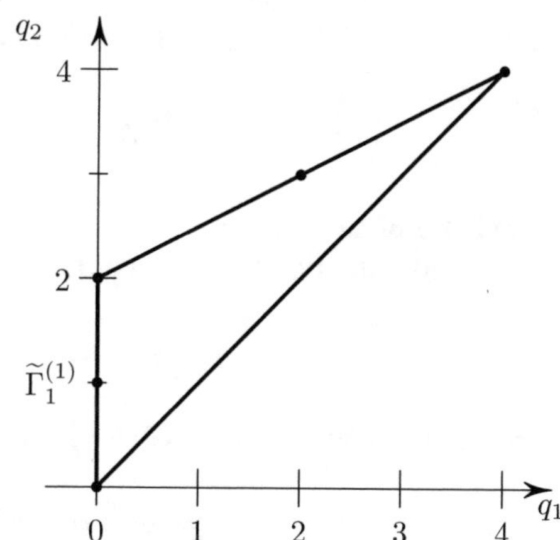

Fig. 2 Support and polygon of the Eq. (10) for $a, b, c, d \neq 0$

$$g_0 \stackrel{def}{=} -x^2zz'' + x^2z'^2 - xzz' + bz + d = 0\,. \tag{11}$$

Support and polygon of Eq. (10) are shown in Fig. 2. Here the truncated Eq. (11) corresponds to the vertical edge $\widetilde{\Gamma}_1^{(1)}$ at the axis $q_1 = 0$. Here $g_2 = az^3$, $g_4 = cz^4$.

After the logarithmic transformation $\xi = \log x$, Eq. (11) takes the form

$$h_0 \stackrel{def}{=} -z\ddot{z} + \dot{z}^2 + bz + d = 0, \tag{12}$$

where $\dot{z} = dz/d\xi$. Support and polygon of Eq. (12) are shown in Fig. 3 in the case $bd \neq 0$. Here $h_2 = az^3$, $h_4 = cz^4$.

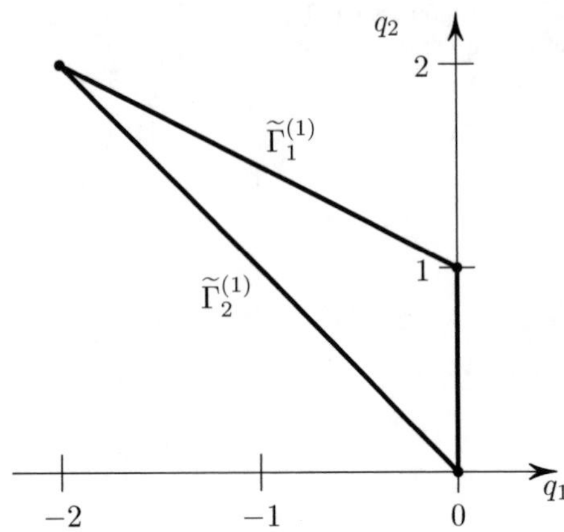

Fig. 3 Support and polygon of the Eq. (12) with $bd \neq 0$

Let $b \neq 0$. The edge $\tilde{\Gamma}_1^{(1)}$ of Fig. 3 corresponds to the truncated equation $\hat{h}_1^{(1)} \stackrel{def}{=} -z\ddot{z} + \dot{z}^2 + bz = 0$. It has the power solution $z = -b\xi^2/2$. According to [1], extending it as expansion in decreasing powers of ξ, we obtain the solutions of Eq. (11)

$$z = -\frac{b}{2}(\log x + \tilde{c})^2 - \frac{d}{2b} = \varphi_0, \tag{13}$$

where $\tilde{c}$ is arbitrary constant.

Let us consider Eq. (11) in the case $b = 0$, $d \neq 0$. Then Eq. (12) has the form

$$h_0 \stackrel{def}{=} -z\ddot{z} + \dot{z}^2 + d = 0.$$

Its polygon coincides with the edge $\tilde{\Gamma}_2^{(1)}$ in Fig. 3. The equation has solutions

$$z = \pm\sqrt{-d}\,(\log x + \tilde{c}) = \varphi_0. \tag{14}$$

Thus, we have proved.

Theorem 3 *All nonconstant solutions to Eq.* (12), *expanded into power series in decreasing powers of* ξ, *form two families:*
the main family (13) *for* $b \neq 0$; *and*
the additional family (14) *for* $b = 0$, $d \neq 0$.

Solutions to Eq. (10) have the form of expansion

$$z = \varphi_0(\xi) + \sum_{k=1}^{\infty} \varphi_{2k}(\xi)x^{2k}, \tag{15}$$

where φ_0 is given by (13) or (14).

In the first case $b \neq 0$, we call family of solutions (15) as **main**, and in the second case $b = 0,\ d \neq 0$, we call the family of solutions (15) as **additional**.

According to Theorem 2, equation for φ_2 is

$$\frac{\delta h_0}{\delta z}(x^2\varphi_2) + x^2 h_2(\varphi_0) = 0. \tag{16}$$

According to (12)

$$\frac{\delta h_0}{\delta z} = -\ddot{z} - z\frac{d^2}{d\xi^2} + 2\dot{z}\frac{d}{d\xi} + b, \quad \frac{\delta^2 h_0}{\delta z^2} = 0. \tag{17}$$

According to (10) $h_2 = az^3$ and according to Corollary 1

$$\frac{d}{d\xi}x^2\varphi_2 = x^2\left[2\varphi_2 + \dot{\varphi}_2\right], \quad \frac{d^2}{d\xi^2}x^2\varphi_2 = x^2\left[4\varphi_2 + 4\dot{\varphi}_2 + \ddot{\varphi}_2\right].$$

So, Eq. (16), after canceling by x^2, takes the form

$$-z\left[4\varphi_2 + 4\dot{\varphi}_2 + \ddot{\varphi}_2\right] + 2\dot{z}\left[2\varphi_2 + \dot{\varphi}_2\right] + (b - \ddot{z})\varphi_2 + az^3 = 0, \tag{18}$$

where $z = \varphi_0$ from (13) or (14).

3.2 The Additional Complicated Family

Let $\xi = \log x + \tilde{c}$, then, according to (14), $z = \varphi_0 = \beta\xi,\ \beta^2 = -d,\ \dot{z} = \beta,\ \ddot{z} = 0$, and Eq. (18) is

$$-\beta\,\xi\left[4\varphi_2 + 4\dot{\varphi}_2 + \ddot{\varphi}_2\right] + 2\,\beta\left[2\varphi_2 + \dot{\varphi}_2\right] + a(\beta\,\xi)^3 = 0.$$

Its support and polygon see in Fig. 4.

Cotangent of the angle of inclination of its right edge equals to -2. So we look for polynomial solution of degree 2. Indeed that equation has a polynomial solution:

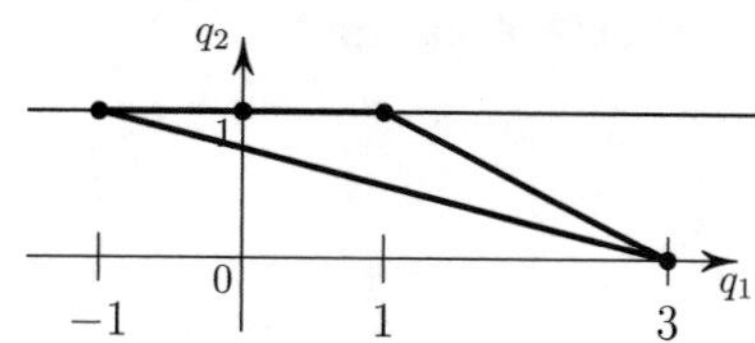

Fig. 4 Support and polygon of equation for φ_2 in additional complicated expansion

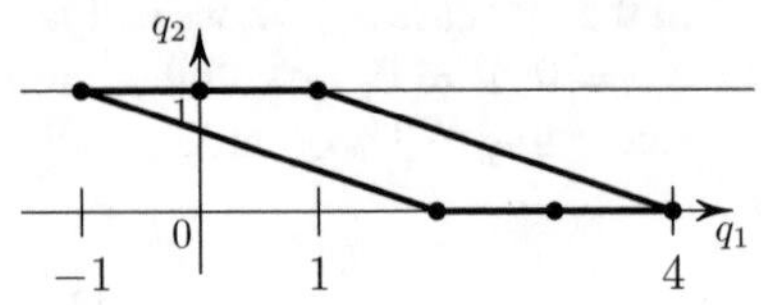

Fig. 5 Support and polygon of the Eq. (20)

$$\varphi_2 = -\frac{ad}{4}\left(\xi^2 - \xi + \frac{1}{2}\right).$$

Here a linear system of 4 algebraic equation is satisfied for 3 constant coefficients. According to Theorem 2, equation for φ_4 is

$$\frac{\delta h_0}{\delta z} x^4 \varphi_4 + x^2 \frac{\delta h_2}{\delta z} x^2 \varphi_2 + x^4 h_4(\varphi_0) = 0\,. \tag{19}$$

According to Corollary 1

$$\frac{d^2}{d\xi^2} x^4 \varphi_4 = x^4[16\varphi_4 + 8\dot{\varphi}_4 + \ddot{\varphi}_4]\,, \quad \frac{d}{d\xi} x^4 \varphi_4 = x^4[4\varphi_4 + \dot{\varphi}_4]\,.$$

Here $\dfrac{\delta h_2}{\delta z} = \dfrac{dh_2}{dz} = 3az^2,\ h_4 = cz^4$.

So after canceling by x^4, Eq. (19) takes the form

$$-\beta\xi[16\varphi_4 + 8\dot{\varphi}_4 + \ddot{\varphi}_4] + 2\beta[4\varphi_4 + \dot{\varphi}_4] + 3a\beta^2\xi^2\varphi_2 + c(\beta\xi)^4 = 0\,. \tag{20}$$

Its support and polygon are shown in Fig. 5.

Cotangent of the angle of inclination of its right edge equals to -3. So the solution to Eq. (14) may be polynomial of order 3

$$\varphi_4 = A\xi^3 + B\xi^2 + C\xi + D\,.$$

Then the sum of two first addends in (20) is

$$-16\beta A\xi^4 + (-16B - 16A)\beta\xi^3 + (-16C - 8B)\beta\xi^2 + (-16D + 2B)\beta\xi + 2(4D + C)\beta\,.$$

Here coefficients near $\xi^2,\ \xi^1$ and $\xi^0 = 1$ for $\beta B,\ \beta C,\ \beta D$ form the matrix

$$\begin{pmatrix} -8 & -16 & 0 \\ 2 & 0 & -16 \\ 0 & 2 & 8 \end{pmatrix}$$

with zero determinant. From the other side, the sum of two last addends in (20) is

$$3a\beta^2\xi^2\varphi_2 + c(\beta\xi)^4 = \frac{3a^2d^2}{4}\left(\xi^4 - \xi^3 + \frac{1}{2}\xi^2\right) + c\beta^4\xi^4 .$$

Coefficients of that sum near ξ^2, ξ^1 and 1 are $\frac{3}{8}a^2d^2$, 0 and 0 correspondingly. Hence, the linear system of equations for A, B, C, D has a solution only if $\frac{3}{8}a^2d^2 = 0$. As $d \neq 0$, then we obtain the condition $a = 0$ for existence A, B, C, D. Under the condition

$$\varphi_4 = \frac{c\beta^3}{16}\left(\xi^3 - \xi^2 + \frac{1}{2}\xi - \frac{1}{8}\right). \tag{21}$$

As $a = 0$, then $g = g_0 + x^4 g_4$, $\varphi_2 = 0$, and the expansion of solution contains powers of x, which are multiple to 4.

Theorem 2 gives for φ_8 the equation

$$\frac{\delta h_0}{\delta z}\, x^8\varphi_8 + x^4\frac{\delta h_4}{\delta z}\, x^4\varphi_4 = 0 . \tag{22}$$

According to (17), here $\frac{\delta h_0}{\delta z} = -z\frac{d^2}{d\xi^2} + \dot{z}\,\frac{d}{d\xi}$, $\frac{\delta h_4}{\delta z} = \frac{dg_4}{dz} = 4cz^3$. According to Corollary 1

$$\frac{d^2}{d\xi^2}x^8\varphi_8 = x^8[64\varphi_8 + 16\dot{\varphi}_8 + \ddot{\varphi}_8] , \quad \frac{d}{d\xi}x^8\varphi_8 = x^8[8\varphi_8 + \dot{\varphi}_8] .$$

As h_4 does not contain derivatives, then variation

$$\frac{\delta h_4}{\delta z} = \frac{dh_4}{dz} = 4c\,(\beta\xi)^3$$

and it commutes with $x^4\varphi_4$. Canceling Eq. (22) by x^4, we obtain equation

$$-\beta\xi\,[64\varphi_8 + 16\dot{\varphi}_8 + \ddot{\varphi}_8] + 2\beta\,[8\varphi_8 + \dot{\varphi}_8] + 4c\beta^3\xi^3\varphi_4 = 0 .$$

It has the polynomial solution

$$\varphi_8 = \frac{c^2\beta^5}{16^2}\left(\xi^5 - 2\xi^4 + \frac{59}{32}\xi^3 - \frac{59}{64}\xi^2 + \frac{59}{4\cdot 64}\xi - \frac{59}{32\cdot 64}\right).$$

According to Theorem 2, we obtain the equation for φ_{12}

$$\frac{\delta h_0}{\delta z}\, x^{12}\varphi_{12} + x^4\frac{\delta h_4}{\delta z}\, x^8\varphi_8 + x^4\,\frac{1}{2}\,\frac{\delta^2 h_4}{\delta z^2}\,\left(x^4\varphi_4\right)^2 = 0 .$$

According to Corollary 1, it has the form

$$-\beta\xi[144\varphi_{12}+24\dot{\varphi}_{12}+\ddot{\varphi}_{12}]+2\beta[12\varphi_{12}+\dot{\varphi}_{12}]+4c(\beta\xi)^3\varphi_8+\frac{1}{2}\cdot 6c(\beta\xi)^2\varphi_4^2=0\,. \tag{23}$$

If to look for solution of the equation as the polynomial of order 7

$$\varphi_{12}=E\xi^7+F\xi^6+G\xi^5+H\xi^4+I\xi^3+J\xi^2+K\xi+L\,,$$

then the sum of terms of small powers of ξ in the first two addends in (23) is

$$\beta(-144K-24J)\xi^2+\beta(-144L+2J)\xi+\beta(24L+2K)\,.$$

Matrix of coefficient near βJ, βK and βL is

$$\begin{pmatrix} -24 & -144 & 0 \\ 2 & 0 & -144 \\ 0 & 2 & 24 \end{pmatrix}.$$

It has zero determinant. From other side, terms of smallest power of ξ in the remaining part of Eq. (23) are

$$3c\beta^2\left(\frac{c\beta^3}{16}\right)^2\left(-\frac{1}{8}\right)^2\xi^2 \tag{24}$$

according to (21). The linear algebraic system of equations for $E,\dots,L$ has a solution, if the coefficient in (24) equals to zero. As $\beta\neq 0$, then $c=0$. In that case the full equation is degenerated into truncated one $g_0=0$, and in expansion $z=\sum\limits_{k=0}^{\infty}\varphi_{4k}(\xi)\,x^{4k}$ all $\varphi_{4k}=0$ for $k>0$. That is the trivially degenerated integrable case with $a=c=0$. So we have proved.

Theorem 4 *In expansion* (15) *of the additional complicated family of solutions to the equation P_3, polynomial coefficients are φ_2 for any values of parameters a and c; also φ_4, $\varphi_6=0$, φ_8 are polynomials for $a=0$. The fifth coefficient φ_8 never is a polynomial, if $|a|+|c|\neq 0$.*

3.3 The Main Complicated Family

Let put $\xi=\log x+\tilde{c}$, then solution (13) is:

$$z=-\frac{b}{2}\xi^2-\frac{d}{2b}=\varphi_0(\xi)\,.$$

Here $\dot{z}=-b\xi$, $\ddot{z}=-b$ and the Eq. (18) has the polynomial solution

$$\varphi_2 = \frac{ab^2}{16}\left[\xi^4 - 2\xi^3 + (2+2\lambda)\xi^2 - (1+2\lambda)\xi + \lambda^2\right],$$

where $\lambda = d/b^2$.

Theorem 5 *In expansion* (15) *of the main complicated family of solutions to the equation* P_3*, the second coefficient* φ_2 *is always a polynomial.*

Farther we consider the main family under the restriction $d = 0$. Then $\lambda = 0$, $z = -\frac{b}{2}\xi^2$, $\dot{z} = -b\xi$, $\ddot{z} = -b$ and

$$\varphi_2 = \frac{ab^2}{16}(\xi^4 - 2\xi^3 + 2\xi^2 - \xi).$$

According to Theorem 2, equation for φ_4 is

$$\frac{\delta h_0}{\delta z}x^4\varphi_4 + x^2\frac{\delta h_2}{\delta z}x^2\varphi_2 + x^4 h_4 = 0.$$

According to (17) and Corollary 1,

$$\frac{\delta h_0}{\delta z}x^4\varphi_4 = x^4\frac{b}{2}\xi^2\left[16\varphi_4 + 8\dot{\varphi}_4 + \ddot{\varphi}_4\right] - x^4 2b\xi\left[4\varphi_4 + \dot{\varphi}_4\right] + x^4\cdot 2b\varphi_4,$$

$$\frac{\delta h_2}{\delta z}x^2\varphi_2 = x^2\frac{3ab^2}{4}\xi^4\varphi_2, \quad h_4 = \frac{c}{16}(b\xi^2)^4.$$

After canceling by x^4, we obtain the equation

$$\frac{b}{2}\xi^2\left[16\varphi_4 + 8\dot{\varphi}_4 + \ddot{\varphi}_4\right] - 2b\xi\left[4\varphi_4 + \dot{\varphi}_4\right] + 2b\varphi_4 + \frac{3}{4}ab^2\xi^4\varphi_2 + \frac{cb^4}{16}\xi^8 = 0.$$

It has the polynomial solution

$$\varphi_4 = a^2b^3\psi_1 + c\,b^3\psi_2,$$

where

$$\psi_1 = \frac{1}{2^9}\left(-3\xi^6 + \frac{15}{2}\xi^5 - \frac{91}{8}\xi^4 + \frac{115}{2}\xi^3 - \frac{115}{4}\xi^2 + \frac{115}{16}\xi\right),$$

$$\psi_2 = \frac{1}{2^7}\left(-\xi^6 + 2\xi^5 - \frac{19}{2^3}\xi^4 + \frac{15}{2^3}\xi^3 - \frac{15}{2^4}\xi^2 + \frac{15}{2^6}\xi\right).$$

According to Theorem 2, we have following equations for φ_6 and φ_8

$$\frac{\delta h_0}{\delta z}\, x^6\varphi_6 + x^2\frac{\delta h_2}{\delta z}\, x^4\varphi_4 + x^2\frac{1}{2}\,\frac{\delta^2 h_2}{\delta z^2}\,(x^2\varphi_2)^2 + x^4\frac{\delta h_4}{\delta z}\, x^2\varphi_2 = 0\,,$$

$$\frac{\delta h_0}{\delta z}\, x^8\varphi_8 + x^2\frac{\delta h_2}{\delta z}\, x^6\varphi_6 + x^2\frac{1}{2}\,\frac{\delta^2 h_2}{\delta z^2}\, 2(x^2\varphi_2)\,(x^4\varphi_4) + x^4\frac{\delta h_4}{\delta z}\, x^4\varphi_4 + \\ + \frac{1}{2}\,\frac{\delta^2 h_4}{\delta z^2}\,(x^2\varphi_2)^2 = 0\,.$$

The equations have polynomial solutions for any parameters $b \neq 0$, a, c, because their parts, containing variations from h_2 and h_4, do not contain ξ^2, ξ and $\xi^0 = 1$.

Hypothesis 1 ([8]) Coefficients $\varphi_{2k}(\xi)$ in expansion (15) of the main complicated family of solutions to the equation P_3 are polynomials in $\log x$, if the parameter of the equation $d = 0$.

3.4 Exotic Expansions for Equation P_3

Now and to the end of the Section, we put $\xi = x^{i\gamma}$, $\gamma \in \mathbb{R}$, $\gamma \neq 0$. Then

$$x = \xi^{1/(i\gamma)}, \quad z' = \frac{i\gamma\dot{z}\xi}{x}, \quad z'' = -\frac{\gamma^2\ddot{z}\,\xi^2 + i\gamma\dot{z}\xi + \gamma^2\dot{z}\xi}{x^2}\,.$$

So the truncated Eq. (11) takes the form

$$\gamma^2 z(\xi^2\ddot{z} + \xi\dot{z}) - \gamma^2\xi^2\dot{z}^2 + bz + d = 0\,.$$

Dividing it by γ^2, we obtain equation

$$h_0 \stackrel{def}{=} z(\xi^2\ddot{z} + \xi\dot{z}) - \xi^2\dot{z}^2 + \tilde{b}z + \tilde{d} = 0\,, \tag{25}$$

where $\tilde{b} = b/\gamma^2$, $\tilde{d} = d/\gamma^2$. In the full (nontruncated) equation $h_2 = \tilde{a}z^3$, $h_4 = \tilde{c}z^4$, where $\tilde{a} = a/\gamma^2$, $\tilde{c} = c/\gamma^2$.

Theorem 6 *All exotic solutions to Eq. (25) in the form of Laurent series*

$$z = A\xi + B + C\xi^{-1} + \cdots\,,$$

where A, B, $C = const \in \mathbb{C}$ are the Laurent polynomials

$$z = A\xi^{-1} + B + C\xi^{-1} = \varphi_0\,, \tag{26}$$

and form one family, where

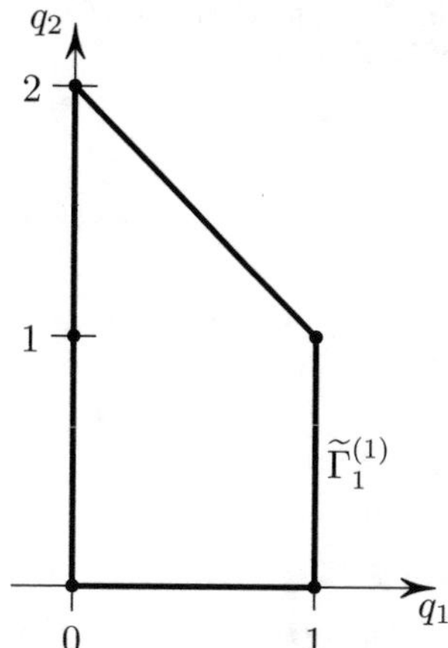

Fig. 6 Support and polygon of Eq. (29)

$$B + \tilde{b} = 0, \quad 4AC - \tilde{b}^2 + \tilde{d} = 0\,. \tag{27}$$

Proof is based on [1]. Polygon Γ of the truncated Eq. (25) is the edge $q_1 = 0,\ 0 \leqslant q_2 \leqslant 2$. Its upper vertex $q_1 = 0,\ q_2 = 2$ corresponds to the truncated equation

$$\hat{h}_0 \stackrel{def}{=} z(\xi^2\ddot{z} + \xi\dot{z}) - \xi^2\dot{z}^2 = 0\,. \tag{28}$$

Its characteristic equation is

$$k(k-1) + k - k^2 \equiv 0\,.$$

So Eq. (28) has power solutions $z = A\xi^\lambda$ with any constants A and λ. In particular, $z = A\xi$ is its solution. We make substitution $z = A\xi + u$ into Eq. (25). Then it takes the form

$$A\xi\left(\xi^2\ddot{u} + u - \xi\dot{u} + \tilde{b}\right) + u\left(\xi^2\ddot{u} + \xi\dot{u}\right) - \xi^2\dot{u}^2 + \tilde{b}u + \tilde{d} = 0\,. \tag{29}$$

Support and the polygon of Eq. (29) are shown in Fig. 6.

It is a quadrangle with the edge $\widetilde{\Gamma}_1^{(1)}$ with normal $P = (1, 0)$, corresponding to the truncated equation

$$A\xi(\xi^2\ddot{u} + u - \xi\dot{u} + \tilde{b}) = 0\,.$$

Its power solution $u = c\xi^2$ with $r = 0$ is $u = -\tilde{b}$. After substitution $u = -\tilde{b} + w$, the Eq. (29) takes the form

$$A\xi(\xi^2\ddot{w} - \xi\dot{w} + w) + (w - \tilde{b})(\xi^2\ddot{w} + \xi\dot{w}) - \xi^2\dot{w}^2 - \tilde{b}^2 + \tilde{b}w + \tilde{d} = 0\,. \tag{30}$$

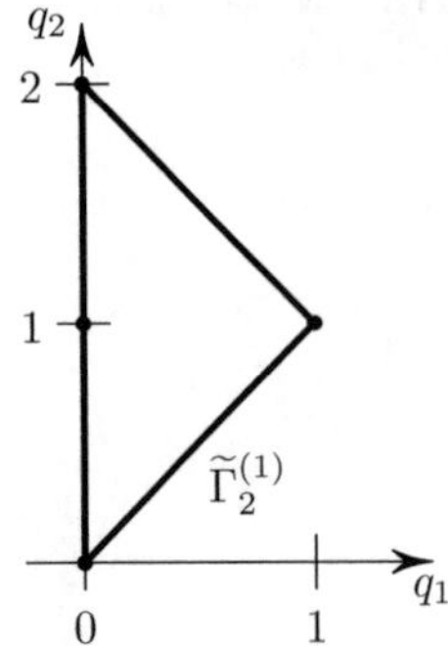

Fig. 7 Support and polygon of Eq. (30)

Its support and polygon Γ are shown in Fig. 7.

Polygon Γ has the edge $\widetilde{\Gamma}_2^{(1)}$ with the normal $P = (1, -1)$, corresponding to the truncated equation

$$A\xi(\xi^2\ddot{w} - \xi\dot{w} + w) - \tilde{b}^2 + \tilde{d} = 0\,.$$

Constant C of its solution $w = C\xi^{(-1)}$ satisfies equation $4AC - \tilde{b}^2 + \tilde{d} = 0$. It is also a solution of equations $w(\xi^2\ddot{w} + \xi\dot{w}) - \xi^2\dot{w}^2 = 0$ and $\tilde{b}(w - \xi^2\ddot{w} - \xi\dot{w}) = 0$. So that solution $w = C\xi^{(-1)}$ is a solution of the Eq. (30). Hence, (26) and (27) are solutions to Eq. (25).

Remark 1 Equation (25) is integrable and Theorem 6 follows from Theorem 1 [9], which describes all solutions of Eq. (25).

Exotic expansion of solutions to the full Eq. (10) again have the form (15). Let us find $\varphi_2(\xi)$. It is a solution to Eq. (16). But now according to (25),

$$\frac{\delta h_0}{\delta z} = z\xi^2\frac{d^2}{d\xi^2} + z\xi\frac{d}{d\xi} - 2\dot{z}\xi^2\frac{d}{d\xi} + \xi\dot{z} + \tilde{b} = z\xi^2\frac{d^2}{d\xi^2} + (z - 2\dot{z}\xi)\xi\frac{d}{d\xi} + \xi\dot{z} + \tilde{b}\,, \tag{31}$$

$$\frac{\delta^2 h_0}{\delta z^2} = -\xi^2\frac{d^2}{d\xi^2} + 2\xi\frac{d}{d\xi}, \quad \frac{\delta^3 h_0}{\delta z^3} = 0, \quad h_2 = \tilde{a}z^3.$$

According to (26) $\xi\dot{z} = A\xi - C\xi^{-1}$, $\xi^2\ddot{z} = 2C\xi^{-1}$. So, applying Corollary 2 to Eq. (16) and dividing it by x^2, we obtain equation

$$\begin{aligned}(A\xi + B + C\xi^{-1})&\left[\frac{2}{i\gamma}\left(\frac{2}{i\gamma} - 1\right)\varphi_2 + \frac{4}{i\gamma}\xi\dot{\varphi}_2 + \xi^2\ddot{\varphi}_2\right] + \\ &+ (-A\xi + B + 3C\xi^{-1})\left[\frac{2}{i\gamma}\varphi_2 + \xi\dot{\varphi}_2\right] + \\ &+ (A\xi - B + C\xi^{-1})\varphi_2 + \tilde{a}(A\xi + B + C\xi^{-1})^3 = 0\,.\end{aligned} \tag{32}$$

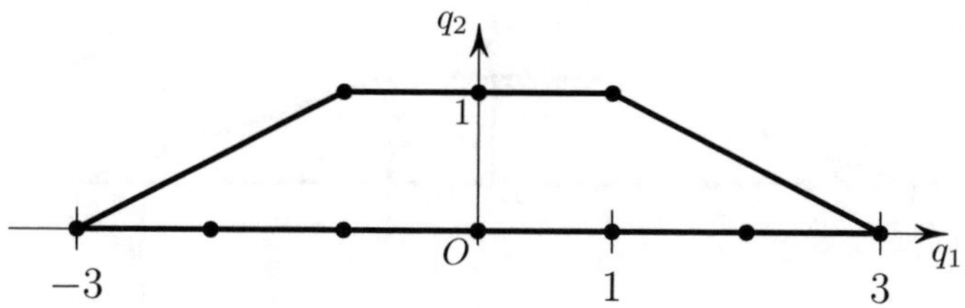

Fig. 8 Support and polygon of Eq. (32)

Its support and the polygon are shown in Fig. 8.

Cotangents of angles of inclination of left and right edges are equal to ± 2. Hence, solution to Eq. (32) in form of a Laurent polynomial must have powers from -2 to $+2$, i.e.

$$\varphi_2 = D\xi^2 + E\xi + F + G\xi^{-1} + H\xi^{-2}\,, \tag{33}$$

where D, E, F, G, H — are constants. Then

$$\xi\dot{\varphi}_2 = 2D\xi^2 + E\xi - G\xi^{-1} - 2H\xi^{-2}\,,$$

$$\xi^2\ddot{\varphi}_2 = 2D\xi^2 + 2G\xi^{-1} + 6H\xi^{-2}\,.$$

Note that

$$\varphi_0^3 = (A\xi + B + C\xi^{-1})^3 = A^3\xi^3 + 3A^2B\xi^2 + 3(AB^2 + A^2C)\xi + B^3 + 6ABC +$$

$$+ 3(AC^2 + B^2C)\xi^{-1} + 3BC^2\xi^{-2} + C^3\xi^{-3}\,.$$

We substitute these expressions into Eq. (32) and nullity coefficients near $\xi^3, \xi^2, \xi, \xi^0, \xi^{-1}, \xi^{-2}, \xi^{-3}$. Then we obtain a system of 7 linear algebraic equations for 5 coefficients D, E, F, G, H. It has the unique solution

$$D = \frac{\tilde{a}A^2\gamma^2}{(2+i\gamma)^2},\ E = \frac{\tilde{a}AB\gamma^2}{2+i\gamma},\ F = \frac{\tilde{a}B^2\gamma^2}{4+\gamma^2} + \tilde{a}AC\frac{(8+6\gamma^2)\gamma^2}{(4+\gamma^2)^2}, \tag{34}$$
$$G = \frac{\tilde{a}BC\gamma^2}{2-i\gamma},\ H = \frac{\tilde{a}C^2\gamma^2}{(2-i\gamma)^2}\,.$$

According to Theorem 2, we have for φ_4 the equation

$$\frac{\delta h_0}{\delta z}x^4\varphi_4 + \frac{1}{2}\frac{\delta^2 h_0}{\delta z^2}(x^2\varphi_2)^2 + x^2\frac{\delta h_2}{\delta z}x^2\varphi_2 + x^4 h_4(\varphi_0) = 0\,, \tag{35}$$

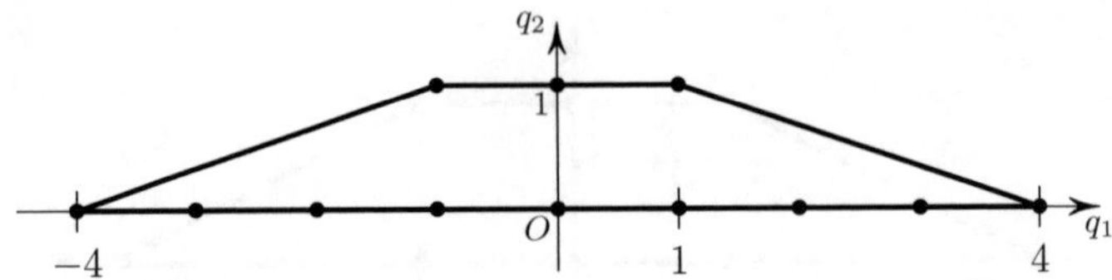

Fig. 9 Support and polygon of Eq. (37)

Let us consider it in the case $a = 0$. Then according to (33) and (34) $\varphi_2 = 0$ and Eq. (35) is

$$\frac{\delta h_0}{\delta z} x^4 \varphi_4 + x^4 h_4 = 0\,, \tag{36}$$

where $\frac{\delta h_0}{\delta z}$ is in (31), $h_4 = \tilde{c} z^4$, $z = \varphi_0 = A\xi + B + C\xi^{-1}$. Using in Eq. (36) Corollary 2 and dividing it by x^4, we obtain equation

$$\begin{aligned}
&(A\xi + B + C\xi^{-1}) \left[\frac{4}{i\gamma} \left(\frac{4}{i\gamma} - 1 \right) \varphi_4 + \frac{8}{i\gamma} \xi \dot{\varphi}_4 + \xi^2 \ddot{\varphi}_4 \right] + \\
&+ (-A\xi + B + 3C\xi^{-1}) \left[\frac{4}{i\gamma} \varphi_4 + \ \xi \dot{\varphi}_4 \right] + \\
&+ (A\xi - B + C\xi^{-1}) \varphi_4 + \tilde{c}(A\xi + B + C\xi^{-1})^4 = 0\,.
\end{aligned} \tag{37}$$

Its support and the Newton polygon are shown in Fig. 9.

Inclinations of its side edges are ± 3. Hence, solution to Eq. (37) in the form of Laurent polynomial must have powers of ξ from -3 up to $+3$.

$$\varphi_4 = I\xi^3 + J\xi^2 + K\xi + L + M\xi^{-1} + N\xi^{-2} + O\xi^{-3}\,. \tag{38}$$

Then

$$\begin{aligned}
\xi \dot{\varphi}_4 &= 3I\xi^3 + 2J\xi^2 + K\xi - M\xi^{-1} - 2N\xi^{-2} - 3O\xi^{-3}\,, \\
\xi^2 \ddot{\varphi}_4 &= 6I\xi^3 + 2J\xi^2 + 2M\xi^{-1} + 6N\xi^{-2} + 12O\xi^{-3}\,.
\end{aligned}$$

Besides,

$$\begin{aligned}
&(A\xi + B + C\xi^{-1})^4 = A^4\xi^4 + 4A^3B\xi^3 + (6A^2B^2 + 4A^3C)\xi^2 + \\
&+ (4AB^3 + 12A^2BC)\xi + B^4 + 6A^2C^2 + 12AB^2C + (4B^3C + 12ABC^2)\xi^{-1} + \\
&+ (6B^2C^2 + 4AC^3)\xi^{-2} + 4BC^3\xi^{-3} + C^4\xi^{-4}\,.
\end{aligned}$$

Substituting these expressions in Eq. (37) and nullifying coefficients near $\xi^4, \xi^3, \xi^2, \xi, \xi^0, \xi^{-1}, \xi^{-2}, \xi^{-3}, \xi^{-4}$, we obtain a system of 9 algebraic equations for 7 coefficients I, J, K, L, M, N, O. The system has solution

$$\begin{aligned}
I &= \frac{\tilde{c}A^3\gamma^2}{4(2+i\gamma)^2}, \quad J = \frac{2\tilde{c}A^2B\gamma^2(3+i\gamma)}{(2+i\gamma)(4+i\gamma)^2},\\
K &= \frac{\tilde{c}AB^2\gamma^2(12+5i\gamma)}{8(2+i\gamma)(4+i\gamma)} + \frac{\tilde{c}A^2C\gamma^2(3+2i\gamma)}{4(2+i\gamma)^2},\\
L &= \frac{\tilde{c}B^3\gamma^2}{16+\gamma^2} + \frac{2\tilde{c}ABC(48+5\gamma^2)}{(16+\gamma^2)^2},\\
M &= \frac{\tilde{c}B^2C\gamma^2(12-5i\gamma)}{8(2-i\gamma)(4-i\gamma)} + \frac{\tilde{c}AC^2\gamma^2(3-2i\gamma)}{4(2-i\gamma)^2},\\
N &= \frac{2\tilde{c}BC^2\gamma^2(3-i\gamma)}{(2-i\gamma)(4-i\gamma)^2}, \quad O = \frac{\tilde{c}C^3\gamma^2}{4(2-i\gamma)^2}.
\end{aligned} \tag{39}$$

Thus, we have proven.

Theorem 7 *In the exotic expansions* (15) *of solutions to equation* P_3*, the second coefficient* $\varphi_2(\xi)$ *is always the Laurent polynomial* (33) *and* (34)*, but the third coefficient* φ_4 *is a Laurent polynomial* (38) *and* (39)*, if the parameter* $a = 0$.

The case $a \neq 0,\ c = 0$ should be studied separately, using Eq. (35).

4 The Fifth Painlevé Equation P_5 in Case I

4.1 Two Cases for Equation P_5

The fifth Painlevé equation P_5 is

$$y'' = \left(\frac{1}{2y} + \frac{1}{y-1}\right)y'^2 - \frac{y'}{x} + \frac{(y-1)^2}{x^2}\left(ay + \frac{b}{y}\right) + \frac{cy}{x} + \frac{dy(y+1)}{y-1}, \tag{40}$$

where a, b, c, d are complex parameters, x and y are independent and dependent variables, $y' = dy/dx$ [5]. To write Eq. (40) as a differential sum, multiply it by $x^2y(y-1)$ and carry all terms into right side. We obtain the equation

$$\begin{aligned}
&-x^2y(y-1)y'' + x^2(3y-1)y'^2/2 - xy(y-1)x' + (y-1)^3(ay^2+b) -\\
&\quad - cxy^2(y-1) + dx^2y^2(y+1)^2 = 0.
\end{aligned} \tag{41}$$

Its support and polygon are shown in Fig. 10.

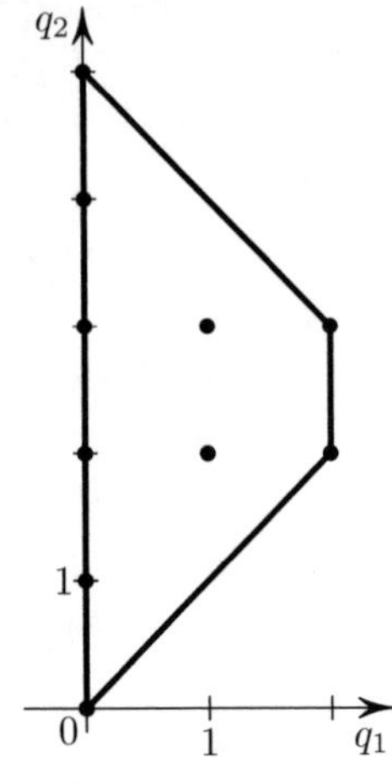

Fig. 10 Support and polygon of Eq. (41)

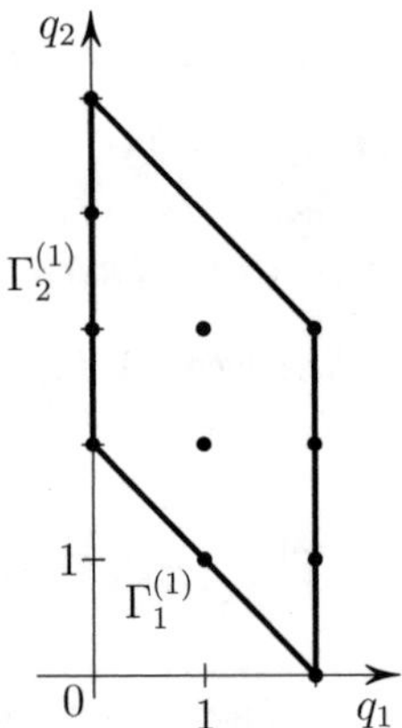

Fig. 11 Support and polygon of Eq. (42)

After substitution $y = 1 + z$ into Eq. (41), we obtain equation

$$-x^2zz''(z+1) + x^2z'^2\left(\frac{3}{2}z+1\right) - xzz'(z+1) + az^3(z+1)^2 + bz^3 + \\ + cxz(z+1)^2 + dx^2(z+1)^2(2+z) = 0\,. \tag{42}$$

Its support and polygon are shown in Fig. 11.

We will differ two cases with different truncated equations:

Case I. Truncated equation corresponds to the low inclined edge $\Gamma_1^{(1)}$ in Fig. 11. It is

$$-z(z''x^2 + z'x) + x^2z/^2 + cxz + 2d = 0$$

and is similar to the truncated Eq. (9) of equation P_3.

Case II. Truncated equation corresponds to the left vertical edge $\Gamma_2^{(1)}$ in Fig. 11.

4.2 Preliminary Transformations in Case I

To transform the edge $\Gamma_1^{(1)}$ in vertical one, we make the power transformation $z = xv$. Then $z' = v + xv'$, $z'' = 2v' + xv''$ and Eq. (42) divided by x^2 takes the form

$$\begin{aligned} g(x, v) \stackrel{def}{=} & -x^2 v v''(1 + xv) + x^2 v'^2 \left(1 + \frac{3}{2} xv\right) - xv'v + \frac{1}{2} xv^3 + \\ & + a(xv^3 + 2x^2v^4 + x^3v^5) + bxv^3 + c(v + 2xv^2 + x^2v^3) + \\ & + d(2 + 5xv + 4x^2v^2 + x^3v^3) = 0 . \end{aligned} \tag{43}$$

Its support and polygon are shown in Fig. 12.

If according to Sect. 2 to write

$$g(x, v) = g_0(x, v) + xg_1(x, v) + x^2 g_2(x, v) + x^3 g_3(x, v),$$

then

$$\begin{aligned} g_0(x, v) &= -x^2 v v'' + x^2 v'^2 - xv'v + cv + 2d , \\ g_1(x, v) &= -x^2 v^2 v'' + \frac{3}{2} x^2 v v'^2 + \left(\tfrac{1}{2} + a + b\right) v^3 + 2cv^2 + 5dv , \\ g_2(x, v) &= 2av^4 + cv^3 + 4dv^2 , \qquad g_3(x, v) = av^5 + dv^3 . \end{aligned} \tag{44}$$

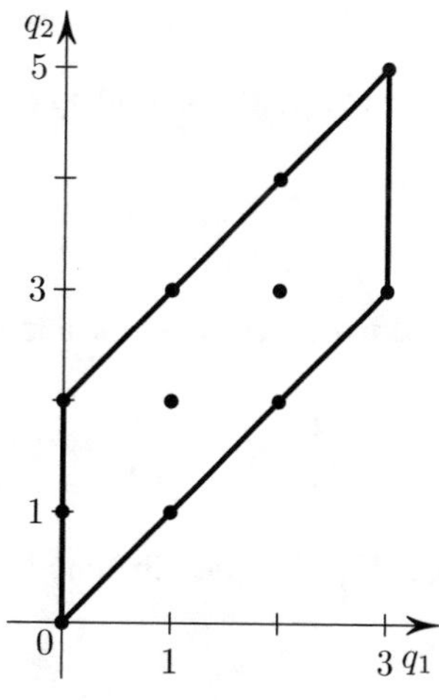

Fig. 12 Support and polygon of Eq. (43)

Complicated and exotic expansions of solutions to Eq. (43) have the form

$$v = \varphi_0(\xi) + x\varphi_1(\xi) + x^2\varphi_2(\xi) + \cdots . \tag{45}$$

According to Theorem 2, equation for the second coefficient φ_1 is

$$\frac{\delta g_0}{\delta v}(x\varphi_1) + x g_1(\varphi_0) = 0 . \tag{46}$$

4.3 Complicated Expansions

In $g_j(x, v)$ from (44), we change the independent variable x by $\xi = \log x + c_0$, where c_0 is arbitrary constant. We obtain

$$\begin{aligned} &g_0^*(\xi, v) = g_0(x, v) = -v\ddot{v} + \dot{v}^2 + cv + 2d , \\ &g_1^*(\xi, v) = g_1(x, v) = -v^2(\ddot{v} - \dot{v}) + \frac{3}{2}v\dot{v}^2 + \omega v^3 + 2cv^2 + 5dv , \\ &g_2^*(\xi, v) = g_2(x, v) = 2av^4 + cv^3 + 4dv^2 , \quad g_3^*(\xi, v) = g_3(x, v) = av^5 + dv^3 , \end{aligned} \tag{47}$$

where $\omega = \dfrac{1}{2} + a + b$.

According to Theorem 3, solutions $v = \varphi_0(\xi)$ to equation $g_0(\xi, v) = 0$, which are the Laurent series in decreasing powers of ξ, form two families:
additional: $\varphi_0 = v = \beta\xi$ for $c = 0$, $\beta^2 = -2d$, $d \neq 0$, and
main: $\varphi_0 = v = -\dfrac{c}{2}\xi^2 - \dfrac{d}{c}$ for $c \neq 0$.

According to (47)

$$\frac{\delta g_0^*}{\delta v} = -v\frac{d^2}{d\xi^2} + 2\dot{v}\frac{d}{d\xi} + c - \ddot{v} .$$

According to Corollary 1

$$\frac{d}{d\xi}(x\varphi_1) = x[\varphi_1 + \dot{\varphi}_1], \quad \frac{d^2}{d\xi^2}(x\varphi_1) = x[\varphi_1 + 2\dot{\varphi}_1 + \ddot{\varphi}_1] .$$

First we consider the additional family. Then

$$\frac{\delta g_0^*}{\delta v} = -\beta\xi\frac{d^2}{d\xi^2} + 2\beta\frac{d}{d\xi}, \quad g_1 = \omega\beta^3\xi^3 + \beta^3\xi^2 + \left(\frac{3}{2}\beta^3 + 5d\beta\right)\xi$$

and Eq. (46) after dividing by x and using $2d = -\beta^2$ takes the form

$$-\beta\xi[\varphi_1 + 2\dot{\varphi}_1 + \ddot{\varphi}_1] + 2\beta[\varphi_1 + \dot{\varphi}_1] + \omega\beta^3\xi^3 + \beta^3\xi^2 - \beta^3\xi = 0 .$$

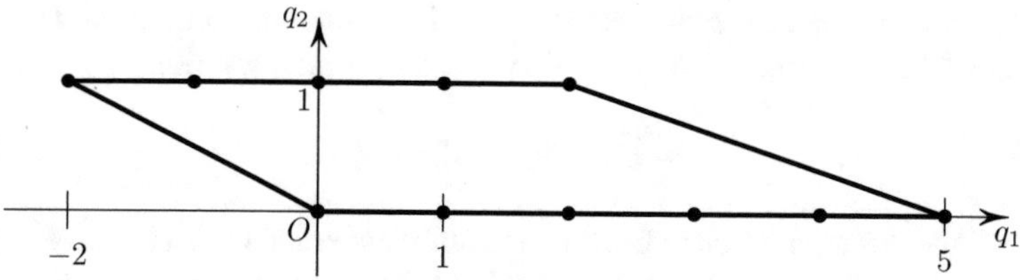

Fig. 13 Support and polygon of Eq. (50)

It has the polynomial solution

$$\varphi_1 = -2\omega d(\xi^2 - 2\xi + 2) - 2d\xi + 2d\,. \tag{48}$$

Now we consider the main family. Then

$$\frac{\delta g_0^*}{\delta v} = \left(\frac{c}{2}\xi^2 + \frac{d}{c}\right)\frac{d^2}{d\xi^2} - 2c\xi\frac{d}{d\xi} + 2c\,,$$
$$g_1^* = \omega v^3 - c\xi v^2 + cv^2 + \frac{3}{2}c^2\xi^2 v + 2cv^2 + 5dv = \omega v^3 - c\xi v^2 + 2dv\,.$$

Equation (46) after division by x is

$$\left(\frac{c}{2}\xi^2 + \frac{d}{c}\right)[\varphi_1 + 2\dot\varphi_1 + \ddot\varphi_1] - 2c\xi[\varphi_1 + \dot\varphi_1] + 2c\varphi_1 + \omega v^3 - c\xi v^2 + 2dv = 0\,. \tag{49}$$

At first we consider auxiliary equation

$$\left(\frac{c}{2}\xi^2 + \frac{d}{c}\right)[\varphi_1 + 2\dot\varphi_1 + \ddot\varphi_1] - 2c\xi[\varphi_1 + \dot\varphi_1] + 2c\varphi_1 + \omega v^3 = 0\,.$$

It has the polynomial solution

$$\varphi_1 = -\omega\frac{c^2}{4}[\xi^4 - 4\xi^3 + (8 + 2\lambda)\xi^2 - (8 + 4\lambda)\xi + \lambda^2]\,,$$

where $\lambda = \dfrac{2d}{c^2}$.

Now we consider Eq. (49) with $\omega = 0$. We divide the equation by $c/2$ and put $\varphi_1 = c^2\psi_1/2$. Then the Eq. (49) takes the form

$$(\xi^2 + \lambda)[\psi_1 + 2\dot\psi_1 + \ddot\psi_1] - 4\xi[\psi_1 + \dot\psi_1] + 4\psi_1 - \xi(\xi^2 + \lambda)^2 + 2\lambda(\xi^2 + \lambda) = 0\,. \tag{50}$$

Its support and polygon are shown in Fig. 13.

As the inclination of the right edge is equal -3, then its solution in decreasing powers of ξ begins from ξ^3. So we look for its polynomial solution

$$\psi = \xi^3 + B\xi^2 + C\xi + D\,.$$

We substitute that expression in Eq. (50) and nullify coefficients for $\xi^5, \xi^4, \xi^3, \xi^2, \xi^1$, ξ^0. We obtain six linear algebraic equations for three coefficients B, C, D. Subsystem of first 4 equations for $\xi^5, \xi^4, \xi^3, \xi^2$ is triangle and has solution

$$B = -2, \quad C = 2 + \lambda, \quad D = -4\lambda.$$

Substituted these values in equation for ξ and ξ^0, we obtain equations $16\lambda = 0$ and $-16\lambda = 0$. Hence, $\lambda = 0$, i.e. $d = 0$. Thus, Eq. (50) has a polynomial solution only for $d = \lambda = 0$, and the solution is

$$\psi_1 = \xi^3 - 2\xi^2 + 2\xi\,.$$

Hence, the Eq. (50) has a polynomial solution only if $d = 0$, and the solution is

$$\varphi_1 = -\omega\frac{c^2}{4}[\xi^4 - 4\xi^3 + 8\xi^2 - 8\xi] + \frac{c^2}{2}[\xi^3 - 2\xi^2 + 2\xi]\,. \tag{51}$$

Theorem 8 *For the equation P_5 in Case I, the second coefficient $\varphi_1(\xi)$ in complicated expansions* (45) *of its solutions is polynomial* (48) *for the additional family always and* (51) *for the main family iff $d = 0$.*

4.4 Exotic Expansions

We introduce new independent variable

$$\xi = x^{i\gamma}, \ \gamma \in \mathbb{R}, \ \gamma \neq 0\,. \tag{52}$$

Then

$$v' = i\gamma\dot{v}\frac{\xi}{x}, \ v'' = \ddot{v}\left(i\gamma\frac{\xi}{x}\right)^2 + \dot{v}(i\gamma)^2\frac{\xi}{x^2} - \dot{v}i\gamma\frac{\xi}{x^2}\,, \tag{53}$$

where $\dot{v} = dv/d\xi$. Then

$$\begin{aligned} xv' &= i\gamma\xi\dot{v}\,, \\ x^2v'' &= -\gamma^2\xi^2\ddot{v} - \gamma^2\xi\dot{v} - i\gamma\xi\dot{v}\,. \end{aligned} \tag{54}$$

Hence, formula (44) give

$$g_0 = \gamma^2 v(\xi^2\ddot{v} + \xi\dot{v}) - \gamma^2\xi^2\dot{v}^2 + cv + 2d\,,$$

$$g_1 = v^2(\gamma^2\xi^2\ddot{v} + \gamma^2\xi\dot{v} + i\gamma\dot{v}) - \frac{3}{2}\gamma^2\xi^2\dot{v}^2 + \omega v^3 + 2cv^2 + 5dv\,.$$

We put

$$\tilde{g}_0 = g_0/\gamma^2,\ \tilde{g}_1 = g_1/\gamma^2,\ \tilde{\omega} = \omega/\gamma^2,\ \tilde{c} = c/\gamma^2,\ \tilde{d} = d/\gamma^2\,.$$

Then these formulas give

$$\tilde{g}_0 = v(\xi^2\ddot{v} + \xi\dot{v}) - \xi^2\dot{v}^2 + \tilde{c}v + 2\tilde{d}\,,$$

$$\tilde{g}_1 = v^2\left[\xi^2\ddot{v} + \xi\dot{v}\left(1 - \frac{1}{i\gamma}\right)\right] - \frac{3}{2}v\xi^2\dot{v}^2 + \tilde{\omega}v^3 + 2\tilde{c}v^2 + 5\tilde{d}v\,. \tag{55}$$

From the first formulae (55) we have

$$\frac{\delta\tilde{g}_0}{\delta v} = v\xi^2\frac{d^2}{d\xi^2} + (v - 2\dot{v}\xi)\xi\frac{d}{d\xi} + \tilde{c} + \xi^2\ddot{v} + \xi\dot{v}\,.$$

According to Theorem 6, all solutions to equation $\tilde{g}_0 = 0$ in the form of Laurent series form one family of solutions

$$\varphi_0 = v = A\xi + B + C\xi^{-1}\,,$$

with following connections

$$B = -\tilde{c}, \qquad 4AC = \tilde{c}^2 - 2\tilde{d}\,.$$

As

$$v - 2\dot{v}\xi = -A\xi + B + 3C\xi^{-1},\quad \tilde{c} + \xi^2\ddot{v} + \xi\dot{v} = A\xi - B + C\xi^{-1}\,,$$

$$\begin{aligned}
\tilde{g}_1 &= \tilde{\omega}v^3 + v^2\left[\ddot{v}\xi^2 + \dot{v}\xi\left(1 - \frac{1}{i\gamma}\right)\right] - \frac{3}{2}v\xi^2\dot{v}^2 + 2\tilde{c}v^2 + 5\tilde{d}v = \\
&= \tilde{\omega}\Big[A^3\xi^3 + 3A^2B\xi^2 + 3(AB^2 + A^2C)\xi + B^3 + 6ABC + 3(AC^2 + B^2C)\xi^{-1} + \\
&\quad + 3BC^2\xi^{-2} + C^3\xi^{-3}\Big] - A^3\frac{2 + i\gamma}{2i\gamma}\xi^3 - A^2B\frac{4 + 3i\gamma}{2i\gamma}\xi^2 + \\
&\quad + \left(-A^2C\frac{2 + 11i\gamma}{2i\gamma} - AB^2\frac{2 + i\gamma}{2i\gamma}\right)\xi + \frac{1}{2}B^3 - 7ABC + \\
&\quad + \left(AC^2\frac{2 - 11i\gamma}{2i\gamma} + B^2C\frac{2 - i\gamma}{2i\gamma}\right)\xi^{-1} + BC^2\frac{4 - 3i\gamma}{2i\gamma}\xi^{-2} + C^3\frac{2 - i\gamma}{2i\gamma}\xi^{-3}\,,
\end{aligned}$$

then equation for $\varphi_1(\xi)$ is

$$(A\xi + B + C\xi^{-1})\left[\frac{1}{i\gamma}\left(\frac{1}{i\gamma} - 1\right)\varphi_1 + \frac{2}{i\gamma}\xi\dot{\varphi}_1 + \xi^2\ddot{\varphi}_1\right] +$$
$$+ (-A\xi + B + 3C\xi^{-1})\left[\frac{1}{i\gamma}\varphi_1 + \xi\dot{\varphi}_1\right] + (A\xi - B + C\xi^{-1})\varphi_1 + \tilde{g}_1 = 0\,.$$

Its solution is the Laurent polynomial

$$\begin{aligned}\varphi_1(\xi) = \tilde{\omega}\gamma^2 \Bigg[& \frac{A^2}{(1+i\gamma)^2}\xi^2 + \frac{2AB}{1+i\gamma}\xi + \frac{B^2}{1+\gamma^2} + \frac{AC(2+6\gamma^2)}{(1+\gamma^2)^2} + \\ & \frac{2BC}{1-i\gamma}\xi^{-1} + \frac{C^2}{(1-i\gamma)^2}\xi^{-2}\Bigg] + \\ + \gamma^2\Bigg[& -\frac{A^2(2+i\gamma)}{2i\gamma(1+i\gamma)^2}\xi^2 - \frac{AB}{i\gamma(1+i\gamma)}\xi + \frac{B^2}{2(1+\gamma^2)} - \\ & \frac{AC(1-\gamma^2)}{(1+\gamma^2)^2} + \frac{BC}{i\gamma(1-i\gamma)}\xi^{-1} + \frac{C^2(2-i\gamma)}{2i\gamma(1-i\gamma)^2}\xi^{-2}\Bigg].\end{aligned} \tag{56}$$

So, we have proved.

Theorem 9 *In exotic expansion* (45) *of solutions to equation P_5 in Case I, coefficient $\varphi_1(\xi)$ is the Laurent polynomial* (56).

5 The Fifth Painlevé Equation P_5 in Case II

5.1 Preliminary Transformations

To obtain polynomial φ_0, we make in Eq. (42) the power transformation $z = \dfrac{1}{w}$. Then

$$z' = -\frac{w'}{w^2}, \quad z'' = \frac{2w'^2 - ww''}{w^3},$$

and Eq. (42), multiplied by x^5, takes the form

$$\begin{aligned}h(x,w) \overset{def}{=} x^2ww''(1+w) - x^2w'^2\left(\frac{1}{2} + w\right) + xww'(1+w) + a(1+w)^2 + \\ + bw^2 + cxw^2(w+1)^2 + dx^2w^2(w+1)^2(1+2w) = 0\,.\end{aligned} \tag{57}$$

Its support and polygon are shown in Fig. 14. If write

$$h(x,w) = h_0(x,w) + xh_1(x,w) + x^2h_2(x,w)\,,$$

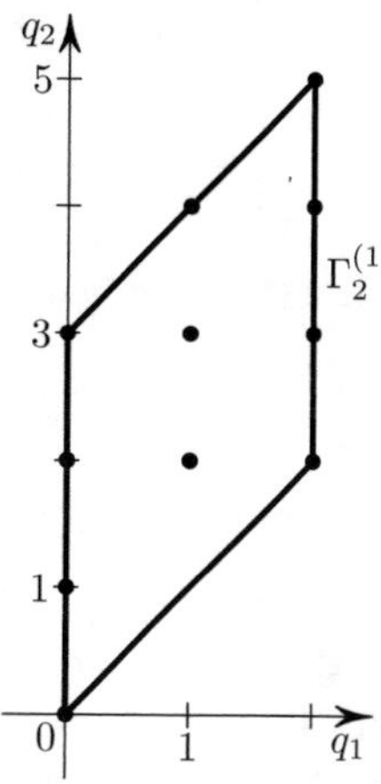

Fig. 14 Support and polygon of Eq. (57)

then

$$
\begin{aligned}
h_0(x, w) &= x^2 w w''(w+1) - x^2 w'^2 \left(w + \frac{1}{2}\right) + x w w'(w+1) + \\
&\quad + a(w+1)^2 + bw^2 , \\
h_1(x, w) &= cw^2(1+w)^2 , \\
h_2(x, w) &= dw^2(w+1)^2(2w+1) .
\end{aligned} \tag{58}
$$

Now formulas (45) and (46) are again correct if we put w instead of v.

5.2 *Complicated Expansions*

In $h_j(x, w)$ from (58), we change the independent variable x by $\xi = \log x + c_0$. We obtain

$$
\begin{aligned}
h_0^*(\xi, w) &= h_0(x, w) = \ddot{w} w(w+1) - \dot{w}^2 \left(w + \frac{1}{2}\right) + a(w+1)^2 + bw^2 , \\
h_1^*(\xi, w) &= h_1(x, w) = cw^2(w+1)^2 , \\
h_2^*(\xi, w) &= h_2(x, w) = dw^2(w+1)(2w+1) .
\end{aligned} \tag{59}
$$

Let us find all solutions of equation $h_0^*(\xi, w) = 0$ in the form of Laurent series.

Theorem 10 *All solutions* $w = \varphi_0(\xi)$ *of equation* $h_0^*(\xi, w) = 0$ *from* (59) *in the form of Laurent series and different from constant form two families:*

main (if $a + b \overset{def}{=} \alpha \neq 0$*)*

$$w = \varphi_0 = \frac{a+b}{2}\,(\xi + c_0)^2 - \frac{a}{a+b} = \frac{\alpha}{2}\,(\xi + c_0)^2 - \frac{a}{\alpha}\,, \tag{60}$$

and additional (if $\alpha = 0,\ a \neq 0$*)*

$$w = \varphi_0 = \beta\,(\xi + c_0)\,, \quad \beta^2 = 2a\,. \tag{61}$$

Here c_0 *is arbitrary constant.*

Proof We will consider 3 cases: (1) $\alpha \neq 0$; (2) $\alpha = 0,\ a \neq 0$; (3) $\alpha = a = 0$.

Case (1) $\alpha \neq 0$. Support and polygon Γ of equation $h_0^*(\xi, w) = 0$ are shown in Fig. 15.

Right side of the boundary $\partial\widetilde{\Gamma}$ of the polygon $\widetilde{\Gamma}$ consists of three vertices $\widetilde{\Gamma}_1^{(0)} = (-2, 3)$, $\widetilde{\Gamma}_2^{(0)} = (0, 2)$, $\widetilde{\Gamma}_3^{(0)} = 0$ and two edges $\widetilde{\Gamma}_1^{(1)}$ and $\widetilde{\Gamma}_2^{(1)}$. Corresponding truncations are

$$\begin{aligned} \hat{h}_1^{*(0)} &= \ddot{w}w^2 - \dot{w}^2 w, \quad \hat{h}_2^{*(0)} = \alpha w^2, \quad \hat{h}_3^{*(0)} = a, \\ \hat{h}_1^{*(1)} &= \ddot{w}w^2 - \dot{w}^2 w + \alpha w^2, \quad \hat{h}_2^{*(2)} = a(w+1)^2 + b w^2\,. \end{aligned}$$

Characteristic equation for truncation $\hat{h}_1^{*(0)}$ is $-r = 0$. It has unique solution $r = 0$. But vector $(1, 0)$ does not belong to the normal cone

$$\mathbf{U}_1^{(0)} = \{P = \lambda_1(0,1) + \lambda_2(1,2), \quad \lambda_1, \lambda_2 \geqslant 0, \quad \lambda_1 + \lambda_2 > 0\}$$

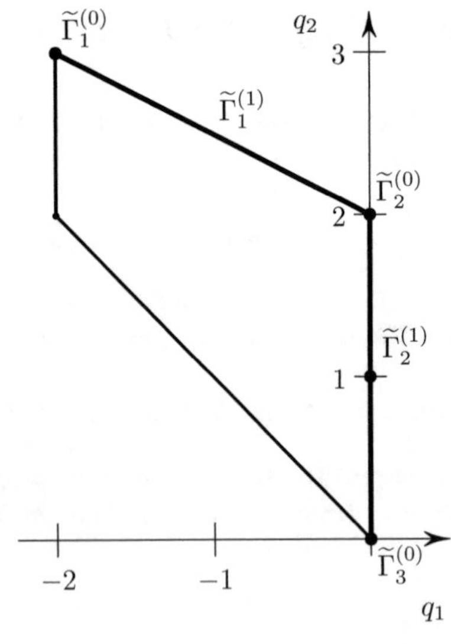

Fig. 15 Support and polygon of equation $h_0^*(\xi, w) = 0$

Truncations $\hat{h}_2^{*(0)}$ and $\hat{h}_3^{*(0)}$ have trivial characteristic equations $\alpha = 0$ and $a = 0$, which have no solutions. Truncated equation $\hat{h}_1^{*(1)} = 0$ has the power solution $w = \alpha\xi^2/2$. According to Sect. 3.3 of [10], we will find critical numbers of that solution. We have

$$\frac{\delta \hat{h}_1^{*(1)}}{\delta w} = w^2 \frac{d^2}{d\xi^2} - 2\dot{w}w\frac{d}{d\xi} + 2\ddot{w}w - \dot{w}^2 + 2\alpha w\,.$$

On the curve $w = \alpha\xi^2/2$, that variation gives operator

$$\mathscr{L}(\xi) = \frac{\alpha^2\xi^4}{4}\frac{d^2}{d\xi^2} - \alpha^2\frac{d}{d\xi} + \alpha^2\xi^2 - \alpha^2\xi^2 + \alpha^2\xi^2\,.$$

Characteristic polynomial of sum $\mathscr{L}(\xi)\xi^k$ is

$$\nu(k) = \frac{\alpha^2}{4}\left[k(k-1) - 4k + 4\right].$$

It has two roots $k_1 = 1$ and $k_2 = 4$. But $k_1 < 2$, and $k_2 > 2$ and it is not a critical number. So we have only one critical number $k_1 = 1$.

According to Sect. 3.4 of [10], the set

$$\mathbf{K}_1^{(1)} = \{2 - 2l,\ l \in \mathbb{N}\} = \{0, -2, -4, \ldots\}\,.$$

Now the critical number $k_1 = 1$ does not belong to the set $\mathbf{K}_1^{(1)}$. Thus, according to Theorem 3 [10], equation $h_0^*(\xi, w) = 0$ has a solution in the form of Laurent series

$$w = \alpha\xi^2/2 + \gamma_0 + \sum_{k=1}^{\infty} \gamma_{2k}\xi^{-2k}\,, \tag{62}$$

where $\gamma_i = const$. To find γ_0, we put $w = \alpha\xi^2/2 + \gamma_0$ into $h_0^*(\xi, w)$. We have $\dot{w} = \alpha\xi$, $\ddot{w} = \alpha$, hence,

$$h_0^*(\xi, \alpha\xi^2/2 + \gamma_0) = \alpha(\alpha\gamma_0 + a)\xi^2 + (2\gamma_0 + 1)(\alpha\gamma_0 + a)\,.$$

Both coefficients near ξ^2 and ξ^0 are zero, iff $\gamma_0 = -a/\alpha$. So, solution (62) is indeed the polynomial

$$w = \alpha\xi^2/2 - a/\alpha.$$

Equation $h_0^*(\xi, w) = 0$ does not contain explicitly the independent variable ξ, so to its solution $w(\xi)$ there correspond solutions $w(\xi + c_0)$, where c_0 is arbitrary constant. Hence, we obtain family (60).

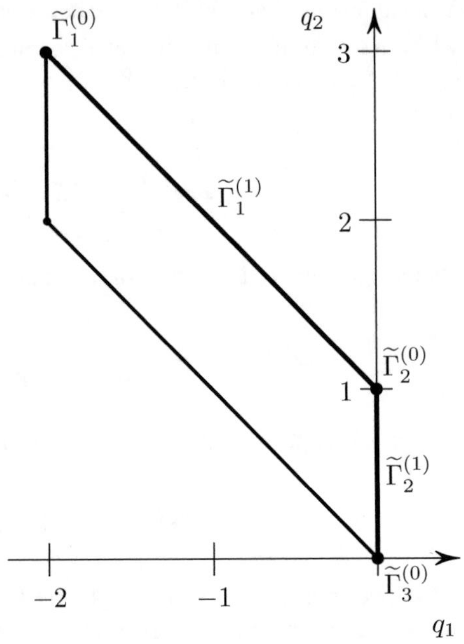

Fig. 16 Support and polygon of equation $h_0^*(\xi, w) = 0$ in case (2)

To finish that case, we must consider the last truncation $\hat{h}_2^{*(1)}$. It is the square polynomial $(a+b)\xi^2 + 2a\xi + a^2$. Its discriminant

$$\Delta = -4ab\,.$$

If $\Delta \neq 0$, then the polynomial has two roots. Each of them is the constant solution of the equation $h_0^*(\xi, w) = 0$ and cannot be continued into power expansion.

If $\Delta = 0$, i.e. $a = 0$ or $b = 0$, then the polynomial has one double solution $w = 0$ or $w = -1$. They are constant double solutions of the full equation $h_0^*(\xi, w) = 0$, and does not give nonconstant solutions to equation $h_0^*(\xi, w) = 0$. But we are looking for nonconstant solutions.

Case (2) $\alpha = 0$, $a \neq 0$. Support and polygon $\widetilde{\Gamma}$ of $h_0^*(\xi, w)$ are shown in Fig. 16.

Right side of the boundary $\partial\widetilde{\Gamma}$ of the polygon $\widetilde{\Gamma}$ consists of three vertices $\widetilde{\Gamma}_1^{(0)} = (-2, 3)$, $\widetilde{\Gamma}_2^{(0)} = (0, 1)$, $\widetilde{\Gamma}_3^{(0)} = 0$ and two edges $\widetilde{\Gamma}_1^{(1)}$ and $\widetilde{\Gamma}_2^{(1)}$. As in case (1), truncated equations, corresponding to all vertices and edge $\widetilde{\Gamma}_2^{(1)}$ do not give us power expansions of solutions to equation $h_0^*(\xi, w) = 0$. So we consider the truncated equation

$$\hat{h}_1^{*(1)} \overset{def}{=} \ddot{w}w^2 - \dot{w}^2 w + 2aw = 0\,.$$

It has power solutions $w = \beta\xi$ with $\beta^2 = 2a$. The solution satisfies the equation

$$h_0^*(\xi, w) - \hat{h}_1^{*(1)}(\xi, w) \stackrel{def}{=} \ddot{w}w - \dot{w}^2/2 + a = 0\,.$$

Hence, the equation has family of solutions (61).

Case (3) $a = b = 0$. Here all solutions of equation $h_0^*(\xi, w) = 0$ belong to two-parameter family $w = \dfrac{[c_2 \exp(c_1\xi) - 1]^2}{4c_2 \exp(c_1\xi)}$, where c_1 and c_2 are arbitrary constants. No one of these solutions has a power expansion. □

Here

$$\frac{\delta h_0^*}{\partial w} = w(w+1)\frac{d^2}{d\xi^2} - 2\left(w + \frac{1}{2}\right)\dot{w}\frac{d}{d\xi} + 2a(w+1) + 2bw + \ddot{w}(2w+1) - \dot{w}^2\,.$$

Let us compute solution φ_1 to Eq. (46) for additional family (61). Here $\dot{w} = \beta$, $\ddot{w} = 0$, $\dot{w}^2 = 2a$ and Eq. (46) divided by x is

$$\beta\xi(\beta\xi + 1)\,[\varphi_1 + 2\dot{\varphi}_1 + \ddot{\varphi}_1] - \beta(2\beta\xi + 1)\,[\varphi_1 + \dot{\varphi}_1] + c\beta^2\xi^2(\beta\xi + 1)^2 = 0\,.$$

It has polynomial solution

$$\varphi_1 = c\beta[-\beta\xi^2 + (2\beta - 1)\xi + 1]\,. \tag{63}$$

For the main family (60), Eq. (46) divided by x is

$$w(w+1)\,[\varphi_1 + 2\dot{\varphi}_1 + \ddot{\varphi}_1] - \alpha\xi(2w+1)\,[\varphi_1 + \dot{\varphi}_1] + \alpha(2w+1)\varphi_1 + cw^2(w+1)^2 = 0\,,$$

where $w = \dfrac{\alpha}{2}\xi^2 - \dfrac{a}{\alpha}$. It has the polynomial solution

$$\varphi_1(\xi) = -c\left[\frac{\alpha^2}{4}\xi^4 - \alpha^2\xi^3 + \left(2\alpha^2 + \frac{\alpha}{2} - a\right)\xi^2 - \right. \\ \left. - \left(2\alpha^2 + \alpha - 2a\right)\xi + \frac{a(a-\alpha)}{\alpha^2}\right]. \tag{64}$$

Thus, we have proven.

Theorem 11 *In Case II of equation P_5 the second coefficient φ_1 of complicated expansions* (45) *is a polynomial* (63) *for the additional family and polynomial* (64) *for the main family.*

5.3 *Exotic Expansions*

Let us introduce new independent variable $\xi = x^{i\gamma}$ according to (52). Then, according to (53), formulas (58), divided by γ^2, take the forms

$$\begin{aligned}\tilde{h}_0(\xi, w) &= \gamma^{-2}h_0(x, w) = -w(w+1)(\xi^2\ddot{w} + \xi\dot{w}) + \left(w + \frac{1}{2}\right)\xi^2\dot{w}^2 + \\ &\quad + \tilde{a}(w+1)^2 + \tilde{b}w^2\,, \\ \tilde{h}_1(\xi, w) &= \gamma^{-2}h_1(x, w) = \tilde{c}w^2(w+1)^2\,,\end{aligned} \tag{65}$$

where $\tilde{a} = a/\gamma^2,\ \tilde{b} = b/\gamma^2,\ \tilde{c} = c/\gamma^2$.

Theorem 12 *All solutions* $w = \varphi_0(\xi)$ *to equation* $\tilde{h}_0(\xi, w) = 0$ *from* (65) *in the form of Laurent series form one family*

$$w = \varphi_0 = A\xi + B + C\xi^{-1}\,, \tag{66}$$

where parameters are connected by equalities

$$B = \tilde{a} + \tilde{b} - \frac{1}{2}\,, \qquad 4AC = (\tilde{a} + \tilde{b})^2 + \tilde{a} - \tilde{b} + \frac{1}{4}\,. \tag{67}$$

Proof First we will show that parameters satisfy to (67) for solution (66) to equation $\tilde{h}_0(\xi, w) = 0$. Let us denote

$$\alpha = A\xi + C\xi^{-1} \quad \text{and} \quad \beta = A\xi - C\xi^{-1}\,.$$

Then $\xi\dot{w} = A\xi - C\xi^{-1}$, $\xi^2\ddot{w} = 2C\xi^{-1}$ and $\xi\dot{w} + \xi^2\ddot{w} = \alpha$. So

$$\begin{aligned}\tilde{h}_0(\xi, w) &= -(\alpha + B)(\alpha + B + 1)\alpha + \left(\alpha + B + \frac{1}{2}\right)\beta^2 + \tilde{a}(\alpha + B + 1)^2 + \tilde{b}(\alpha + B)^2 = \\ &= -\alpha^3 + \alpha\beta^2 - \alpha^2(2B+1) + \left(B + \frac{1}{2}\right)\beta^2 + \tilde{a}\alpha^2 + \tilde{b}\alpha^2 - B(B+1)\alpha + 2\tilde{a}(B+1)\alpha + \\ &\quad + 2\tilde{b}B\alpha + \tilde{a}(B+1)^2 + \tilde{b}B^2 = \alpha[\beta^2 - \alpha^2] + \alpha^2[\tilde{a} + \tilde{b} - (2B+1)] + \left(B + \frac{1}{2}\right)\beta^2 + \\ &\quad + \alpha[2\tilde{a}(B+1) + 2\tilde{b}B - B(B+1)] + \tilde{a}(B+1)^2 + \tilde{b}B^2\,.\end{aligned}$$

We have

$$\beta^2 - \alpha^2 = (\beta - \alpha)(\beta + \alpha) = 2A\xi(-2C\xi^{-1}) = -4AC\,.$$

Hence, $\beta^2 = \alpha^2 - 4AC$ and

$$\begin{aligned}\tilde{h}_0(\xi, w) &= -4AC\alpha + \alpha^2\left[\tilde{a} + \tilde{b} - (2B+1) + B + \frac{1}{2}\right] - 4AC\left(B + \frac{1}{2}\right) + \\ &\quad + \alpha[2\tilde{a}(B+1) + 2\tilde{b}B - B(B+1)] + \tilde{a}(B+1)^2 + \tilde{b}B^2 = \alpha^2\left[\tilde{a} + \tilde{b} - B - \frac{1}{2}\right] + \\ &\quad + \alpha[2\tilde{a}(B+1) + 2\tilde{b}B - B(B+1) - 4AC] + \tilde{a}(B+1)^2 + \tilde{b}B^2 - 4AC\left(B + \frac{1}{2}\right).\end{aligned}$$

But $\alpha^2 = A^2\xi^2 + 2AC + C^2\xi^{-2}$, hence,

$$\tilde{h}_0(\xi, w) = (A^2\xi^2 + C^2\xi^{-2})\left(\tilde{a} + \tilde{b} - B - \frac{1}{2}\right) + \alpha[2\tilde{a}(B+1) + 2\tilde{b}B - B(B+1) - 4AC] +$$
$$+\tilde{a}(B+1)^2 + \tilde{b}B^2 - 4AC\left(B + \frac{1}{2}\right) + 2AC\left(\tilde{a} + \tilde{b} - B - \frac{1}{2}\right) \equiv 0\,.$$

It means that coefficients for $\xi^{\pm 2}$, α and ξ^0 are zero. Exactly $\tilde{a} + \tilde{b} - B - \frac{1}{2} = 0$, i.e.

$$\tilde{a} + \tilde{b} = B + \frac{1}{2}\,; \tag{68}$$

$$0 = 2\tilde{a}(B+1) + 2\tilde{b}B - B^2 - B - 4AC = 2\left(B + \frac{1}{2}\right)B + 2\tilde{a} - B^2 - B - 4AC =$$
$$= B^2 + 2\tilde{a} - 4AC$$

according to (68), i.e.

$$4AC = B^2 + 2\tilde{a}\,. \tag{69}$$

Finally,

$$\tilde{a}(B+1)^2 + \tilde{b}B^2 - 4AC\left(B + \frac{1}{2}\right) = (\tilde{a} + \tilde{b})B^2 + 2\tilde{a}B + \tilde{a} - 4AC\left(B + \frac{1}{2}\right) =$$
$$= (\tilde{a} + \tilde{b})[B^2 + 2\tilde{a} - 4AC] = 0$$

according to (68) and (69).

Now we will show that, for any solution

$$w = A\xi^1 + B + C\xi^{-1} + D\xi^{-l} + \cdots, \quad l \geqslant 2 \tag{70}$$

to equation $\tilde{h}_0(\xi, w) = 0$, coefficient $D = 0$. We insert (70) in $\tilde{h}_0(\xi, w)$ and find in it a term with maximal power ξ, containing D. Terms of the third order in $\tilde{h}_0(\xi, w)$ are

$$-w^2(\xi^2\ddot{w} + \xi\dot{w}) + w\xi^2\dot{w}^2 \stackrel{def}{=} \Omega_3\,.$$

We assume that $w = A\xi + D\xi^{-l}$, then $\xi^2\ddot{w} + \xi\dot{w} = A\xi + l^2D\xi^{-l}$ and

$$\Omega_3 = -(A + D\xi^{-l})^2(A\xi + l^2D\xi^{-l}) + (A\xi + D\xi^{-l})(A\xi - lD\xi^{-l})^2 =$$
$$= -(l+1)^2A^2D\xi^{2-l} + \cdots$$

Coefficient before the ξ^{2-l} must be zero. But $(l+1)^2 \neq 0$, $A^2 \neq 0$, hence $D = 0$. □

According to (65)

$$\frac{\delta\tilde{h}_0}{\delta w} = -w(w+1)\xi^2\frac{d^2}{d\xi^2} - w(w+1)\xi\frac{d}{d\xi} + 2\left(w+\frac{1}{2}\right)\dot{w}\xi^2\frac{d}{d\xi} +$$

$$+ 2\tilde{a}(w+1) + 2\tilde{b}w - (2w+1)\left(\xi^2\ddot{w} + \xi\dot{w}\right) + \xi^2\dot{w}^2 .$$

Equation (46) for $\varphi_1(\xi)$ is

$$a_1\left[\frac{1}{i\gamma}\left(\frac{1}{i\gamma}-1\right)\varphi_1 + \frac{2}{i\gamma}\xi\dot{\varphi}_1 + \xi^2\ddot{\varphi}_1\right] + a_2\left[\frac{1}{i\gamma}\varphi_1 + \xi\dot{\varphi}_1\right] + a_3\varphi_1 + \tilde{h}_1 = 0, \tag{71}$$

where

$$\begin{aligned}
a_1 &= -w(w+1) = -A^2\xi^2 - A(2B+1)\xi - 2AC - B(B+1) - \\
&\quad - (2B+1)C\xi^{-1} - C^2\xi^{-2}, \\
a_2 &= (2w+1)\dot{w}\xi - w(w+1) = A^2\xi^2 - [2AC + B(B+1)] - 2(2B+1)C\xi^{-1} - \\
&\quad - 3C^2\xi^{-2}, \\
a_3 &= 2\tilde{a}(w+1) + 2\tilde{b}w - (2w+1)(\xi^2\ddot{w} + \xi\dot{w}) + \xi^2\dot{w}^2 = -A^2\xi^2 - 2AC + \\
&\quad + B(B+1) - C^2\xi^{-2}, \\
\frac{\tilde{h}_1}{\tilde{c}} &= w^2(w+1)^2 = A^4\xi^4 + 2A^3(2B+1)\xi^3 + [4A^3C + A^2\beta]\xi^2 + \\
&\quad + [6A^2(2B+1)C + 2AB(B+1)(2B+1)]\xi + 6A^2C^2 + 2A\beta C + \\
&\quad + B^2(B+1)^2 + [6A(2B+1)C^2 + 2B(B+1)(2B+1)C]\xi^{-1} + \\
&\quad + [4AC^3 + \beta C^2]\xi^{-2} + 2(2B+1)C^3\xi^{-3} + C^4\xi^{-4} \stackrel{def}{=} \tilde{h}_{15}, \\
\beta &= 6B(B+1) + 1.
\end{aligned}$$

Support and the Newton polygon Γ for Eq. (69) are shown in Fig. 17. As inclinations of side edges of the polygon Γ are ± 2, then polynomial solutions to Eq. (71) should be as

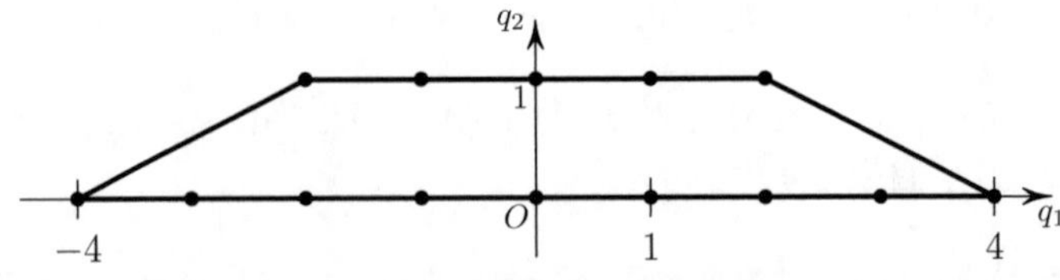

Fig. 17 Support and polygon of Eq. (69)

$$\varphi_1 = D\xi^2 + E\xi + F + G\xi^{-1} + H\xi^{-2}. \tag{72}$$

Inserting that φ_1 into Eq. (71), we obtain a linear system of 9 algebraic equations for 5 coefficients D, E, F, G, H. Equations correspond to vanish of coefficients near $\xi^4, \xi^3, \xi^2, \xi, \xi^0, \xi^{-1}, \xi^{-2}, \xi^{-3}, \xi^{-4}$. From coefficients near ξ^4, ξ^3, ξ^2, we find

$$\begin{aligned} D &\stackrel{def}{=} D_1 = -c\frac{A^2}{(1+i\gamma)^2}, \quad E \stackrel{def}{=} E_1 = -c\frac{A(2B+1)}{1+i\gamma}, \\ F &\stackrel{def}{=} F_1 = -c\frac{2AC(1-\gamma^2)}{(1+\gamma^2)^2} - c\frac{B(B+1)(1-3\gamma^2)}{(1+\gamma^2)^2}. \end{aligned} \tag{73}$$

From coefficients near ξ^{-2}, ξ^{-3}, ξ^{-4}, we find

$$\begin{aligned} H = H_2 &= -c\frac{C^2}{(1-i\gamma)^2}, \quad G = G_2 = -c\frac{(2B+1)C}{1-i\gamma}, \\ F = F_2 &= -c\frac{2AC(1+7\gamma^2)}{(1+\gamma^2)^2} - c\frac{B(B+1)(1+5\gamma^2)}{(1+\gamma^2)^2}. \end{aligned} \tag{74}$$

According to (73) and (74), equality $F_1 = F_2$ is possible, iff

$$2AC + B(B+1) = 0. \tag{75}$$

Then

$$F = -c\frac{4AC\gamma^2}{(1+\gamma^2)^2} = c\frac{2B(B+1)\gamma^2}{(1+\gamma^2)^2}. \tag{76}$$

Inserting found values (73), (74) and (76) of coefficients D, E, F, G, H into equations near ξ and ξ^{-1}, we obtain, that for $A(2B+1)C \neq 0$ they are fulfilled, if $\gamma^4 = 1$, i.e. $\gamma^2 = \pm 1$. As $\gamma^2 > 0$, it means that $\gamma^2 = 1$. We have obtain the second condition

$$A(2B+1)C(\gamma^2 - 1) = 0. \tag{77}$$

Equation near ξ^0 is satisfied under substitution of find coefficients and condition (77). Thus, we have proven.

Theorem 13 *In the exotic expansion* (45) *of solutions to equation* P_5 *in Case II, the second coefficient* $\varphi_1(\xi)$ *is a Laurent polynomial* (72), (73), (74) *and* (76), *iff* 2 *conditions* (75) *and* (77) *are fulfilled.*

6 The Sixth Painlevé Equation P_6

6.1 Preliminary Transformations

Usually the sixth Painlevé equation [6] is

$$y'' = \frac{y'^2}{2}\left(\frac{1}{y} + \frac{1}{y-1} + \frac{1}{y-x}\right) - y'\left(\frac{1}{x} + \frac{1}{x-1} + \frac{1}{y-x}\right) +$$
$$+ a\frac{y(y-1)(y-x)}{x^2(x-1)^2} + b\frac{(y-1)(y-x)}{x(x-1)^2 y} + c\frac{y(y-x)}{x^2(x-1)(y-1)} + d\frac{y(y-1)}{x^2(x-1)^2(y-x)}\,.$$

We put $z = -y$, multiply the equation by its common denominator $x^2(x-1)^2 y(y-1)(y-x)$ and translate all terms into the right side of equation. So we obtain the equation

$$\begin{aligned} g(x, z) \stackrel{def}{=} &- z'' x^2(x-1)^2 z(z+1)(z+x) + \\ &+ \frac{1}{2} z'^2 x^2(x-1)^2[(z+1)(z+x) + z(z+x) + z(z+1)] - \\ &- z' z(z+1)[x(x-1)^2(z+x) + x^2(x-1)(z+x) + x^2(x-1)(z+x) - x^2(x-1)^2] + \\ &+ a z^2(z+1)^2(z+x)^2 + b x(z+1)^2(z+x)^2 + c(x-1)z^2(z-x)^2 + \\ &+ d x(x-1)^2 z^2(z+1)^2 = 0\,. \end{aligned} \tag{78}$$

Support and polygon of the equation are shown in Fig. 18.

If we write

$$g(x, z) = g_0(x, z) + x g_1(x, z) + x^2 g_2(x, z) + x^3 g_3(x, z)$$

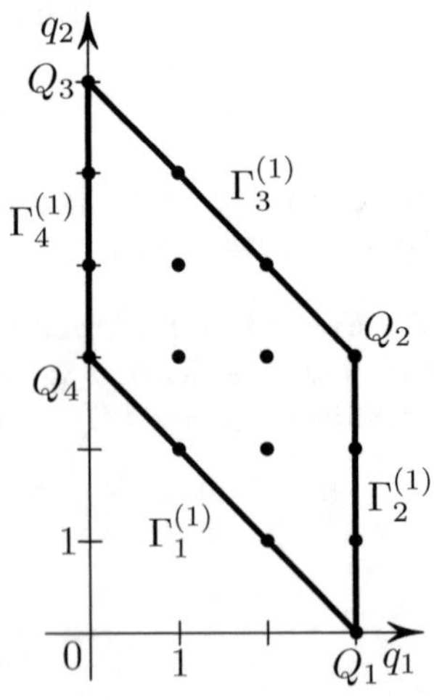

Fig. 18 Support and polygon of Eq. (78)

according to Sect. 2.2, then

$$g_0(x,z) = -z''x^2z^2(z+1) + z'^2x^2z\left(\frac{3}{2}z+1\right) - z'xz^2(z+1) + az^4(z+1)^2 - cz^4\,,$$
$$g_1(x,z) = z''x^2(z+1)(2z-1) - z'^2x^2z\left(3z^2+z-\frac{1}{2}\right) + 3z'xz^2(z+1) +$$
$$+ 2az^3(z+1)^2 + bz^2(z+1)^2 + cz^3(z-2) - dz^2(z+1)^2\,,$$
$$g_2(x,z) = -z''x^2z(z+1)(z-2) + z'^2x^2(3z^2-2z-2) - z'xz(z+1)(2z-1) -$$
$$- az^2(z+1)^2 - bz(z-1)^2 - cz^2(2z-1) - dz^2(z+1)^2\,,$$
$$g_3(x,z) = -2z''x^2z(z+1) + z'^2x^2(2z+1) - z'xz(z+1) + b(z+1)^2 - cz^2\,.$$

Note, that $g_0(x,z)$ coincides with the upper line of formula (42), multiplied by z, if $-c$ change by b. Now in Eq. (78) we make the power transformation $z = \dfrac{1}{w}$. Then

$$z' = -\frac{w'}{w^2}, \quad z'' = \frac{2w'^2 - ww''}{w^3}\,,$$

Denote $h_i(x,w) = g_i\left(x, \dfrac{1}{w}\right) \cdot w^6,\ i = 0, 1, 2, 3$. Then

$$h_0(x,w) = ww''x^2(1+w) - w'^2x^2\left(w+\frac{1}{2}\right) + w'xw(w+1) + a(w+1)^2 - cw^2\,,$$
$$h_1(x,w) = ww''x^2(w+1)(w-2) + w'^2x^2\left(-\frac{3}{2}w^2+w+1\right) - 3w'xw(w+1) +$$
$$+ 2aw(w+1)^2 + cw^2(1-2w) + (b-d)w^2(w+1)^2\,. \tag{79}$$

After change $-c$ by b, $h_0(x,w)$ coincides with $h_0(x,w)$ from (58), but in $h_1(x,w)$ here only one term $(b-d)w^2(w+1)^2$ coincides with $h_1(x,w)$ in (58), but now h_1 has several other terms.

6.2 *Complicated Expansions*

In $h_i(x,w)$ from (79), we change independent variable $\xi = \log x + c_0$ and obtain

$$h_0^*(\xi,w) = h_0(x,w) = \ddot{w}w(w+1) - \dot{w}^2\left(w+\frac{1}{2}\right) + a(w+1)^2 - cw^2\,,$$
$$h_1^*(\xi,w) = h_1(x,w) = \ddot{w}w(w+1)(w-2) - \dot{w}^2\left(\frac{3}{2}w^2 - w - 1\right) - \dot{w}w(w+1)^2 +$$
$$+ 2aw(w+1)^2 - cw^2(2w-1) + \omega w^2(w+1)^2\,,$$

where $\omega = b - d$.

According to Theorem 10 all nonconstant power series solutions to equation $h_0^*(\xi, w) = 0$ form two families:
main (if $\alpha \overset{def}{=} a - c \neq 0$)

$$w = \varphi_0 = \frac{\alpha}{2}(\xi + c_0)^2 - \frac{a}{\alpha}, \tag{80}$$

and additional (if $\alpha = 0,\ a \neq 0$)

$$w = \varphi_0 = \beta(\xi + c_0), \quad \beta^2 = 2a, \tag{81}$$

where c_0 is arbitrary constant. Let us compute the second coefficient $\varphi_1(\xi)$ of expansion (45), using Eq. (46). Here

$$\frac{\delta h_0^*}{\delta z} = w(w+1)\frac{d^2}{d\xi^2} - (2w+1)\dot{w}\frac{d}{d\xi} + 2a(w+1) - 2cw + \ddot{w}(2w+1) - \\ - \dot{w}^2 \overset{def}{=} a_1\frac{d^2}{d\xi^2} + a_2\frac{d}{d\xi} + a_3 .$$

According to Corollary 1 Eq. (46) for φ_1 is equivalent to equation

$$a_1[\varphi_1 + 2\dot{\varphi}_1 + \ddot{\varphi}_1] + a_2[\varphi_1 + \dot{\varphi}_1] + a_3\varphi_1 + h_1^* = 0 . \tag{82}$$

Denote $\xi = \log x + c_0$. For the additional family (81)

$$a_1 = \beta\xi(\beta\xi + 1), \quad a_2 = -\beta(2\beta\xi + 1), \quad a_3 = 0, \\ h_1 = 2a(w+1)^2 - \beta w(w+1)^2 + \omega w^2(w+1)^2 ,$$

because here $a = c$. Equation (82) has polynomial solution

$$\varphi_1 = 2\omega a\xi^2 + [\omega(4a - \beta) + 2a]\xi + \omega(\beta - 4a) + \beta - 2a . \tag{83}$$

Calculation of φ_2 see in [9].
For the main family (80)

$$a_1 = w(w+1), \quad a_2 = -\alpha\xi(2w+1), \quad a_3 = \alpha(2w+1) , \\ h_1^* = \omega w^2(w+1)^2 - \alpha\xi w(w+1)^2 + 2a(w+1)^2 .$$

If in Eq. (82) $h_1^* = \omega w^2(w+1)^2$, then according to Theorem 11, it has polynomial solution (64) with ω instead of c. Now we consider Eq. (82) for $\omega = 0$. We look for its polynomial solution in the form

$$\varphi_1 = A\xi^4 + B\xi^3 + C\xi^2 + D\xi + E . \tag{84}$$

For 5 coefficients A, B, C, D, E we obtain a system of 9 linear algebraic equations. They correspond to vanishing coefficients near $\xi^8, \xi^7, \ldots, \xi^0$, which arrive after substitution of expression (84) into Eq. (82). From coefficients near $\xi^8, \xi^7, \ldots, \xi^4$, we obtain

$$A = 0, \quad B = \alpha^2/2, \quad C = -\alpha^2, \quad D = \alpha^2 + \alpha - a, \quad E = 0.$$

Inserting these values into coefficient near $\xi^3, \xi^2, \xi^1, \xi^0$, we obtain the zeroes. And polynomial solution (84) of the full Eq. (82) has

$$\begin{aligned} &A = -\omega\frac{\alpha^2}{4}, \; B = \omega\alpha^2 + \frac{\alpha^2}{2}, \; C = -\omega\left(2\alpha^2 + \frac{\alpha}{2} - a\right) - \alpha^2, \\ &D = \omega\left(2\alpha^2 + \alpha - 2a\right) + \alpha^2 + \alpha - a, \; E = -\omega\frac{a(a-\alpha)}{\alpha^2}. \end{aligned} \tag{85}$$

Thus, we have proven.

Theorem 14 *The second coefficient φ_1 of the complicated expansion* (45) *of solution to equation P_6 is a polynomial* (84) *and* (85) *for the main family and is a polynomial* (83) *for the additional family.*

Calculation of φ_2 see in [9].

6.3 Exotic Expansions

Let us introduce new independent variable $\xi = x^{i\gamma}$ according to (52), (53) and (54). Then expressions (79) after division by γ^2 take forms

$$\begin{aligned} &\tilde{h}_0(\xi, w) = \gamma^{-2}h_0(x, w) = -(\dot{w}\xi + \ddot{w}\xi^2)w(w+1) + \dot{w}^2\xi^2(w + \tfrac{1}{2}) + \tilde{a}(w+1)^2 - \tilde{c}w^2, \\ &\tilde{h}_1(\xi, w) = \gamma^{-2}h_1(x, w) = -(\dot{w}\xi + \ddot{w}\xi^2)w(w+1)(w-2) + \dot{w}^2\xi^2(\tfrac{3}{2}w^2 - w - 1) + \\ &\quad + \frac{1}{i\gamma}\xi\dot{w}w(w+1)^2 + 2\tilde{a}w(w+1)^2 - \tilde{c}w^2(2w-1) + \tilde{\omega}w^2(w+1)^2, \end{aligned} \tag{86}$$

where

$$\tilde{a} = a/\gamma^2, \; \tilde{b} = b/\gamma^2, \; \tilde{c} = c/\gamma^2, \; \tilde{\omega} = \omega/\gamma^2, .$$

In (86) $\tilde{h}_0(\xi, w)$ coincides with $\tilde{h}_0(\xi, w)$ from (65), if $-c$ is changed by b. So according to Theorem 12, all power series solutions to equation $\tilde{h}_0(\xi, w) = 0$ from (86) are

$$w = \varphi_0 = A\xi + B + C\xi^{-1},$$

where

$$B = \tilde{a} - \tilde{c} - \frac{1}{2}, \quad 4AC = (\tilde{a} - \tilde{c})^2 + \tilde{a} + \tilde{c} + \frac{1}{4}\,.$$

According to (86),

$$\frac{\delta \tilde{h}_0}{\delta w} = -w(w+1)\xi^2 \frac{d^2}{d\xi^2} + [(2w+1)\dot{w}\xi - w(w+1)]\frac{d}{d\xi} + 2\tilde{a}(w+1) - \\ - 2\tilde{c}w - (2w+1)(\xi\dot{w} + \xi^2\ddot{w}) + \xi^2\dot{w}^2\,.$$

According to Corollary 2, Eq. (46) for $\varphi_1(\xi)$ is Eq. (71) with following changes: a_1, a_2 and a_3 are the same as in Sect. 5.3, with $-2\tilde{c}$ instead of $2\tilde{b}$, $\tilde{h}_1 = \tilde{h}_{16} + \tilde{\omega}\tilde{h}_{15}$, where $\tilde{h}_{15}$ is from Sect. 5.3 and

$$\begin{aligned}
\tilde{h}_{16} = {} & \frac{2+i\gamma}{2i\gamma}A^4\xi^4 + \left(\frac{3+2i\gamma}{i\gamma}B + \frac{2+i\gamma}{i\gamma}\right)A^3\xi^3 + \\
& + \left(10AC + \frac{2+4i\gamma}{i\gamma}B^2 + \frac{3+5i\gamma}{i\gamma}B + \frac{2+i\gamma}{2i\gamma}\right)A^2\xi^2 + \\
& + \left(20ABC + 4AC - \frac{6i\gamma-1}{2i\gamma}B^3 + \frac{i\gamma-1}{2i\gamma}B^2 - \frac{5}{2}B\right)A\xi + \\
& + 20AB^2C + 14ABC + \frac{1}{4}B(B+1)(9B^2+13B+2) + \\
& + \left(20ABC + 4AC - \frac{1+6i\gamma}{2i\gamma}B^3 + \frac{1+i\gamma}{2i\gamma}B^2 - \frac{5}{2}B\right)C\xi^{-1} + \\
& + \left(10AC - \frac{2-4i\gamma}{i\gamma}B^2 - \frac{3-5i\gamma}{i\gamma}B - \frac{2-i\gamma}{2i\gamma}\right)C^2\xi^{-2} + \\
& + \left(-\frac{3-2i\gamma}{i\gamma}B - \frac{2-i\gamma}{i\gamma}\right)C^3\xi^{-3} - \frac{2-i\gamma}{2i\gamma}C^4\xi^{-4}\,.
\end{aligned}$$

Polynomial solution φ_1 to new Eq. (71) we look for in the form (72). Again we obtain a system of 9 linear algebraic equations for 5 coefficients. Let us consider case $\tilde{\omega} = 0$. From vanishing coefficients near ξ^4, ξ^3, ξ^2, we find

$$\begin{aligned}
D &= -\frac{(2+i\gamma)\gamma^2}{2i\gamma(1+i\gamma)^2}A^2\,, \\
E &= -\left[\frac{B}{i\gamma(1+i\gamma)} + \frac{2+i\gamma}{2i\gamma(1+i\gamma)}\right]\gamma^2 A = -\left[\frac{\Omega}{2i\gamma(1+i\gamma)} + \frac{1}{2i\gamma}\right]\gamma^2 A\,, \\
F_1 &= 2AC\gamma^2\left[\frac{(2+i\gamma)(1+4i\gamma-\gamma^2)}{2i\gamma(1+i\gamma^2)^2} - \frac{5}{(1-i\gamma)^2}\right] + \\
& + B(B+1)\gamma^2\left[\frac{(2+i\gamma)(1+4i\gamma+\gamma^2)}{2i\gamma(1+i\gamma^2)^2} - \frac{4}{(1-i\gamma)^2}\right],
\end{aligned} \tag{87}$$

where $\Omega = 2B + 1$.

From vanishing coefficients near $\xi^{-4}, \xi^{-3}, \xi^{-2}$, we obtain

$$
\begin{aligned}
H &= -\frac{(2-i\gamma)\gamma^2}{2i\gamma(1-i\gamma)^2}C^2\,, \\
G &= \left[\frac{B}{i\gamma(1-i\gamma)} + \frac{2-i\gamma}{2i\gamma(1-i\gamma)}\right]\gamma^2 C = \left[\frac{\Omega}{2i\gamma(1-i\gamma)} + \frac{1}{2i\gamma}\right]\gamma^2 C\,, \\
F_2 &= -2AC\gamma^2\left[\frac{(2-i\gamma)(1-4i\gamma-9\gamma^2)}{2i\gamma(1+i\gamma^2)^2} + \frac{5}{(1+i\gamma)^2}\right] - \\
&\quad - B(B+1)\gamma^2\left[\frac{(2-i\gamma)(1-4i\gamma-7\gamma^2)}{2i\gamma(1+i\gamma^2)^2} + \frac{4}{(1+i\gamma)^2}\right].
\end{aligned}
\tag{88}
$$

Equality $F_1 = F_2$ is possible, iff $2AC + B(B+1) = 0$, see (75). Then

$$
F = -2AC\frac{\gamma^2}{(1+\gamma^2)^2} = B(B+1)\frac{\gamma^2}{(1+\gamma^2)^2}\,. \tag{89}
$$

Coefficients near ξ^1 and ξ^{-1} vanish for values (87)–(89). Coefficient near ξ^0 vanishes if

$$
AC(6B^2 - B - 3) = 0\,. \tag{90}
$$

If $\tilde{\omega} \neq 0$, we have additional condition (77) for polynomiality of $\varphi_1(\xi)$, i.e.

$$
\omega A(2B+1)C(\gamma^2 - 1) = 0\,. \tag{91}
$$

Thus, we have proven.

Theorem 15 *In the exotic expansion* (45) *of solutions to equation P_6, the second coefficient $\varphi_1(\xi)$ is a Laurent polynomial* (72), (73) + (87), (74) + (88), (76) + (89) *with $\omega = b - d$ instead of c, iff 3 conditions* (75), (90) *and* (91) *are fulfilled.*

Usually the equation for $\varphi_k(\xi)$ has two solutions: with increasing and with decreasing powers of ξ. But they coincide if the solution is an usual or Laurent polynomial. If all coefficients $\varphi_k(\xi)$ are polynomials then there is one family of exotic expansions. In another case there are two different families. Details see in [11].

7 Conclusion

In both cases: complicated and exotic expansions we have its own alternative. In complicated expansion the coefficient $\varphi_k(\xi)$ is either a polynomial or a divergent Laurent series. In exotic expansion the coefficient $\varphi_k(\xi)$ is either a Laurent polynomial, in that case it is unique, or a Laurent series, then there are two different coefficients in form of convergent series. The convergence follows from [12].

In all considered cases, when coefficient $\varphi_k(\xi) = D\xi^m + E\xi^{m-1} + F\xi^{m-2} + \cdots$ of the complicated or exotic expansion is an usual or Laurent polynomial, its coefficients $D, E, F, \ldots$, satisfy to a system of linear algebraic equations. And number of equations is more then number of these coefficients. Such linear systems have solutions only in degenerated cases when rank of the extended matrix of the system is less then the maximal possible. Existence of such situations in the Painlevé equations shows their degeneracy or their inner symmetries.

We have considered 4 cases: equations P_3, Case I of P_5, Case II of P_5, P_6. In each of them there are 3 families: additional complicated, main complicated and exotic. Among these 12 families, 9 have polynomial second coefficient, but 3 families demand for that some conditions on parameters. Namely, main complicated family for Case I of P_5 demands one condition; exotic families for Case II of P_5 and for P_6 demand 2 conditions and 3 conditions correspondingly. In all cases number of conditions is less than difference between number of equations and number of unknowns.

All these calculations were made by hands. Further computations should be made using Computer Algebra.

Acknowledgements This work was supported by RFBR, grant Nr. 18-01-00422, and by Program of the Presidium of RAS Nr. 01 “Fundamental Mathematics and its Applications” (Grant PRAS-18-01).

References

1. Bruno, A.D.: Asymptotics and expansions of solutions to an ordinary differential equation. Uspekhi Matem. Nauk **59**(3), 31–80 (2004) (in Russian). Russ. Math. Surv. **59**(3), 429–480 (2004) (in English)
2. Bruno, A.D.: Complicated expansions of solutions to an ordinary differential equation. Doklady Akademii Nauk **406**(6), 730–733 (2006) (in Russian). Dokl. Math. **73**(1), 117–120 (2006) (in English)
3. Bruno, A.D.: Exotic expansions of solutions to an ordinary differential equation. Doklady Akademii Nauk **416**(5), 583–587 (2007) (in Russian). Dokl. Math. **76**(2), 714–718 (2007) (in English)
4. Bruno, A.D.: On complicated expansions of solutions to ODE. Keldysh Institute Preprints. No. 15. Moscow, 2011, 26 p. (in Russian). http://library.keldysh.ru/preprint.asp?id=2011-15 (2011)
5. Bruno, A.D., Parusnikova, A.V.: Local expansions of solutions to the fifth Painlevé equation. Doklady Akademii Nauk **438**(4), 439–443 (2011) (in Russian). Dokl. Math. **83**(3), 348–352 (2011) (in English)
6. Bruno, A.D., Goryuchkina, I.V.: Asymptotic expansions of solutions of the sixth Painlevé equation. Trudy Mosk. Mat. Obs. **71**, 6–118 (2010) (in Russian). Transactions of Moscow Mathematical Society **71**, 1–104 (2010) (in English)
7. Hazewinkel, M. (ed.): Multinomial coefficient (http://www.encyclopediaofmath.org/index.php?title=p/m065320). Encyclopedia of Mathematics. Springer, Berlin (2001)

8. Bruno, A.D.: Calculation of complicated asymptotic expansions of solutions to the Painlevé equations. Keldysh Institute Preprints, No. 55, Moscow, 2017, 27 p. (Russian). https://doi.org/10.20948/prepr-2017-55, http://library.keldysh.ru/preprint.asp?id=2017-55
9. Bruno, A.D.: Calculation of exotic expansions of solutions to the third Painlevé equation. Keldysh Institute Preprints, No. 96, Moscow, 2017, 22 p. (in Russian). https://doi.org/10.20948/prepr-2017-96, http://library.keldysh.ru/preprint.asp?id=2017-96
10. Bruno, A.D.: Elements of Nonlinear Analysis (in that book)
11. Bruno, A.D.: Complicated and exotic expansions of solutions to the fifth Painlevé equation. Keldysh Institute Preprints, No. 107, Moscow, 2017 (in Russian). https://doi.org/10.20948/prepr-2017-107, http://library.keldysh.ru/preprint.asp?id=2017-107
12. Bruno, A.D., Goryuchkina, I.V.: On the convergence of a formal solution to an ordinary differential equation. Doklady Akademii Nauk **432**(2), 151–154 (2010) (in Russian). Dokl. Math. **81**(3), 358–361 (2010) (in English)

Part III
Summability of Divergent Solutions of ODEs

The Borel Transform of Canard Values and Its Singularities

P. Pavis d'Escurac

Abstract Canards were discovered in the early 80s by É. Benoît, F. and M. Diener, and J.-L. Callot in the study of the famous van der Pol equation (Benoît et al. [3]). Given a real differential equation, singularly perturbed by ε small, a canard solution — if it exists — has the particularity to follow partially or totally a slow curve from its attractive part to its repulsive part for a certain value of the control parameter a, named a *canard value*. A generalization to complex ODEs leads to overstable solutions, bounded in a neighbourhood of a turning point, i.e. a point where the slow curve presents an inversion of stability. It is known (Benoît et al. [5] and Canalis-Durand et al. [7]) that canard values admit a unique asymptotic expansion of Gevrey order 1 denoted by $\hat{a}$, so that the Borel transform $\tilde{a}(t)$ of $\hat{a}(\varepsilon)$ is analytic near the origin. Using the recent theory of composite asymptotic expansions due to A. Fruchard and R. Schäfke (Fruchard and Schäfke [11]), we study and describe the first singularity of this Borel transform $\tilde{a}$. This article focuses on a Riccati equation

$$\varepsilon \frac{dy}{dx} = (x(1-x))^2 - y^2 - a$$

where $x, y, \varepsilon, a \in \mathbb{C}$. For this equation, the formal series $\hat{a}$ is Borel summable in every direction except the real positive axis which constitutes a Stokes line. We first obtain an estimate of the difference of the canard values. This estimate contains an exponentially small term and a Gevrey asymptotic expansion. Then this result is translated into the Borel plane. It follows that the Borel transform $\tilde{a}$ can be analytically continued to $\mathbb{C} \setminus [1/3, +\infty[$ and has an isolated singularity at $t = 1/3$ on the first sheet.

Keywords Singular perturbation · Composite asymptotic expansions · Turning point · Canard solution · Overstability

MSC Primary 34E17 · Secondary 34E20, 34E10

P. Pavis d'Escurac (✉)
UHA Mulhouse, Mulhouse, France
e-mail: philippe.pavis-descurac@uha.fr

G. Filipuk et al. (eds.), *Formal and Analytic Solutions of Diff. Equations*, Springer Proceedings in Mathematics & Statistics 256,
https://doi.org/10.1007/978-3-319-99148-1_8

1 Introduction

The Riccati equation

$$\varepsilon y' = (x(1-x))^2 - y^2 - a \tag{1}$$

falls within the scope of singularly perturbed ordinary differential equations of order 1 of the form

$$\varepsilon y' = \Phi(x, y, \varepsilon, a), \tag{2}$$

where Φ is analytic in some domain of $\mathbb{C}^4$, ε is a small parameter and a is a control parameter. The prime denotes the derivation with respect to the variable x. These equations are of particular interest when they admit a value a_0 of the parameter a and a curve — called *slow curve* — given by an equation $y = y_0(x)$ such that $\Phi(x, y_0(x), 0, a_0) = 0$.

Related to a slow curve, we consider the functions f and F respectively defined by

$$f(x) := \frac{\partial \Phi}{\partial y}(x, y_0(x), 0, a_0) \quad \text{and} \quad F(x) := \int_{x_0}^{x} f(t)dt$$

with $x_0 \in \mathbb{C}$ arbitrary whose choice will be precised later. An important function is the so-called *relief function* associated to a direction $d \in \mathbb{R}$, denoted by R_d and given by

$$R_d(x) := \Re(F(x)e^{-id}).$$

J.-L. Callot showed its relevance in [6], Theorem 2 p. 156: given a solution $\tilde{y}$ of (2) defined in a neighbourhood of $x^* \in \mathbb{C}$ such that $|\tilde{y}(x^*) - y_0(x^*)| = O(|\varepsilon|)$ when ε tends to 0 will exist and be defined for all x that can be reached from x^* by a path where R_d decreases, $d = \arg \varepsilon$. Moreover, for all such x, $|\tilde{y}(x) - y_0(x)| = O(|\varepsilon|)$ when ε tends to 0.

We are interested in the case where there exists a point x_0 such that $f(x_0) = 0$. Such a point is called a *turning point* and constitutes a saddle point of the relief function R_d. We will only consider simple turning points, i.e. $f'(x_0) \neq 0$.

For some values of the control parameter a close to a_0, the existence of a solution which remains close to the slow curve y_0 in a neighbourhood of a turning point x_0, is remarkable: it is an *overstable solution* that generalizes the real *canard solution.* Roughly speaking, a canard solution is a solution that follows partially or totally a slow curve from its repulsive part to its attractive part. In [7], the authors showed that these values of a are exponentially close to each other and that they have a common Gevrey asymptotic expansion $\hat{a} = \sum_{n \geq 0} a_n \varepsilon^n$ when ε tends to 0. This formal series generally diverges but is Borel summable. This last property was first proven on the example of the van der Pol equation. In [10], A. Fruchard and R. Schäfke constructed a solution — called *exceptional solution* — associated to an analytic function $a(\varepsilon)$

for the control parameter, defined for ε on a sectorial domain of opening 3π. (A planar domain whose boundary lies on finitely many rays leaving the origin is called a sectorial domain). This result allowed them to deduce the Borel summability of the asymptotic expansion $\hat{a}$ for every direction of the complex plane, except the positive real axis. Moreover ([13, 14]), É. Matzinger computed an asymptotic equivalence for the monodromy $a(\varepsilon) - a(\varepsilon e^{-2\pi i})$ and then for the coefficients a_n of $\hat{a}$.

In this article we will go further than [8], in which the Riccati equation (1) is also studied with the same perspectives as those of the van der Pol equation: the authors showed the existence of exceptional solutions for (1) with $\varepsilon > 0$ and then computed the monodromy associated using the *matching* method. Actually we use here the recent theory of *Composite Asymptotic Expansions* due to A. Fruchard and R. Schäfke ([11]) in order to get a Gevrey asymptotic expansion of the product of some exponential term by the monodromy. We will construct exceptional solutions defined on domains slightly different from [8]: we will remain in a neighbourhood of the origin of order η where $\eta^2 = \varepsilon$. We need to evaluate the difference of our exceptional solutions somewhere where they are not exponentially close to each other i.e. near the origin which is the second turning point. Indeed, composite asymptotic expansions are well suited to describe the behaviour of a solution of a singularly perturbed ordinary differential equations near a turning point.

Finally we will translate this monodromy in the Borel plane and show that the Borel transform of the formal series $\hat{a}$ can be analytically continued to $\mathbb{C} \setminus [1/3, +\infty[$ and has an isolated singularity at $t = 1/3$ on the first sheet.

2 Statements

2.1 Canard Phenomenon

The real canard phenomenon can first be observed graphically for the Riccati equation (1), considering x, y, a and ε as real numbers. The Riccati equation is symmetric with respect to the transformations $x \mapsto 1 - x$ and $y \mapsto -y$, and it admits two slow curves of equation $y(x) = y_0(x) := x(1 - x)$ and $y(x) = -y_0(x)$. Moreover the points $x = 0$ and $x = 1$ are both turning points of these two slow curves. Thus it can appear both canard solutions near $x = 1$ for the slow curve y_0 and near $x = 0$ for the slow curve $-y_0$. Furthermore, we call *long* canard a solution that follows the slow curve y_0 on $]0, +\infty[$. See Fig. 1. In the following, we will only focus on canard solutions near $x = 1$ for the slow curve $y_0(x) = x(1 - x)$.

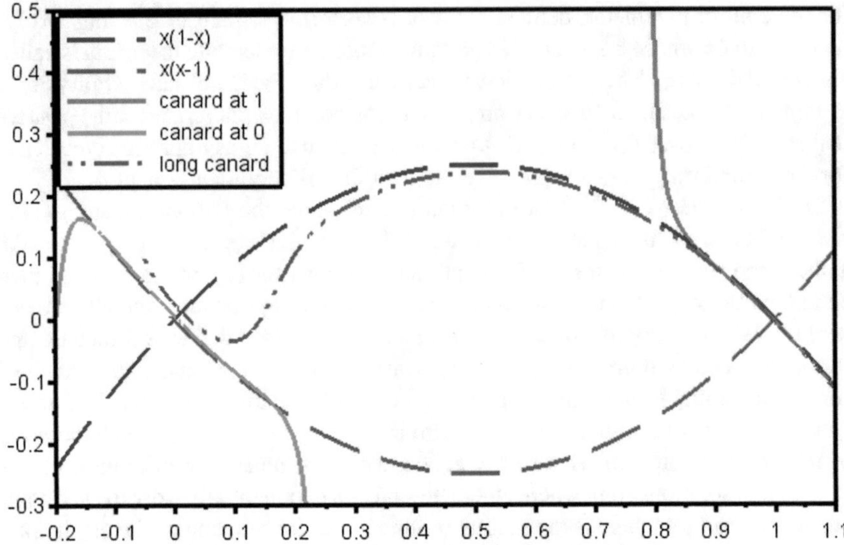

Fig. 1 Real canard phenomenon for the Riccati equation (1) near the turning points $x = 1$ and $x = 0$ associated to the slow curves $y(x) = y_0(x) := x(1 - x)$ and $y(x) = -y_0(x)$

2.2 Formal Solution

We are interested in a formal solution of (1) that does not have poles at $x = 1$. First we can see that Eq. (1) admits a formal solution $(\hat{a}, \hat{y})$:

$$\begin{cases} \hat{a}(\varepsilon) = \sum_{n=0}^{\infty} a_n \varepsilon^n & (3) \\ \hat{y}(x, \varepsilon) = \sum_{n=0}^{\infty} y_n(x) \varepsilon^n & (4) \end{cases}$$

completely determined by the condition that the functions y_n do not have poles at $x = 1$. It is recursively given by $a_0 = 0$, $y_0(x) = x(1 - x)$, $a_1 = 1$, $y_1(x) = -\frac{1}{x}$ and for $n \geq 2$,

$$a_n = -y'_{n-1}(1) - \sum_{k=1}^{n-1} y_k(1) y_{n-k}(1),$$

$$y_n(x) = -\frac{1}{2y_0(x)} \left(y'_{n-1}(x) + \sum_{k=1}^{n-1} y_k(x) y_{n-k}(x) + a_n \right). \tag{5}$$

Computing the first terms of these two series gives:

$$\hat{a}(\varepsilon) = \varepsilon - 2\varepsilon^2 - 9\varepsilon^3 + O(\varepsilon^4),$$
$$\hat{y}(x,\varepsilon) = x(1-x) - \frac{\varepsilon}{x} - \frac{1+x}{x^3}\varepsilon^2 - \frac{5+9x+9x^2+9x^3}{2x^5}\varepsilon^3 + O(\varepsilon^4), \quad \varepsilon \to 0. \tag{6}$$

Moreover, we can easily check recursively that for $n \geq 2$, y_n is a rational function of the form

$$y_n(x) = x^{1-2n} P_{2n-3}(x) \tag{7}$$

where P_{2n-3} is a polynomial of degree at most $2n-3$.

The Gevrey theory is a fundamental tool in this study. Let us recall the definition of a Gevrey asymptotic expansion for a *sector* $S(\beta_1, \beta_2, r)$, i.e. the set:

$$S(\beta_1, \beta_2, r) := \{\eta \in \mathbb{C};\quad \beta_1 < \arg \eta < \beta_2 \quad \text{and} \quad 0 < |\eta| < r\}.$$

Definition 2.1 Given $p \in \mathbb{N}^*$ and $\rho > 0$, a series $\hat{b} = \sum_{n\geq 0} b_n \varepsilon^n$ is said to be *Gevrey of order* $\frac{1}{p}$ *and type* ρ if there exists a constant $C > 0$ such that for all $n \in \mathbb{N}$, $|b_n| \leq C\left(\frac{1}{\rho}\right)^n \Gamma\left(\frac{n}{p}+1\right)$ where Γ is the usual Gamma function.

Definition 2.2 Let $S = S(\beta_1, \beta_2, r)$ be a sector, b be a holomorphic and bounded function on S and p be a positive integer. We say that b *admits* $\sum_{n\geq 0} b_n \varepsilon^n$ *as Gevrey asymptotic expansion of order* $\frac{1}{p}$ *and type* ρ *on* S if there exists a constant $C > 0$ such that for all $N \in \mathbb{N}^*$ and for all $\varepsilon \in S$,

$$\left| b(\varepsilon) - \sum_{n=0}^{N-1} b_n \varepsilon^n \right| \leq C\left(\frac{1}{\rho}\right)^N \Gamma\left(\frac{N}{p}+1\right) |\varepsilon|^N.$$

Finally, thanks to the literature ([7]), we get the following result.

Proposition 2.3 *Every pair* (a, y)*, where* y *is a canard or an overstable solution of* (1) *for the control parameter* a *and related to the slow curve* y_0*, is Gevrey-1 asymptotic to the formal solution* $(\hat{a}, \hat{y})$ *whose coefficients are given by* (5).

2.3 Borel Summability

First let us recall the definition of Borel summability.

Definition 2.4 A formal series $\hat{a} = \sum_{n=1}^{\infty}$ is said to be *Borel summable in the direction* θ if there exists $\phi > 0$ for which the function $\tilde{a}(t) = \sum_{n=1}^{\infty} \frac{a_n}{(n-1)!} t^{n-1}$ can be analytically continued to a sector $S = S(\theta - \phi/2, \theta + \phi/2, \infty)$ with at most exponential growth on S.

As soon as we will have proved the existence of exceptional solutions of (1) associated to an analytic function $a(\varepsilon)$ which is defined for ε on a sectorial domain of opening 3π and Gevrey-1 asymptotic to the formal solution $\hat{a}$, we will deduce the Borel summability of $\hat{a}$ thanks to a famous result that can be found in [1], Theorem 3.1.

Proposition 2.5 *The formal solution $\hat{a}$ of* (1) *given by* (5) *is Borel summable in every direction of the complex plane, except the real positive axis.*

2.4 Exceptional Solutions

Now, for $\delta > 0$ sufficiently small, we will consider $\arg\varepsilon \in]\theta-\delta, \theta+\delta[$ where $\theta \in \left]-\frac{5\pi}{2}+3\delta, \frac{\pi}{2}-3\delta\right[$. However it will be more convenient to use the variable η related to ε by $\eta^2 = \varepsilon$. Indeed ε comes from the Riccati equation studied at the beginning and the functions that we construct are defined for ε on a sectorial domain of opening 3π. This means to work with the Riemann surface of the logarithm function. Then, we prefer to work with η on a sectorial domain of opening $\frac{3}{2}\pi < 2\pi$. Let us introduce the complex domain defined as follows.

Given $0 < \delta < \pi/4$, consider $\eta_0 > 0$ satisfying

$$\sqrt{3}\eta_0 < \delta\sqrt{\delta}\sin\delta.$$

Let $r := (2\delta\sin\delta)^{-2}$ and $\frac{\theta}{2} \in \left]-\frac{5\pi}{4}+\frac{3\delta}{2}, \frac{\pi}{4}-\frac{3\delta}{2}\right[$. Denote also A_θ and B_θ the points near 0 such that $F(A_\theta) = F(0) - 2r|\eta|^2 e^{i(\theta+2\delta)}$ respectively $F(B_\theta) = F(0) - 2r|\eta|^2 e^{i(\theta-2\delta)}$.

Then $D_\theta(\delta, |\eta|^2)$ is the simply connected complex domain containing the turning point $x = 1$ and satisfying for all $x \in D_\theta(\delta, |\eta|^2)$, $|F(x) - F(0)| > 2r|\eta|^2$ and whose boundaries are the curves given by the equations $\arg\,(F(x) - F(A_\theta)) = -\frac{3\pi}{2}+\theta+2\delta$, $\arg\,(F(x) - F(B_\theta)) = \frac{3\pi}{2}+\theta-2\delta$ and $|F(x)| = 2r|\eta|^2$. We give a representation of the domain $D_0(\delta, |\eta^2|)$ and its image by F on Fig. 2.

The endpoints $\theta = -5\pi/2$ and $\theta = \pi/2$ must be eliminated because, in those cases, the point $x = 1$ is exactly on the level curves of the associated relief function $R_\theta = 0$. See Fig. 4. We also give an illustration in Fig. 3 of the evolution of the mountain excluded from the domain $D_\theta(\delta, |\eta|^2)$ as θ varies.

Denote by S and Ω the sets defined by

$$S := \left\{\eta \in \mathbb{C}; 0 < |\eta| < \eta_0 \quad \text{and} \quad \arg\eta \in \left]-\frac{5\pi}{4}+\frac{3\delta}{2}, \frac{\pi}{4}-\frac{3\delta}{2}\right[\right\}$$

and

$$\Omega := \left\{(x,\eta); \eta \in S, x \in D_{2\arg\eta}(\delta, |\eta|^2)\right\}.$$

We give here the result concerning exceptional solutions of (1) that will be proved in Sect. 3.

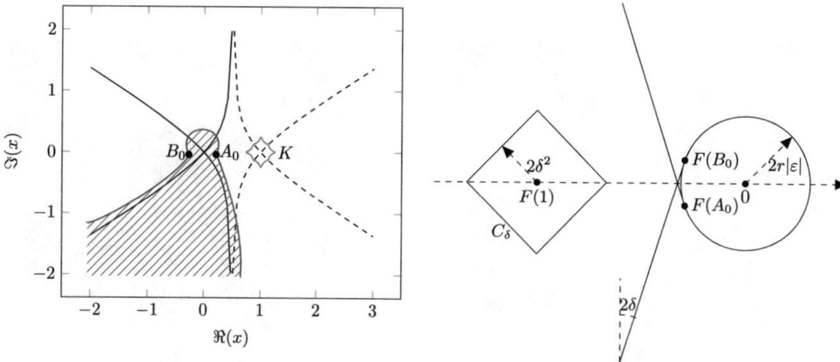

Fig. 2 On the left: Representation of the domain $D_0(\delta, |\eta^2|)$. The hatched region corresponds to the complementary of $D_0(\delta, |\eta^2|)$. Near 1 is sketched the compact set K of Sect. 3.1. We add the level curves of the relief function R_0 passing through 0 (solid lines) respectively 1 (dotted lines). On the right: Representation of the image of the boundaries of the domain $D_0(\delta, |\varepsilon|)$ by the application F and the square C_δ of Sect. 3.1, image of K by F

Theorem 2.6 *For all $\delta > 0$ sufficiently small, there exist $\eta_0 > 0$ and two unique analytic functions $a : S \to \mathbb{C}$, $y : \Omega \to \mathbb{C}$ such that*

- *$y(\cdot, \eta)$ is a solution of* (1) *with the control parameter $a(\eta)$,*
- *$y(\cdot, \eta)$ tends to $y_0(x) := x(1 - x)$ when $\eta \to 0$.*

For $\arg \eta = 0$, let denote by (a^+, y^+) respectively (a^-, y^-) the restrictions of the functions $a(x, \eta)$, $y(x, \eta)$ respectively $a(x, -\eta)$, $y(x, -\eta)$ provided by Theorem 2.6 and defined for $x \in D_0(\delta, |\eta|^2)$ respectively $x \in D_{-2\pi}(\delta, |\eta|^2) = \overline{D_0(\delta, |\eta|^2)}$ and $\eta \in S_0 := S(-\delta/2, \delta/2, \eta_0)$. Here $\overline{A}$ is the image of $A \subset \mathbb{C}$ by the complex conjugation. More precisely,

- for all $x \in D_0(\delta, |\eta|^2)$ and all $\eta \in S_0$,

$$\begin{aligned} y^+(x, \eta) &= y(x, \eta), \\ a^+(\eta) &= a(\eta), \end{aligned} \tag{8}$$

- and for all $x \in D_{-2\pi}(\delta, |\eta|^2)$ and all $\eta \in S_0$,

$$\begin{aligned} y^-(x, \eta) &= y(x, -\eta), \\ a^-(\eta) &= a(-\eta). \end{aligned} \tag{9}$$

Furthermore, using the complex conjugate one has also $a^-(\eta) := \overline{a^+(\overline{\eta})}$ and

$$y^- : (x, \eta) \mapsto \overline{y^+(\overline{x}, \overline{\eta})} \tag{10}$$

for $x \in \overline{D_0(\delta, |\eta|^2)}$ and $\eta \in S_0$.

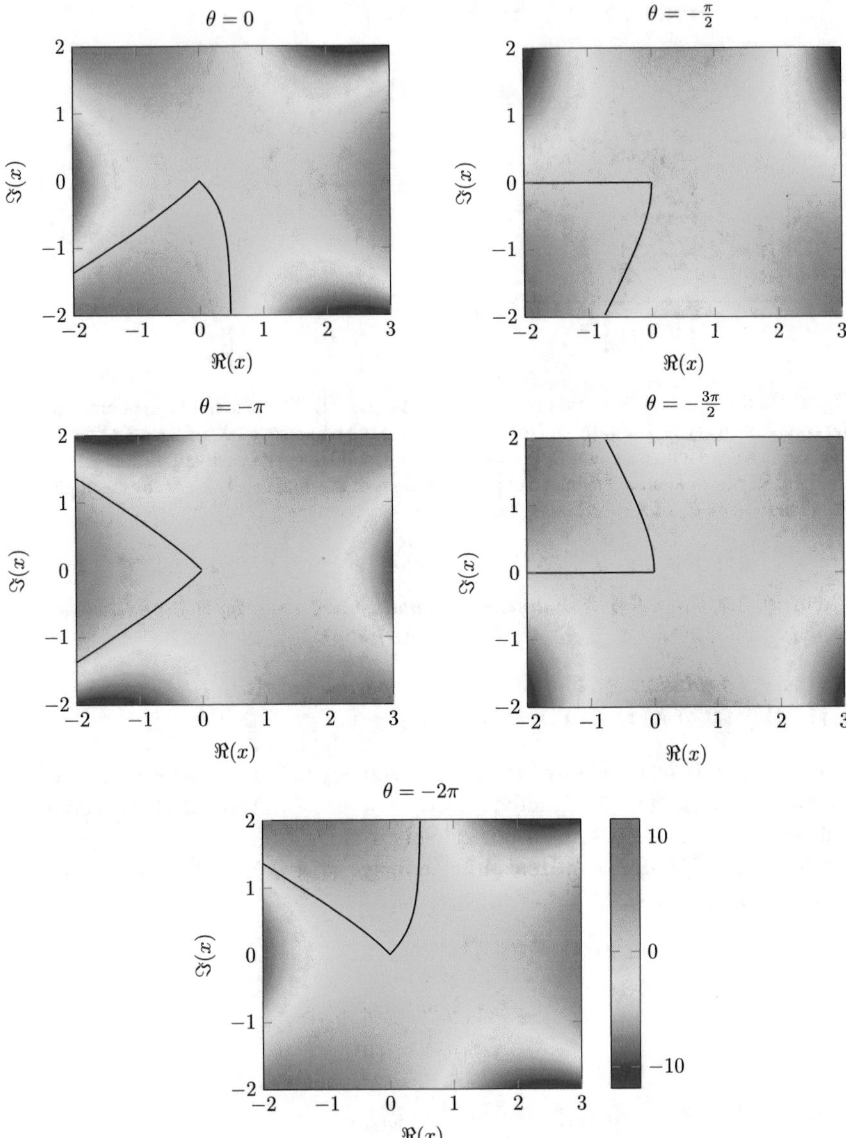

Fig. 3 Representation of the relief function R_θ for different values of θ. The connected set delimited by the level curves $R_\theta = 0$ constitutes the mountain excluded from the domain $D_\theta(\delta, |\eta|^2)$ associated

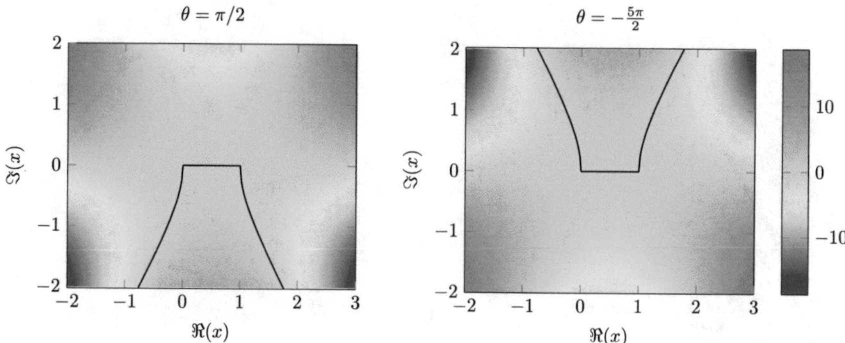

Fig. 4 Representation of R_θ and level curves $R_\theta = 0$ passing through the points $x = 0$ and $x = 1$ for $\theta = -5\pi/2$ and $\pi/2$

Later, we will be interested in giving an asymptotic expansion of the monodromy $(a^+(\eta) - a^-(\eta))e^{1/(3\eta^2)}$ for $\arg \eta = 0$.

2.5 Composite Asymptotic Expansions

Before giving the definition of a Gevrey composite asymptotic expansion due to A. Fruchard and R. Schäfke (cf. [11]), we introduce some notations and definitions. Let $\mathcal{D}(x_0, \mu)$ denotes the disk of center x_0 and radius $\mu > 0$.

Definition 2.7 For $\mu > 0$, $\mathcal{V}(\beta_1, \beta_2, r, \mu)$ is the union of the sector $S(\beta_1, \beta_2, r)$ and the disk $\mathcal{D}(0, \mu)$:

$$\mathcal{V}(\beta_1, \beta_2, r, \mu) := \{x \in \mathbb{C};\ (|x| < r \quad \text{and} \quad \beta_1 < \arg x < \beta_2) \quad \text{or} \quad |x| < \mu\}$$

and for $\mu < 0$,

$$\mathcal{V}(\beta_1, \beta_2, r, \mu) := \{x \in \mathbb{C};\ -\mu < |x| < r \quad \text{and} \quad \beta_1 < \arg x < \beta_2\}\,.$$

These two sets are called *quasi-sectors* (Fig. 5).

Definition 2.8 Let $r_0 > 0$, $\mu \in \mathbb{R}$ and $\mathcal{V} = \mathcal{V}(\alpha, \beta, \infty, \mu)$ be an infinite quasi-sector. Let also $(c_n)_{n\in\mathbb{N}}$ respectively $(g_n)_{n\in\mathbb{N}}$ be two families of holomorphic and bounded functions on the disk $\mathcal{D}(0, r_0)$, respectively the quasi-sector $\mathcal{V}$. A composite formal series associated to $\mathcal{D}(0, r_0)$ and $\mathcal{V}$ is an expression of the form

$$\hat{y}(x, \eta) = \sum_{n\geq 0} \left(c_n(x) + g_n\left(\tfrac{x}{\eta}\right)\right) \eta^n$$

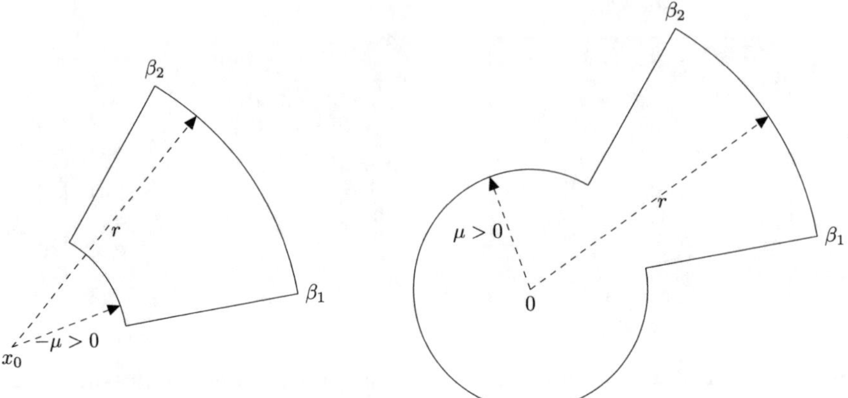

Fig. 5 Representations of a quasi-sector $\mathcal{V}(\beta_1, \beta_2, r, \mu)$ depending on $\mu < 0$ or $\mu > 0$

where $g_n(X) \sim \sum_{m>0} g_{nm} X^{-m}$ as $\mathcal{V} \ni X \to \infty$. We say that $\hat{y}$ *is Gevrey of order* $\frac{1}{p}$ $(p \in \mathbb{N}^*)$ *and type* (L_1, L_2), if there is a constant $C > 0$ such that for all $n \in \mathbb{N}$, one has $\sup_{|x|<r_0} |c_n(x)| \leq C L_1^n \Gamma\left(\frac{n}{p}+1\right)$ and for all $n, M \in \mathbb{N}$ and all $X \in \mathcal{V}$, one has

$$\left| g_n(X) - \sum_{m=1}^{M-1} g_{nm} X^{-m} \right| \leq C L_1^n L_2^M \Gamma\left(\frac{M+n}{p}+1\right) |X|^{-M}.$$

Definition 2.9 Let $\mathcal{V} = \mathcal{V}(\alpha, \beta, \infty, \mu)$ be an infinite quasi-sector, $S_2 = S(\alpha_2, \beta_2, \eta_0)$ a finite sector and $\alpha_1 < \beta_1$ such that $\alpha \leq \alpha_1 - \beta_2 < \beta_1 - \alpha_2 \leq \beta$. Let $y(x, \eta)$ be a holomorphic function defined for $\eta \in S_2$ and $x \in \mathcal{V}(\alpha_1, \beta_1, r_0, \mu|\eta|)$. Let $(c_n)_{n\in\mathbb{N}}$ respectively $(g_n)_{n\in\mathbb{N}}$ two families of holomorphic and bounded functions on the disk $\mathcal{D}(0, r_0)$, respectively the quasi-sector $\mathcal{V}$. Let finally $\hat{y}(x, \eta) = \sum_{n\geq 0} \left(c_n(x) + g_n(\frac{x}{\eta})\right) \eta^n$ be a composite formal series associated to $\mathcal{D}(0, r_0)$ and $\mathcal{V}$. We say that y *has* $\hat{y}$ *as composite asymptotic expansion of Gevrey order* $\frac{1}{p}$ $(p \in \mathbb{N}^*)$ and we write $y(x, \eta) \sim_{\frac{1}{p}} \hat{y}(x, \eta)$, when $S_2 \ni \eta \to 0$ and $x \in \mathcal{V}(\alpha_1, \beta_1, r_0, \mu|\eta|)$ if $\hat{y}(x, \eta)$ is Gevrey of order $\frac{1}{p}$ and type (L_1, L_2) in the sense of Definition 2.8 for some $L_1, L_2 > 0$ and if there exists a constant C such that, for all N, for all $\eta \in S_2$ and all $x \in \mathcal{V}(\alpha_1, \beta_1, r_0, \mu|\eta|)$,

$$\left| y(x, \eta) - \sum_{n=0}^{N-1} \left(c_n(x) + g_n\left(\frac{x}{\eta}\right) \right) \eta^n \right| \leq C L_1^N \Gamma\left(\frac{N}{p}+1\right) |\eta|^N.$$

Our goal is to prove the following proposition in Sect. 3.

Proposition 2.10 *For all* $0 < \delta$ *sufficiently small, there exist* $\mu, \eta_0 > 0$ *and a sequence* $(g_n^+)_{n\in\mathbb{N}}$ *of holomorphic functions tending to 0 at infinity such that*

$$y^+(x,\eta) \sim_{\frac{1}{2}} x(1-x) + \eta \left(-2\frac{x}{\eta} + \frac{e^{\left(\frac{x}{\eta}\right)^2}}{\int_{i\infty}^{\frac{x}{\eta}} e^{T^2} dT} \right) + \sum_{n \geq 1} g_{n+1}^+ \left(\frac{x}{\eta}\right) \eta^{n+1} \tag{11}$$

for $\eta \in S(-\frac{\delta}{2}, \frac{\delta}{2}, \eta_0)$ *and* $x \in \mathcal{V}^+ := \mathcal{V}(-\frac{\pi}{6}, \frac{7\pi}{6}, 1+\delta, \mu|\eta|)$.

Remark 2.11 Section 3.2 contains some information about the functions g_n^+.

Remark 2.12 An analogous result holds for the function y^- using the symmetry (10).

2.6 Asymptotic Expansion Versus Monodromy of Canard Values

Thanks to the previous result concerning composite asymptotic expansions, the evaluation of the monodromy $a^+ - a^-$ leads to the following theorem that will be proved in Sect. 3.

Theorem 2.13 *The difference of the canard values* a^+ *and* a^- *associated respectively to the exceptional solutions* y^+ *and* y^- *provided by Theorem 2.6 is exponentially small. More precisely, for* $\varepsilon > 0$, $(a^+(\varepsilon) - a^-(\varepsilon))e^{1/(3\varepsilon)}$ *has a Gevrey-1 asymptotic expansion:*

$$(a^+(\varepsilon) - a^-(\varepsilon))e^{1/(3\varepsilon)} \sim_1 4i + \sum_{n \geq 1} b_n \varepsilon^n$$

with $b_1 = -\frac{106}{3}i$.

Remark 2.14 The constant $4i$ was already established in [8] using the matching. This method consists in finding an accurate approximation to the solution of a perturbed ODE combining the inner and the outer expansions together. The use of composite asymptotic expansions contributes to an improvement. Indeed the latter are well suited to describe the behaviour of a solution near a turning point. Moreover it provides us the Gevrey property which is essential to study of the first singularity of the Borel transform defined below.

2.7 First Singularity of the Borel Transform

The evaluation of the monodromy $a^+ - a^-$ computed in the previous Theorem 2.13 allows us to study the Borel transform of the formal solution $\hat{a}$ introduced in Sect. 2.2 by (5). Writing $\hat{a}(\varepsilon) = \sum_{n \geq 1} a_n \varepsilon^n$, we define its Borel transform as

$$\tilde{a}(t) := \sum_{n\geq 1} \frac{b_n}{(n-1)!} t^{n-1}.$$

Since the serie $\hat{b}(\varepsilon) := \sum_{n\geq 1} a_n \varepsilon^n$ provided by Theorem 2.13 is Gevrey of order 1, its Borel transform, $\tilde{b}(t) = \sum_{n\geq 1} \frac{b_n}{(n-1)!} t^{n-1}$ converges in a disk of radius — say strictly greater than $M > 0$—into which its sum is also denoted by $\tilde{b}$ for convenience. Then we get the following theorem that we will prove in Sect. 3.

Theorem 2.15 *The Borel transform $\tilde{a}$ of the formal series $\hat{a}(\varepsilon) = \sum_{n\geq 1} a_n \varepsilon^n$ given by* (5) *has a ramified singularity at $t = 1/3$, isolated on the first sheet: there exists $0 < \tilde{M} \leq M$ such that $\tilde{a}$ can be analytically continued on* $]1/3, 1/3 + \tilde{M}[$ *in two different ways depending on* $\arg t \to -2\pi^+$ *or* $\arg t \to 0^-$. *Moreover, the growth of $\tilde{a}(t)$ as $t \to 1/3$ is at most logarithmic. Furthermore, the difference of these determinations is $\tilde{b}$. More precisely, there exists $0 < \tilde{M} \leq M$ and an analytic function $\tilde{d}$ on the half-plane*

$$\mathcal{H} := \left\{ t \in \mathbb{C}; \Re(t) < 1/3 + \tilde{M} \right\}$$

such that for all $t \in \mathcal{H}$, the Borel transform $\tilde{a}$ of the formal series $\hat{a}(\varepsilon) = \sum_{n\geq 1} a_n \varepsilon^n$ given by (5) *can be written as*

$$\tilde{a}(t) = \frac{2}{\pi(t - 1/3)} + \frac{1}{2\pi i} \int_{1/3}^{1/3+M} \frac{\tilde{b}(\tau - 1/3)}{\tau - t} d\tau + \tilde{d}(t). \tag{12}$$

Remark 2.16 The term $\frac{1}{2\pi i} \int_{1/3}^{1/3+M} \frac{\tilde{b}(\tau-1/3)}{\tau-t} d\tau$ presented in (12) is called the *Cauchy-Heine transform of $\tilde{b}$* in the sense of [2].

2.8 Perspectives and Remarks

This work constitutes one part of my PhD thesis. The reader will find more details in the memoir [15]. In this memoir, we also give analogous results for the famous van der Pol equation. However, we chose to present only the result for the Riccati equation because they are more explicit. Besides, the Riccati equation is still greatly studied because of the link with its application in quantum mechanics. See the works of J. Zinn-Justin in [16] or [17] for instance.

We also emphasize the proofs of Theorem 2.13 and Theorem 2.15 because they are new results compared to [8]. In Proposition 2.10 we add here a related result which is not presented but can be found in [15]. Indeed, one can give an explicit expression of the composite asymptotic expansion of the exceptional solution y^+. More precisely, all functions g_n^+ are given by:

$$g_n^+(X) = \sum_{k=0}^{n} P_{n,k}(X)\left(\mathcal{U}^+(X)\right)^k$$

where $P_{n,k}$ are polynomials and $\mathcal{U}^+(X) = \frac{e^{X^2}}{\int_{i\infty}^{X} e^{T^2}dT}$.

Besides, in [15] we answer a conjecture of [4]. As for perspectives, the main and most natural one is the resurgence of the Borel transform of the canard value function which still remains to be studied.

3 Proofs

In the whole Sect. 3, $\delta > 0$ is fixed small enough.

3.1 Proof of Theorem 2.6

We first restrict η to some sector $S_\theta := S(\frac{\theta-\delta}{2}, \frac{\theta+\delta}{2}, \eta_0)$ with $\theta \in \left]-\frac{5\pi}{2}+3\delta, \frac{\pi}{2}-3\delta\right[$. Then we prove the following theorem applying the same method as in [10]: we show that some operator — that will be precised later — is contracting and finally we use the famous fixed point theorem.

Theorem 3.1 *There exists $\eta_0 > 0$ such that for all $\eta \in S_\theta := S(\frac{\theta-\delta}{2}, \frac{\theta+\delta}{2}, \eta_0)$, there exists a unique value a_θ for which equation* (1) *admits, for $a = a_\theta$, a unique solution y_θ such that for all x in $D_\theta(\delta, |\eta|^2)$, $|y_\theta(x,\eta) - x(1-x)| = O(|\eta|)$ when $\eta \to 0$. Moreover, $a_\theta(\eta) = O(\eta^2)$.*

However, to prove this theorem it will be more convenient to use the variable $\varepsilon = \eta^2$. We proceed to the change of variable $y(x) = y_0(x) - \frac{\varepsilon}{x} + \varepsilon u(x)$ in Eq. (1). This leads to

$$\varepsilon\frac{du}{dx} = -2x(1-x)u + \left(\left(\frac{2}{x} - u\right)u - \frac{2}{x^2}\right)\varepsilon - b \tag{13}$$

where b is defined by $a = \varepsilon + \varepsilon b$.

For an arbitrary function h analytic and bounded on $D_\theta(\delta, |\varepsilon|)$, we study the first order singularly perturbed ordinary differential equation

$$\varepsilon\frac{du}{dx} = -2x(1-x)u + h - b \tag{14}$$

and then we define $\tilde{u}$ by

$$\tilde{u}(x) := \frac{1}{\varepsilon}\int_{\gamma_x} e^{(F(x)-F(t))/\varepsilon}(h(t) - b)dt. \tag{15}$$

This function $\tilde{u}$ is given by the variation of constant formula for Eq. (14). To ensure that $\tilde{u}$ is well defined on $D_\theta(\delta, |\varepsilon|)$, that is to say the integral in (15) does not depend on the path, we choose b such that :

$$\int_{\gamma_\infty} e^{-F(t)/\varepsilon}\,(h(t)-b)\,dt = 0,$$

i.e.

$$b = \frac{\int_{\gamma_\infty} e^{-F(t)/\varepsilon} h(t)dt}{\int_{\gamma_\infty} e^{-F(t)/\varepsilon} dt}$$

where γ_∞ is an infinite path which links the two mountains of the domain $D_\theta(\delta, |\varepsilon|)$.

In the complex plane of the variable $z = F(x)$, consider the closed square C_δ centered at $F(1) = -1/3$, whose half side is $2\delta^2$ and whose one diagonal is included in the real axis. The choice $0 < \delta \leq \frac{1}{20}$ ensures that C_δ is included in $F(D_\theta(\delta, |\varepsilon|))$. See Fig. 2 for an illustration of C_δ. The inverse image $F^{-1}(C_\delta)$ has three connected components. Denote by K the connected component containing the point $x = 1$. K is a compact set of $D_\theta(\delta, |\varepsilon|)$ included in the disk $D(1, \delta)$. See Fig. 2 for an illustration of K in the case $\theta = 0$. We must distinguish the case where $x \in D_\theta(\delta, |\varepsilon|) \setminus K$ from the case where $x \in K$. Indeed we will introduce some *ascending path* (cf. property (16) below) and will deduce upper bounds in the case when $x \in D_\theta(\delta, |\varepsilon|) \setminus K$. The maximum modulus principle will let us deduce the same upper bounds when $x \in K$. Thus, inequalities will stay valid for all $x \in D_\theta(\delta, |\varepsilon|)$.

The domain $D_\theta(\delta, |\varepsilon|)$ has the following property: for all $x \in D_\theta(\delta, |\varepsilon|) \setminus K$, there exists a path $\gamma_x : [0, +\infty[\to D_\theta(\delta, |\varepsilon|)$ with $\gamma_x(0) = x$, $|\gamma_x(\tau) - 1| \geq \delta$, $|\gamma_x'(\tau)| = 1$ and

$$\frac{d(R_d \circ \gamma)}{d\tau}(\tau) \geq \sqrt{|\varepsilon|} \tag{16}$$

for all $\tau \geq 0$ where $R_d(x) = \Re(F(x)e^{-id})$ and $d = \arg\varepsilon \in]\theta - \delta, \theta + \delta[$. Such a path is called an *ascending path.*

Thus, for $x \in D_\theta(\delta, |\varepsilon|) \setminus K$, one can choose for γ_x an ascending path in (15). Then we define E_∞ as the set of functions h analytic on $D_\theta(\delta, |\varepsilon|)$ and bounded for the norm

$$||h||_\infty := \sup_{x \in D_\theta(\delta, |\varepsilon|)} |h(x)|.$$

We also define the space $E := \{(b, u) \in \mathbb{C} \times E_\infty\}$ endowed with the norm

$$||(b, u)|| = |b| + \frac{\sqrt{|\varepsilon|}}{2} ||u||_\infty$$

and the operator $\mathcal{G}$ defined on E by

$$\mathcal{G}(b,u) = (g_1, g_2) \tag{17}$$

with

$$g_1 = \frac{\varepsilon \int_{\gamma_\infty} e^{-F(t)/\varepsilon} \left(\left(\frac{2}{t} - u\right) u - \frac{2}{t^2} \right) dt}{\int_{\gamma_\infty} e^{-F(t)/\varepsilon} dt}$$

and

$$g_2(x) = \int_{\gamma_x} e^{(F(x)-F(t))/\varepsilon} \left(\left(\frac{2}{t} - u \right) u - \frac{2}{t^2} - \frac{b}{\varepsilon} \right) dt$$

where γ_x is an ascending path (cf. (16)) when $x \in D_\theta(\delta, |\varepsilon|) \setminus K$.

A quite long calculation shows that for all $\delta > 0$ sufficiently small, there exists $\eta_0 > 0$ such that for all $\eta \in S_\theta = S(\frac{\theta-\delta}{2}, \frac{\theta+\delta}{2}, \eta_0)$, there exists $\rho > 0$ such that the operator $\mathcal{G}$ defined by (17) is a $\frac{1}{2}$-contraction of the closed ball $B_\rho := \{(b,u) \in E, ||(b,u)|| \leq \rho\}$. This calculation can be found in [15]. The fixed point theorem allows us to conclude the proof of Theorem 3.1.

Since for all $\delta > 0$, the sector

$$S := \left\{\eta \in \mathbb{C};\ |\eta| < \eta_0 \quad \text{and} \quad \arg \eta \in \left] -\tfrac{5\pi}{4} + \tfrac{3\delta}{2}, \tfrac{\pi}{4} - \tfrac{3\delta}{2} \right[\right\}$$

can be covered by a finite number of sectors of opening δ, Theorem 3.1 provides solutions which coincide in the intersections of their domains by the uniqueness property. This defines the analytic functions y and a of Theorem 2.6.

3.2 *Proof of Proposition 2.10*

First recall that $\delta > 0$ has been fixed small enough. Let us consider $\eta \in S(-\delta/2, \delta/2, \eta_0)$ as in Theorem 2.6 and fix a to the value a^+ in (1). Recall that η is linked to ε by $\eta^2 = \varepsilon$. The relief function R_0 is composed of three mountains: one at the east of $x = 1$, one at the northwest of $x = 0$ and the last one — which is excluded from the domain $D_{2\arg\eta}(\delta, |\eta|^2)$ — at the southwest of $x = 0$. Our main tool is Corollary 5.16 from [11]. One has to do the change of variable $v = y - y_0$ where $y_0(x) = x(1-x)$ in Eq. (1) to obtain

$$\varepsilon v' = -2x(1-x)v - \varepsilon(1-2x) - a^+ - v^2. \tag{18}$$

Indeed, this new Eq. (18) satisfies the required hypotheses. Thus, the mentioned corollary provides us a solution $\widetilde{v^+}$ which has a composite asymptotic expansion of order 1/2 and defined for x in some arbitrary quasi-sector included in the northwest mountain near $x = 0$ and its adjacent valleys but not containing $x = 1$. For example, one can assume that $\widetilde{v^+}$ is defined for x in the quasi-sector $\mathcal{V}(-\frac{\pi}{6}, \frac{7\pi}{6}, 1-\delta, \mu_0|\eta|)$ with $\mu_0 < 0$. Reducing η_0 if necessary, one can assume that the composite asymptotic expansion of $\widetilde{v^+}$ is also valid for $\eta \in S(-\delta/2, \delta/2, \eta_0)$. More precisely, we show the following proposition.

Proposition 3.2 *There exist $\eta_0 > 0$ and $\mu_0 < 0$ such that the solution $v^+ := y^+ - y_0$ of Eq.* (18) *obtained from Theorem 2.6 is exponentially close to $\widetilde{v^+}$ and has the same Gevrey composite asymptotic expansion of order 1/2 as $\widetilde{v^+}$ when $S(-\frac{\delta}{2}, \frac{\delta}{2}, \eta_0) \ni \eta \to 0$ and $x \in \mathcal{V}(-\frac{\pi}{6}, \frac{7\pi}{6}, \frac{1}{2}, \mu_0|\eta|) \subset D_0(\delta, |\eta^2|)$.*

Proof Let us first consider x_0 on the boundary of the quasi-sector $\mathcal{V}(-\frac{\pi}{6}, \frac{7\pi}{6}, 1-\delta, \mu_0|\eta|)$ satisfying:

$$R_0(x_0) = \sup_{x \in \mathcal{V}(-\frac{\pi}{6}, \frac{7\pi}{6}, 1-\delta, \mu_0|\eta|)} R_0(x).$$

To show Proposition 3.2 we need to compute the difference $w := v^+ - \widetilde{v^+}$ and to see that it is exponentially small. Actually, w satisfies the equation

$$\eta^2 w' = \left(-2x(1-x) + v^+ + \widetilde{v^+}\right) w$$

whose solution can be written:

$$w(x, \eta) = w(x_0, \eta) \exp\left\{\frac{1}{\eta^2} \int_{x_0}^{x} \left(-2t(1-t) + v^+(t) + \widetilde{v^+}(t)\right) dt\right\}.$$

Denote by

$$M := \inf_{\substack{x \in \mathcal{V}(-\frac{\pi}{6}, \frac{7\pi}{6}, \frac{1}{2}, \mu_0|\eta|),\\ \eta \in S(-\delta/2, \delta/2, \eta_0)}} \left(R_{2\arg\eta}(x_0) - R_{2\arg\eta}(x)\right).$$

Since $\mathcal{V}\left(-\frac{\pi}{6}, \frac{7\pi}{6}, \frac{1}{2}, \mu_0|\eta|\right) \subset \mathcal{D}\left(0, \frac{1}{2}\right)$ and $|x_0| = 1-\delta$, we have $M > 0$ for δ sufficiently small. Then for all $x \in \mathcal{V}\left(-\frac{\pi}{6}, \frac{7\pi}{6}, \frac{1}{2}, \mu_0|\eta|\right)$ and all $\eta \in S\left(-\delta/2, \delta/2, \eta_0\right)$, we have

$$w(x, \eta) = O\left(e^{-M/|\eta|^2}\right).$$

By application of [11], Proposition 3.13, we conclude that w has a Gevrey composite asymptotic expansion when $S(-\delta/2, \delta/2, \eta_0) \ni \eta \to 0$ and $x \in \mathcal{V}(-\frac{\pi}{6}, \frac{7\pi}{6}, \frac{1}{2}, \mu_0|\eta|)$ which is identically zero. In other words, v^+ and $\widetilde{v^+}$ have the same composite asymptotic expansion of Gevrey order 1/2. □

Now it only remains to extend the quasi-sector $\mathcal{V}(-\frac{\pi}{6}, \frac{7\pi}{6}, \frac{1}{2}, \mu_0|\eta|)$ as in the statement of Proposition 2.10. In order to do this, we need to compute the first terms of this composite asymptotic expansion. Roughly speaking, thanks to Proposition 2.6

of [11], a composite asymptotic expansion is the sum of the outer expansion for which one has eliminated the non-positive powers of x and the inner expansion for which one has eliminated the non-negative powers of $X := \frac{x}{\eta}$. Since the outer expansion of y^+ is given by (5), property (7) reduces the terms provided by the outer expansion to only one term: $x(1-x)$. Concerning the inner expansion, we do the change of variables $(X, V^+) := (\frac{x}{\eta}, \frac{v^+}{\eta})$ in equation (18):

$$\frac{dV^+}{dX} = -2XV^+ - 1 - \frac{a^+}{\eta^2} + 2\eta X(1 + XV^+) - (V^+)^2.$$

Denoting by $\sum_{n\geq 0} V_n^+(X)\eta^n$ the inner expansion of v^+, one has:

$$(V_0^+)' = -2XV_0^+ - 2 - (V_0^+)^2 \tag{19}$$

$$(V_1^+)' = -2(X + V_0^+)V_1^+ + 2X(1 + XV_0^+) \tag{20}$$

and for $n \geq 2$,

$$(V_n^+)' = -2(X + V_0^+)V_n^+ + 2X^2 V_{n-1}^+ - \sum_{k=1}^{n-1} V_k^+ {V_{n-k}}^+ - \mathfrak{a}_n \tag{21}$$

with $\mathfrak{a}_n = \begin{cases} a_{k+1} & \text{if there exists} \quad k \geq 1 \quad \text{such that} \quad n = 2k \\ 0 & \text{otherwise.} \end{cases}$

One can remark that the unique solution of (19) asymptotic to $\left(-\frac{1}{X}\right)$ when $X \to i\infty$ is given by

$$V_0^+(X) = -2X + \frac{e^{X^2}}{\int_{i\infty}^X e^{T^2} dT}. \tag{22}$$

Then, we set $g_1^+ := V_0^+$.

We define V_n^+ as the unique solution with, at most, polynomial growth when $X \to i\infty$. Denote by P_n the polynomial part of V_n^+. Thus, the function g_{n+1}^+ of Proposition 2.10 is defined by $g_{n+1}^+ := V_n^+ - P_n$ and tends to 0 as X tends to $i\infty$.

Then, on the one hand because the outer expansion is analytic near $x = 1$, Proposition 2.20 from [11] allows us to extend our quasi-sector $\mathcal{V}(-\frac{\pi}{6}, \frac{7\pi}{6}, \frac{1}{2}, \mu_0|\eta|)$ beyond $x = 1$. On the other hand, because V_0^+ is defined for $X = 0$, there exists $\mu > 0$ such that the Gevrey composite asymptotic expansion of y^+ is valid for $x \in \mathcal{V}(-\frac{\pi}{6}, \frac{7\pi}{6}, 1 + \delta, \mu|\eta|)$. This is due to Theorem 5.17 from [11].

A straightforward computation detailed in [15] yields for the next terms:

$$V_1^+(X) = \left(2X^2+1\right) + \left(-\frac{2X^3}{3} - 2X\right)\mathcal{U}^+(X) + \left(\frac{X^2}{3} + \frac{2}{3}\right)\mathcal{U}^+(X)^2, \tag{23}$$

$$\begin{aligned} V_2^+(X) =& X + \left(\frac{2}{9}X^6 + \frac{4}{3}X^4 + X^2 - \frac{1}{12}\right)\mathcal{U}^+(X) + \left(-\frac{1}{3}X^5 - \frac{3}{2}X^3 - \frac{5}{4}X\right)\left(\mathcal{U}^+(X)\right)^2 \\ &+ \left(\frac{1}{3}X^2 + \frac{2}{3}\right)^2 \left(\mathcal{U}^+(X)\right)^3, \end{aligned} \tag{24}$$

with $\mathcal{U}^+(X) = \frac{e^{X^2}}{\int_{i\infty}^X e^{T^2} dT}$. Since $\mathcal{U}^+(X)$ has an asymptotic expansion when $X \to \infty$ (starting with $2X$), the asymptotic expansion of V_1^+ at infinity has no positive powers of X. Thus, g_2^+ coincides with V_1^+. Similarly, $g_3^+ = V_2^+$.

3.3 Proof of Theorem 2.13

We first need a formula in order to get an estimation of the difference of the canard values a^+ and a^- associated with the exceptional solutions y^+ respectively y^-. Let $0 < \varepsilon < \varepsilon_0$ where ε_0 is provided by Theorem 2.6. Since these two functions are solutions of (1), the difference $z := y^+ - y^-$ satisfies the equation:

$$\varepsilon \frac{dz}{dx} = -(y^+ + y^-)z - (a^+ - a^-).$$

By the variation of constant formula, one obtains

$$z(x) = \left(y^+(x_0) - y^-(x_0)\right) e^{-\frac{1}{\varepsilon}\int_{x_0}^x (y^+ + y^-)} - \frac{a^+ - a^-}{\varepsilon} \int_{x_0}^x e^{-\frac{1}{\varepsilon}\int_t^x (y^+ + y^-)} dt \tag{25}$$

with $x_0 \in \mathbb{C}$ to be chosen. More precisely, x_0 must be chosen such that the difference $y^+(x_0) - y^-(x_0)$ is not exponentially small. Actually, it is only near the origin that we can observe a sensitive difference between y^+ and y^-. Moreover, thanks to the composite asymptotic expansions, we saw in the previous subsection that y^+ (and also $y^- := \overline{y^+}$) can be analytically continued until $x = 0$. That is the reason why we can choose $x_0 = 0$ in (25).

Because z vanishes at infinity, one has

$$a^+ - a^- = \varepsilon \left(y^+(0) - y^-(0)\right) \frac{E}{D} \tag{26}$$

with

$$E = \exp\left\{-\frac{1}{\varepsilon}\int_0^1 (y^+ + y^-)\right\} \quad \text{and} \quad D = \int_0^\infty \exp\left\{\frac{1}{\varepsilon}\int_1^t (y^+ + y^-)\right\} dt. \tag{27}$$

To prove Theorem 2.13, it remains to compute each term of this formula (26): the difference $y^+(0) - y^-(0)$, D and E. We begin with D.

3.3.1 Calculation of D

We will use the following Gevrey version of the Laplace method. The proof can be found in [15].

Theorem 3.3 *Let be $L, \delta > 0$. Let $f(x, \varepsilon)$ and $g(x, \varepsilon)$ be two holomorphic functions defined for $x \in \mathcal{D}(0, L + \delta)$ and $\varepsilon \in S(-\delta, \delta, \varepsilon_0)$ having a Gevrey asymptotic expansion of order 1 as $\varepsilon \to 0$ uniformly with respect to $x \in \mathcal{D}(0, L + \delta)$:*

$$f(x, \varepsilon) \sim_1 \sum_{k\geq 0} f_k(x)\varepsilon^k,$$

$$g(x, \varepsilon) \sim_1 \sum_{k\geq 0} g_k(x)\varepsilon^k,$$

with for all $k \in \mathbb{N}$, f_k and g_k analytic on $\mathcal{D}(0, L + \delta)$. Moreover suppose that $f(x, \varepsilon)$ is such that $f_0(x) = \sum_{l\geq 2} f_{0l}x^l$ with $f_{02} > 0$ and f_0 has real values on $[0, L + \delta[$. Then there exists $r > 0$ such that for all $\varepsilon > 0$, the integral

$$\int_0^r e^{-f(x,\varepsilon)/\varepsilon} g(x, \varepsilon)dx$$

has a Gevrey asymptotic expansion of order 1 in powers of $\sqrt{\varepsilon}$, as $\varepsilon \to 0$, without constant term.

The function $t \mapsto \int_1^t (y^+ + y^-)$ admits a local maximum denoted by c near $t = 1$. More precisely, using the outer expansion $\hat{y}$ of $y^\pm$ given by (5), we find that $c = 1 - \varepsilon + O(\varepsilon^2)$. Recall that $y^\pm$ is Gevrey-1 asymptotic to $\hat{y}$ (cf. Proposition 2.3) and the functions y_n in the expansion $\hat{y}$ are analytic for $x > 0$. It is sufficient to study

$$\tilde{D} := \int_{1/2}^{3/2} \exp\left\{\frac{1}{\varepsilon}\int_1^t (y^+ + y^-)\right\} dt$$

because the difference $D - \tilde{D}$ is negligible with respect to $\tilde{D}$: This difference is exponentially small, say $O(e^{-M/\varepsilon})$, with some $M > 0$ and $\tilde{D}$ turns out to be of order $\sqrt{\varepsilon}$. The change of variable $x = t - c$ leads to:

$$\tilde{D} = \int_{1/2-c}^{0} \exp\left\{\frac{1}{\varepsilon}\int_1^{x+c} (y^+ + y^-)\right\} dx + \int_0^{3/2-c} \exp\left\{\frac{1}{\varepsilon}\int_1^{x+c} (y^+ + y^-)\right\} dx.$$

Then we can apply Theorem 3.3 to the term $\int_0^{3/2-c} \exp\left\{\frac{1}{\varepsilon}\int_1^{x+c}(y^+ + y^-)\right\} dx$. The same holds for the integral $\int_{1/2-c}^{0} \exp\left\{\frac{1}{\varepsilon}\int_1^{x+c}(y^+ + y^-)\right\} dx$ thanks to the transformation $x \mapsto -x$. As a consequence, only the even powers of η will remain, i.e. the expansion of $D/\sqrt{\varepsilon}$ only contains powers of ε.

One finally obtains the next result for D. The details of the computation can be found in [15].

Lemma 3.4 *The term D introduced in formula* (27) *is such that $D/\sqrt{\varepsilon}$ has an asymptotic expansion of Gevrey order 1 in terms of ε when $\varepsilon \to 0$, $\varepsilon > 0$. Moreover, the calculation of the first term gives:*

$$\frac{D}{\sqrt{\pi\varepsilon}} = 1 + \frac{35}{12}\varepsilon + O(\varepsilon^2).$$

Before detailing the computation of the other two terms E and $y^+(0) - y^-(0)$ of formulae (26) and (27), we would like to present a symmetry property. This property explains why the expansion of Theorem 2.13 only contains powers of ε, i.e. even powers of η.

3.3.2 A Symmetry Property

Let us start from equations (9) and (10). Combining them, we obtain

$$y(x, -\eta) = \overline{y(\overline{x}, \overline{\eta})} \tag{28}$$

valid for all $\eta \in S_0 = S(-\delta/2, \delta/2, \eta_0)$ and all $x \in D_{-2\pi}(\delta, |\eta|^2) \cap D_0(\delta, |\eta|^2)$.

In Proposition 2.10 we show the existence of a Gevrey composite asymptotic expansion of order 1/2 for y^+:

$$y^+(x, \eta) \sim_{\frac{1}{2}} x(1 - x) + \sum_{n\geq 1} g_n^+\left(\tfrac{x}{\eta}\right)\eta^n \tag{29}$$

valid in particular for $\eta \in S_0$ and x in the quasi-sector $\mathcal{V}^+ := \mathcal{V}(-\frac{\pi}{6}, \frac{7\pi}{6}, 1 + \delta, \mu|\eta|)$, where $\mu > 0$.

Applying (28) to the composite asymptotic expansion (29), one has for all $n \geq 1$, all $\eta \in S_0$ and all $x \in \mathcal{V}^+ \cap \overline{\mathcal{V}^+}$,

$$g_n^+\left(\tfrac{x}{-\eta}\right)(-1)^n = \overline{g_n^+\left(\tfrac{\overline{x}}{\overline{\eta}}\right)}. \tag{30}$$

For $n \geq 1$, let $A_n := \Re(g_n^+)$ and $B_n := \Im(g_n^+)$, such that $g_n^+ = A_n + iB_n$. In the particular case where $\eta > 0$ and $x \in \mathbb{R}$, Eq. (30) is equivalent to:

$$\begin{cases} A_n\left(\frac{x}{-\eta}\right)(-1)^n = A_n\left(\frac{x}{\eta}\right) \\ B_n\left(\frac{x}{-\eta}\right)(-1)^n = -B_n\left(\frac{x}{\eta}\right). \end{cases}$$

In other words, for $k \geq 0$, the restriction of A_{2k+1} to $\mathbb{R}$ is an odd function while the restriction of B_{2k+1} to $\mathbb{R}$ is an even function and for $k \geq 1$, the restriction of A_{2k} to $\mathbb{R}$ is even while the restriction of B_{2k} to $\mathbb{R}$ is odd.

3.3.3 Calculation of E

Concerning E of formula (27), we show the following result.

Lemma 3.5 *The term $\varepsilon e^{1/(3\varepsilon)} E$ admits a Gevrey asymptotic expansion of order 1 in powers of ε when $\varepsilon \to 0$, $\varepsilon > 0$. Moreover,*

$$\varepsilon e^{1/(3\varepsilon)} E = \pi + \left(\frac{16}{9} - \frac{35}{6}\pi\right)\varepsilon + O(\varepsilon^2).$$

Proof Note first that

$$\int_0^1 (y^+(x) + y^-(x))dx = 2\int_0^1 \Re(y^+(x))dx \to \frac{1}{3} \quad \text{as} \quad \varepsilon \to 0.$$

This is due to $2\int_0^1 x(1-x)dx = 1/3$. Because we use (29), we need to study the integral $\int_0^1 \eta^n A_n(\frac{x}{\eta})dx = \eta^{n+1}\int_0^{1/\eta} A_n(X)dX$ where A_n is defined in the previous Sect. 3.3.2.

When n is even, say $n = 2k$, A_{2k} is even. If G_{2k} denotes the primitive of A_{2k} such that $G_{2k}(0) = 0$, then G_{2k} is odd and $G_{2k}(X)$ has an asymptotic expansion in odd powers of X when $X \to \infty$. As a consequence,

$$\eta^{2k+1}\int_0^{1/\eta} A_{2k}(X)dX = \eta^{2k+1}G_{2k}(1/\eta)$$

has an asymptotic expansion in even powers of η when $\eta \to 0$, that is to say in powers of ε.

The same argument holds for the case n odd. We just need to pay attention to the case $n = 1$ because $g_1^{\pm}(X) \sim -\frac{1}{X}$ when $X \to \infty$, which provides the logarithmic term $\varepsilon \log \varepsilon$ after integration. Anyway, in any case $\eta^n \int_0^1 A_n(\frac{x}{\eta})dx$, $n \geq 2$, has an expansion in powers of ε, which means that $\frac{1}{3} + \varepsilon \log \varepsilon - \int_0^1 (y^+ - y^-)$ has an expansion in powers of ε without constant term. Therefore, $\varepsilon e^{1/(3\varepsilon)} E$ has an expansion in powers of ε only. Thanks to Proposition 2.9 of [11] concerning the integration of a composite asymptotic expansion, one deduces that $\frac{1}{\varepsilon}\left(\frac{1}{3} + \varepsilon \log \varepsilon - \int_0^1 (y^+ + y^-)\right)$ admits a Gevrey asymptotic expansion of order 1 in powers of ε. Finally, the composition with the exponential function gives the result. More details of the computation can be found in [15]. Mainly, the computation uses (22), (23) and (24). □

3.3.4 Calculation of $y^+(0) - y^-(0)$

Using the symmetry property described in previous Sect. 3.3.2, one deduces that for $\eta > 0$, $y^+(0) - y^-(0)$ has a Gevrey asymptotic expansion of order 1/2 as follows:

$$y^+(0) - y^-(0) = 2i\Im(y^+(0)) \sim_{1/2} 2i \sum_{n\geq 1} B_n(0)\eta^n = 2i\eta \sum_{k\geq 0} B_{2k+1}(0)\eta^{2k}.$$

With $0 < \eta = \sqrt{\varepsilon}$, one can also write

$$y^+(0) - y^-(0) \sim_{\frac{1}{2}} 2i\sqrt{\varepsilon} \sum_{k\geq 0} B_{2k+1}(0)\varepsilon^k. \tag{31}$$

The computation of the first terms of this expansion uses (22) and (24). More details can be found in [15]. Finally, one gets

Lemma 3.6 *The difference* $\frac{1}{\sqrt{\varepsilon}}(y^+(0) - y^-(0))$ *where* y^+ *and* y^- *denote the exceptional solutions provided by Theorem 2.6 has a Gevrey asymptotic expansion of order 1 in powers of* ε *when* $\varepsilon \to 0$, $\varepsilon > 0$. *Moreover,*

$$-i\sqrt{\frac{\pi}{\varepsilon}}\big(y^+(0) - y^-(0)\big) = 4 - \frac{1}{3}\left(1 + \frac{64}{3\pi}\right)\varepsilon + O(\varepsilon^2).$$

3.3.5 End of the Proof of Theorem 2.13

Now we use some classical results of the Gevrey theory that can be found in [12] or also in [15]. More precisely, for two functions f and g admitting each one a Gevrey-1 asymptotic expansion on some sector S:

- The product fg also admits a Gevrey-1 asymptotic expansion on S.
- Suppose that f is bounded below, i.e. there exists $\sigma > 0$ such that for all $\varepsilon \in S$, $|f(\varepsilon)| \geq \sigma$. Then, the inverse $1/f$ also admits a Gevrey-1 asymptotic expansion on S.

In Lemma 3.4, we got a Gevrey asymptotic expansion of order 1 for $D/\sqrt{\varepsilon}$ starting with a non-zero constant term. Thus, one can deduce that $\frac{\sqrt{\varepsilon}}{D}$ has also a Gevrey asymptotic expansion of order 1 in powers of ε.

Then, thanks to Lemmas 3.5 and 3.6, the product of $\frac{\sqrt{\varepsilon}}{D}$, $\frac{1}{\sqrt{\varepsilon}}(y^+(0) - y^-(0))$ and $\varepsilon e^{1/(3\varepsilon)}E$ have a Gevrey asymptotic expansion of order 1 in powers of ε when $\varepsilon \to 0$.

According to (26) this means that $(a^+ - a^-)e^{1/(3\varepsilon)}$ has a Gevrey asymptotic expansion of order 1 in powers of ε.

3.4 Proof of Theorem 2.15

We introduce the truncated Laplace transform.

Definition 3.7 Given an analytic function g defined on a disk $\mathcal{D}(0, r)$ with finite radius $r > 0$ and $T \in \mathcal{D}(0, r)$, the *truncated Laplace transform of* g *at* T is:

$$\mathcal{L}_T g(\varepsilon) = \int_0^T e^{-T/\varepsilon} g(t)dt.$$

Denote by $\mathcal{T}$ the isomorphism which maps a function holomorphic in a neighbourhood of 0 to its Taylor series:

$$\mathcal{T} : \phi \mapsto \sum_{n\geq 0} \tfrac{\phi^{(n)}(0)}{n!} t^n.$$

Then the interest of the truncated Laplace transform is given by the following lemma; see [15] for a proof.

Lemma 3.8 *Let g be a function holomorphic in a neighbourhood of the origin and $T = |T|e^{i\psi}$ in the disk of convergence of $\mathcal{T}g$. Then $\mathcal{L}_T g$ is analytic in $\mathbb{C}^*$. Moreover, for any $\phi \in \left]0, \frac{\pi}{2}\right[$, $\mathcal{L}_T g$ admits a Gevrey asymptotic expansion of order 1, obtained by termwise integration of the expansion of g, in the sector $S(\psi - \phi, \psi + \phi, \infty)$. Precisely, if $\mathcal{T}g(t) = \sum_{n\geq 0} c_n t^n$ then*

$$\mathcal{L}_T g \sim_{1,|T|\cos\phi} \sum_{n\geq 0} n! c_n \varepsilon^{n+1}$$

on the infinite sector $S(\psi - \phi, \psi + \phi, \infty)$.

Theorem 2.13 provides us a Gevrey-1 asymptotic expansion for $(a^+(\varepsilon) - a^-(\varepsilon))$ $e^{1/(3\varepsilon)}$. An immediate consequence of Lemma 3.8 is the following result: For $\delta, \phi \in \left]0, \frac{\pi}{2}\right[$ and $T = M\cos\delta$, one has

$$(a^+(\varepsilon) - a^-(\varepsilon))e^{1/(3\varepsilon)} \sim_{1,M\cos\delta\cos\phi} 4i + \mathcal{L}_{M\cos\delta}\tilde{b}(\varepsilon)$$

with $\mathcal{L}_{M\cos\delta}\tilde{b}(\varepsilon) := \int_0^{M\cos\delta} \tilde{b}(t)e^{-t/\varepsilon}dt$ and $[0, M\cos\delta]$ as path of integration.

In other words,

$$(a^+(\varepsilon) - a^-(\varepsilon))e^{1/(3\varepsilon)} - 4i - \mathcal{L}_{M\cos\delta}\tilde{b}(\varepsilon) \sim_{1,M\cos\delta\cos\phi} \hat{0}.$$

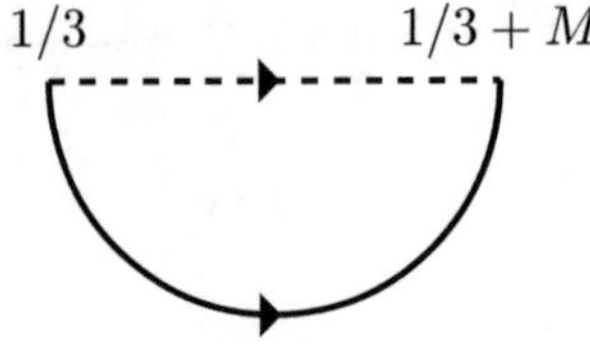

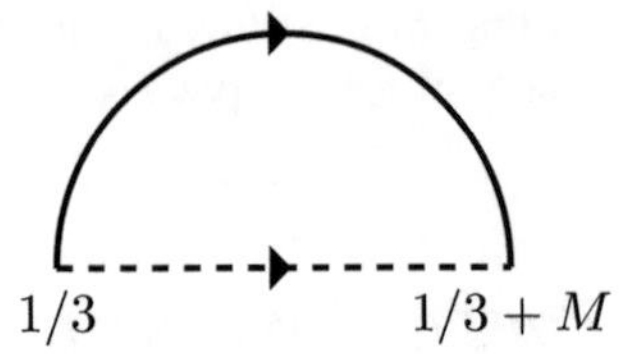

Fig. 6 Deformation of the path of integration for $\tilde{B}$

Using that a function which is Gevrey-1 asymptotic to the identically zero expansion decreases exponentially, one has

$$a^+(\varepsilon) - a^-(\varepsilon) = e^{-1/(3\varepsilon)}\left(4i + \mathcal{L}_{M\cos\delta}\tilde{b}(\varepsilon) + O\left(\varepsilon^{-1/2}e^{-M\cos\delta\cos\phi/\varepsilon}\right)\right), \quad \varepsilon \to 0.$$

Choosing a smaller $M > 0$, one can always write:

$$a^+(\varepsilon) - a^-(\varepsilon) = e^{-1/(3\varepsilon)}\left(4i + \mathcal{L}_{M\cos\delta}\tilde{b}(\varepsilon) + O\left(e^{-M\cos\delta\cos\phi/\varepsilon}\right)\right), \quad \varepsilon \to 0. \tag{32}$$

Now introduce, for $t \in \mathbb{C} \setminus [1/3, 1/3 + M]$,

$$\tilde{f}(t) := \frac{2}{\pi(t - 1/3)} + \tilde{B}(t)$$

where

$$\tilde{B}(t) := \frac{1}{2\pi i}\int_{1/3}^{1/3+M} \frac{\tilde{b}(\tau - 1/3)}{\tau - t}d\tau.$$

From $\tilde{f}$, we are going to construct two functions f_S and f_N whose difference $f_S - f_N$ satisfies an equation similar to (32). By a deformation of the path of integration, $\tilde{B}$ (and thus $\tilde{f}$) can be analytically continued to $]1/3, 1/3 + M[$ by two ways depending on $\arg t \to -2\pi^+$ or $\arg t \to 0^-$ (cf. Fig. 6). Denote by $\tilde{B}_S$, respectively $\tilde{B}_N$, these two continuations of $\tilde{B}$. The letter S respectively N refers to the deformation of the path from the *South* respectively from the *North*. For all $t \in]1/3, 1/3 + M[$, one also has

$$\tilde{B}_S(t) - \tilde{B}_N(t) = \tilde{b}(t - 1/3). \tag{33}$$

Moreover, $\tilde{B}(t) = O(\log(t - 1/3))$ when $t \to 1/3$, hence $\tilde{B}$ is integrable at $t = 1/3$. Denote $\tilde{f}_N(t) := \frac{2}{\pi(t-1/3)} + \tilde{B}_N(t)$ and $\tilde{f}_S(t) := \frac{2}{\pi(t-1/3)} + \tilde{B}_S(t)$.

Then define

$$f_N(\varepsilon) := \int_0^{1/3+Me^{i\delta}} \tilde{f}_N(t)e^{-t/\varepsilon}dt$$

and

$$f_S(\varepsilon) := \int_0^{1/3+Me^{-i\delta}} \tilde{f}_S(t)e^{-t/\varepsilon}dt$$

with the segments $\left[0, 1/3 + Me^{\pm i\delta}\right]$ as paths of integration.

Lemma 3.9 *For $\varepsilon > 0$, the difference $f_S - f_N$ satisfies:*

$$f_S(\varepsilon) - f_N(\varepsilon) = e^{-1/(3\varepsilon)}\left(4i + \mathcal{L}_{M\cos\delta}\tilde{b}(\varepsilon) + O\left(e^{-M\cos\delta/\varepsilon}\right)\right).$$

Proof First, on the closed path γ defined by the concatened segments $\left[1/3 + Me^{i\delta}, 0\right]$, $\left[0, 1/3 + Me^{-i\delta}\right]$ and $\left[1/3 + Me^{-i\delta}, 1/3 + Me^{i\delta}\right]$, one has

$$\int_\gamma \frac{2}{\pi(t-1/3)}e^{-t/\varepsilon}dt = 4ie^{-1/(3\varepsilon)}$$

thanks to the residue theorem. Then, on $\left[1/3 + Me^{-i\delta}, 1/3 + Me^{i\delta}\right]$, we have

$$\int_{1/3+Me^{-i\delta}}^{1/3+Me^{i\delta}} \frac{2}{\pi(t-1/3)}e^{-t/\varepsilon}dt = O\left(e^{-(1/3+M\cos\delta)/\varepsilon}\right).$$

Thus,

$$\left(\int_0^{1/3+Me^{-i\delta}} \frac{2}{\pi(t-1/3)}e^{-t/\varepsilon}dt\right) - \left(\int_0^{1/3+Me^{i\delta}} \frac{2}{\pi(t-1/3)}e^{-t/\varepsilon}dt\right) \\ = e^{-1/(3\varepsilon)}\left(4i + O\left(e^{-M\cos\delta/\varepsilon}\right)\right). \tag{34}$$

We decompose each path of integration $\left[0, 1/3 + Me^{\pm i\delta}\right]$ in two segments

$$[0, 1/3 + M\cos\delta] \quad \text{and} \quad \left[1/3 + M\cos\delta, 1/3 + Me^{\pm i\delta}\right].$$

Thanks to (33), one gets:

$$\begin{aligned}
&\int_0^{1/3+Me^{-i\delta}} e^{-t/\varepsilon}\tilde{B}_S(t)dt - \int_0^{1/3+Me^{i\delta}} e^{-t/\varepsilon}\tilde{B}_N(t)dt \\
&= \int_{1/3}^{1/3+M\cos\delta} e^{-t/\varepsilon}\tilde{b}(t-1/3)dt + \int_{1/3+Me^{i\delta}}^{1/3+Me^{-i\delta}} e^{-t/\varepsilon}\tilde{B}(t)dt \\
&= \mathcal{L}_{M\cos\delta}(\tilde{b})(\varepsilon) + O\left(e^{-M\cos\delta/\varepsilon}\right).
\end{aligned} \tag{35}$$

Finally, (34) and (35) conclude the proof of the lemma. □

End of proof of Theorem 2.15. Our canard value function a can be decomposed as follows. For arbitrary $0 < \mu < \pi/2$ and $\varepsilon_0 > 0$, let $S_N := S(-2\pi - \mu, -\pi + \mu, \varepsilon_0)$ and $S_S := S(-\pi - \mu, \mu, \varepsilon_0)$. Define $a_N := a|_{S_N}$ and $a_S := a|_{S_S}$. We will need the following result.

Theorem 3.10 ([9], Theorem 1) *Suppose that sectors $S_j = S(\alpha_j, \beta_j, r)$, $1 \leq j \leq m$ are such that $\alpha_1, \dots, \alpha_m, \alpha_1 + 2\pi$ and $\beta_1, \dots, \beta_m, \beta_1 + 2\pi$ are increasing sequences satisfying $\alpha_{j+1} < \beta_j$ for $j = 1, \dots, m-1$ and $\alpha_1 + 2\pi < \beta_m$. For $j = 1, \dots, m$ suppose that $d_j : S_j \to \mathbb{C}$ is bounded, analytic and that there exists constants $r_j \in \mathbb{C}$ and $\phi_j \in]\alpha_{j+1}, \beta_j[$ such that for $\varepsilon \in S_j \cap S_{j+1}$ with* $\arg\varepsilon = \phi_j$ *one has*

$$|d_{j+1}(\varepsilon) - d_j(\varepsilon)| = O\left(e^{-r_j/|\varepsilon|}\right).$$

Then the Borel transform $\tilde{d}$ of the common asymptotic expansions of the d_j can be analytically continued to the connected set

$$\bigcap_{j=1}^{m} \left\{t \in \mathbb{C}; \Re(te^{-i\phi_j}) < r_j\right\}.$$

Now we apply Theorem 3.10 to the functions d_N and d_S defined by $d_N : S_N \to \mathbb{C}, \varepsilon \mapsto a_N(\varepsilon) - f_N(\varepsilon)$ and $d_S : S_S \to \mathbb{C}, \varepsilon \mapsto a_S(\varepsilon) - f_S(\varepsilon)$.

On the one hand, on the sector $S(-\pi - \mu, -\pi + \mu, \varepsilon_0)$, $d_S - d_N = 0$. On the other hand, on $S(-\mu, \mu, \varepsilon_0)$, for $\varepsilon > 0$, $d_S(\varepsilon) - d_N(\varepsilon) = O\left(e^{-(1/3+M\cos\delta\cos\phi)/\varepsilon}\right)$. Therefore the Borel transform $\tilde{d}$ of the common asymptotic expansions of d_N and d_S can be analytically continued to

$$\mathcal{H} := \{t \in \mathbb{C}; \Re(t) < 1/3 + M\cos\delta\cos\phi\}.$$

We deduce that $\tilde{a} - \tilde{f}$ is analytic on $\mathcal{H}$, which ends the proof of Theorem 2.15.

Acknowledgements I would like to express my gratitude to Prof. A. Fruchard and Prof. R. Schäfke for the supervision of this work.

References

1. Balser, W.: From Divergent Power Series to Analytic Functions : Theory and Application of Multisummable Power Series. Lecture Notes in Mathematics. Springer, Berlin (1994)
2. Balser, W.: Formal Power Series and Linear Systems of Meromorphic Ordinary Differential Equations. Universitext, Springer (2000)
3. Benoît, É., Callot, J.-L., Diener, F., Diener, M.: Chasse au canard. Collectanea Mathematica **32**, 37–119 (1981)
4. Benoît, É., El Hamidi, A., Fruchard, A.: On combined asymptotic expansions in singular perturbations. Electron. J. Differ. Equ. (EJDE) [electronic only], 2002: Paper No. 51, 27 p., (2002)
5. Benoît, É., Fruchard, A., Schäfke, R., Wallet, G.: Solutions surstables des équations différentielles complexes lentes-rapides à point tournant. Annales de la Faculté des Sciences de Toulouse. Mathématiques. Série VI **7**(4), 627–658 (1998)
6. Callot, J.-L.: Champs lents-rapides complexes à une dimension lente. Annales scientifiques de l'École Normale Supérieure **26**(2), 149–173 (1993)
7. Canalis-Durand, M., Ramis, J.-P., Schäfke, R., Sibuya, Y.: Gevrey solutions of singularly perturbed differential equations. Crelles J. **518**. (Journal für die Reine und Angewandte Mathematik, 1999)
8. Fruchard, A., Matzinger, É.: Matching and singularities of canard values. In: Costin, O., Kruskal, M.D., Macintyre, A. (eds.), Analyzable functions and applications: International Workshop on Analyzable Functions and Applications, June 17–21, 2002, International Centre for Mathematical Sciences, Edinburgh, Scotland. Contemporary mathematics, vol. 373, pp. 317–335. American Mathematical Society (2005)
9. Fruchard, A., Schäfke, R.: On the Borel transform. C. R. Acad. Sci. Paris Sér. I Math. **323**(9), 999–1004 (1996)
10. Fruchard, A., Schäfke, R.: Exceptional complex solutions of the forced van der Pol equation. Funkcialaj Ekvacioj **42**(2), 201–223 (1999)
11. Fruchard, A., Schäfke, R.: Composite Asymptotic Expansions. Lecture Notes in Mathematics, vol. 2066. Springer, Berlin (2013)
12. Loday-Richaud, M.: Divergent Series, Summability and Resurgence II. Lecture Notes in Mathematics. Springer, Berlin (2016)
13. Matzinger, É.: Étude d'équations différentielles ordinaires singulièrement perturbées au voisinage d'un point tournant. Thèse, Strasbourg 1 (2000)
14. Matzinger, É.: Étude des solutions surstables de l'équation de van der Pol. Annales de la faculté des sciences de Toulouse **10**(4), 713–744 (2001)
15. Pavis d'Escurac, P.: Étude des singularités de la fonction valeur à canard de certaines équations différentielles complexes singulièrement perturbées. Preprint, Doctoral Dissertation, UHA, Mulhouse (2018)
16. Zinn-Justin, J.: Multi-instanton contributions in quantum mechanics. Nucl. Phys. B **192**(1), 125–140 (1981)
17. Zinn-Justin, J., Jentschura, U.D.: Multi-instantons and exact results i: conjectures, WKB expansions, and instanton interactions. Ann. Phys. **313**(1), 197–267 (2004)

Quantization Conditions on Riemannian Surfaces and Spectral Series of Non-selfadjoint Operators

Andrei Shafarevich

Abstract In the paper, the review of the papers [26–30, 32–34] devoted to the semi-classical asymptotic behavior of the eigenvalues of some nonself-adjoint operators important for applications is given. These operators are the Schrödinger operator with complex periodic potential and the operator of induction. It turns out that the asymptotics of the spectrum can be calculated using the quantization conditions, which can be represented as the condition that the integrals of a holomorphic form over the cycles on the corresponding complex Lagrangian manifold, which is a Riemann surface of constant energy, are integers. In contrast to the real case (the Bohr–Sommerfeld–Maslov formulas), to calculate a chosen spectral series, it is sufficient to assume that the integral over only one of the cycles takes integer values, and different cycles determine different parts of the spectrum.

Keywords Non-selfadjoint operators · Quantization conditions · Riemannian surfaces

MSC Primary 47F05

A. Shafarevich (✉)
"M.V. Lomonosov" Moscow State University, Leninskie Gory,1, Moscow, Russia
e-mail: shafarev@yahoo.com

A. Shafarevich
Moscow Institute of Physics and Technology, Institutskii Pereulok, 9, Dolgoprudny, Russia

A. Shafarevich
Institute for Problems in Mechanics of the Russian Academy of Sciences, Prospekt Vernadskogo, 101, Moscow, Russia

A. Shafarevich
Russian National Scientific Centre "Kurchatov Institute", Ploshad' Akademika Kurchativa, 1, Moscow, Russia

G. Filipuk et al. (eds.), *Formal and Analytic Solutions of Diff. Equations*, Springer Proceedings in Mathematics & Statistics 256,
https://doi.org/10.1007/978-3-319-99148-1_9

1 Introduction

One of the main problems of the semiclassical theory (see, for example, [1]) — is the description of the asymptotic behavior of the spectrum of operators of the form $\hat{H} = H(x, -ih\frac{\partial}{\partial x}), \quad h \to 0$. In this case, the problem can naturally be divided into two following subproblems.

1. To solve the spectral equation approximately, i.e., to find numbers λ and functions ψ, satisfying the following equation for some $N > 1$:

$$\widehat{H}\psi = \lambda\psi + O(h^N). \tag{1}$$

2. To choose numbers of the form λ that approach spectral points of the operator $\hat{H}$, i.e., to choose points λ such that

$$|\lambda - \lambda_0| = O(h^N) \tag{2}$$

for some point λ_0 of the spectrum of the operator $\hat{H}$.

If the operator $\hat{H}$ is self-adjoint, then the estimate (2) automatically follows from Eq. (1) (see, e.g., [1–3]). At the same time, the first problem is highly nontrivial and is related to the study of invariant sets of the corresponding classical Hamiltonian system. Recall how to solve this problem (1) in the integrable case. Let

$$H(x, p) : R^{2n} \to R$$

be a smooth function, and let the Hamiltonian system defined by the function H be Liouville integrable. Let $f_1 = H, \ldots, f_n$ be the commuting first integrals; consider the domain of the phase space smoothly fibered into Liouville tori Λ which are the compact connected components of the common level sets of the form $f_j = c_j$. We assume that the Weyl operator $\hat{H}$ is self-adjoint in $L^2(R^n_x)$. The following theorem is due to V. P. Maslov.

Theorem 1 *Suppose that a Liouville torus Λ satisfies the following conditions (the so-called Bohr–Sommerfeld–Maslov quantization rules, see [1, 2, 4, 5]):*

$$\frac{1}{2\pi h}\int_{\gamma}(p, dx) = m + \frac{\mu(\gamma)}{4}, \tag{3}$$

where $m = O(1/h) \in Z$, γ is an arbitrary cycle on Λ, and $\mu(\gamma)$ is the Maslov index of the cycle. Then there is a function $\psi \in L^2(R^n)$, $||\psi|| = 1$, such that

$$\hat{H}\psi = \lambda\psi + O(h^2), \quad \lambda = H|_\Lambda.$$

Remark 1 The function ψ mentioned in the theorem can be described in a computable way, namely, it is of the form $K(1)$, where K stands for the Maslov canonical operator on the Liouville torus Λ. Integer m can be chosen in the form $m = [1/h]_{int} + m_0$, where $[1/h]_{int}$ stands for the integral part of the real number $1/h$ and m_0 does not depend on h.

Remark 2 As was already noted above, it follows automatically from the statement of the theorem that the point λ is at a distance of the order of $O(h^2)$ from the spectrum of the operator $\hat{H}$.

Remark 3 We stress that the topological condition (3) must be satisfied for *all* cycles of the torus Λ (in other words, the quantization condition is the condition that the cohomology class

$$\frac{1}{2\pi h}[\theta] + \frac{1}{4}[\mu]$$

is integer, where $[\theta]$ stands for the class of the form (p, dx) and $[\mu]$ for the Maslov class).

Remark 4 In action–angle variables $(I_1, \ldots, I_n, \varphi_1, \ldots, \varphi_n)$, the quantization conditions and formula for the spectrum have simple form (see e.g. [2])

$$I_j = h(m_j + \frac{\mu_j}{4}), \quad \lambda = H(I_1, \ldots, I_n).$$

The nonself-adjoint case is less investigated, and quite incompletely; however, spectral problems for nonself-adjoint operators arise in many important physical applications (like the theory of hydrodynamic stability, a description of magnetic fields of the Earth and of galaxies, the PT-symmetric quantum theory, statistical mechanics of Coulomb gases and many other problems; see, for example, [6–11]).

In the paper, we consider two classes of nonself-adjoint operators, namely, the one-dimensional Schrödinger operator with complex potential and the operator of magnetic induction on a two- dimensional symmetric surface. The spectrum of these operators, in the semiclassical limit, is concentrated in the $O(h^2)$-neighborhood of some curves in the complex plane E; these curves form the so-called *spectral graph*. It turns out that each edge of the spectral graph corresponds to a certain cycle on the Riemann surface defined by the classical complex Hamiltonian system (this is a surface of constant energy).

The asymptotics of the eigenvalues can be calculated by using complex equations which are similar to the Bohr–Sommerfeld–Maslov quantization conditions on the Riemann surface. However, in contrast to the self-adjoint case, in order to evaluate the eigenvalues, it is required to satisfy the corresponding condition on *only one cycle*, and it turns out that different cycles determine different parts of the spectrum (and different edges of the spectral graph).

2 Schrödinger Equation with a Complex Potential

The spectral problem for the Schrödinger equation on the circle with a purely imaginary potential

$$-h^2\psi'' + \imath V(x)\psi = \lambda\psi, \quad \psi(x+2\pi) = \psi(x) \tag{4}$$

arises, in particular, as a model problem for the Orr–Sommerfeld operator in the theory of hydrodynamic stability (see, e.g. [12–25]) Close problem appears in the statistical mechanics of the Coulomb gas (see [11]). Here $h \to 0$ is a small parameter and $V(x)$ is a trigonometric polynomial. The asymptotic behavior of the spectrum of this operator for different trigonometric polynomials V as $h \to 0$ was calculated in [26–31]; it turns out here that the numbers λ satisfying (1) fill a half-strip in the complex plane entirely, while the actual spectrum is discrete and concentrates near some graph. The results of these papers can be reformulated in terms of the quantization rules on Riemann surfaces as follows (for the proof see [30, 31]). Consider a Riemann surface Λ in the complex phase space $\Phi = (C/2\pi Z) \times C$ with coordinates (x, p), where Λ is given by the equation

$$p^2 + \imath V(x) = \lambda;$$

this surface is obtained by gluing together two cylinders of the variable x, $\Re x \in S^1$, $\Im x \in R$ along finitely many cuts, namely, the zeros of the trigonometric polynomial $\imath V - \lambda$ are joined to one another and to the points at infinity. The results of the papers mentioned above imply the following assertion. Consider the set $\Gamma \in C$, defined by quantization conditions

$$\frac{1}{2\pi h}\int_\gamma p dx = m + \frac{\mu}{4}, \tag{5}$$

where γ is an element of a certain finite set of cycles on the surface Λ, $\mu(\gamma) \in \{0, 2\}$, and $m = O(1/h)$ is an integer.

Theorem 2 *Let K be a compact subset of the complex plane, independent of h. Then there exists a constant C such that the following holds when $h > 0$ is small enough: If $\lambda \in K$ is an eigenvalue of the Schrödinger operator, then the distance d between λ and Γ satisfies the estimate*

$$d \leq Ch^2.$$

Conversely, for each point of the intersection $\Gamma \cap K$, except maybe the points corresponding to singular surfaces Λ, there exits an eigenvalue with the same estimate.

Remark 5 In contrast to the self-adjoint case (see Theorem 1), for each λ the quantization condition must hold on *at least one cycle* in the given number of cycles, and different cycles determine different parts of the spectrum. So the complex quantization conditions do not imply that a certain cohomology class of Λ should be integral.

Remark 6 The choice of the set of cycles depends on the potential V only; there exists an analytic algorithm selecting the cycles.

Remark 7 Separating the real and imaginary parts in Eq. (5) we obtain the system

$$Im \int_{\gamma} pdx = 0, \tag{6}$$

$$Re \frac{1}{2\pi h} \int_{\gamma} pdx = m + \frac{\mu}{4}. \tag{7}$$

The first equation does not depend on h. The combination of these equations for different cycles defines a set of analytical curves in the complex plane λ, the so-called *spectral graph*. The second equation defines a discrete set of asymptotic eigenvalues; for a fixed cycle γ, these eigenvalues are concentrated near the corresponding edge of the spectral graph.

Remark 8 In [26–30] examples of spectral graphs for specific surfaces Λ are presented. In particular, if $V(x) = \cos x$, then the surface Λ is homeomorphic to a torus with two punctures; the corresponding spectral graph consists of three edges corresponding to the three cycles in the surface and has the shape shown in Fig. 1. If

$$V = \cos x + \cos 2x,$$

then the surface is homeomorphic to a pretzel with two punctures (a sphere with two handles and with two disks removed); the corresponding spectral graph is shown in Fig. 2 and consists of five edges (note that the one-dimensional homology of Λ is the five-dimensional in this case).

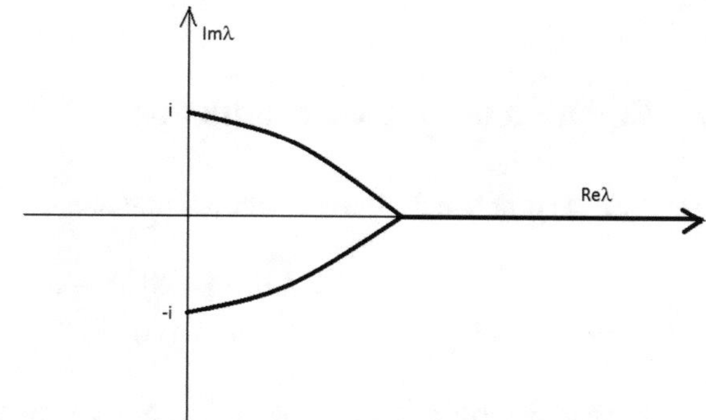

Fig. 1 Spectral graph for the case $V = \cos x$

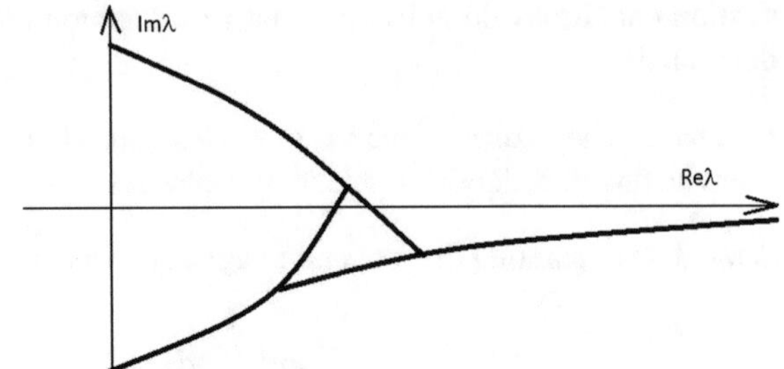

Fig. 2 Spectral graph for the case $V = \cos x + \cos 2x$

Remark 9 Evidently, the set $\Gamma \cap K$ is a discrete subset of a finite graph (a union of finite number of analytic curves).

Remark 10 The equations for the asymptotic eigenvalues can be represented by explicit formulas

$$\int_{x_j}^{x_k} \sqrt{\lambda - \imath V(x)}dx = \pi h(m_{kj} + \mu/4) \tag{8}$$

where m_{kj} are integers, $\mu \in \{0, 2\}$, and x_k and x_j are zeros of the integrand. In this case, the equation

$$\Im \int_{x_j}^{x_k} \sqrt{\lambda - \imath V(x)}dx = 0 \tag{9}$$

defines the edges of the spectral graph, and the spectral points are defined by the equations:

$$\Re \int_{x_j}^{x_i} \sqrt{\lambda - \imath V(x)}dx = \pi h(m_{ij} + \mu/4) \tag{10}$$

Remark 11 Integer μ is the analog of the Maslov index; however, the definition of this number is quite different. Namely, $\mu(\gamma)$ equals the index of intersection of the cycle γ with the pull-back of the real circle $\Im x = 0$ with respect to the projection $(x, p) \to x$.

3 Equation of Magnetic Induction

The spectral problem for the operator of induction,

$$h^2 \Delta B - \{v, B\} = -\lambda B, \tag{11}$$
$$div B = 0 \tag{12}$$

arises when describing the magnetic field in a conductive liquid (in particular, the magnetic fields of planets, stars, and galaxies, see, e.g., [9]). Here, v stands for a given smooth divergence-free field on a Riemannian manifold M, Δ for the Laplace–Beltrami operator, $\{,\}$ for the commutator of vector fields and B is the desired vector field (the magnetic field). The parameter h characterizes the resistance in the liquid, and the passage to the limit as $h \to 0$ corresponds to a high conductivity.

Clearly, the spectrum of the operator of induction substantially depend on the manifold M and on the field v and can be computed efficiently in special situations only. Below we consider special case of this kind, namely, a two-dimensional surface of revolution with the flow along the parallels. This case was discussed in detail in [29] (see also [32, 33]); we present the main results only. Recall that a two-dimensional compact surface of revolution is diffeomorphic either to a torus or to a sphere.

3.1 Torus

The torus is obtained by rotating a smooth closed curve around an axis that does not intersect the curve, and the metric is of the form

$$ds^2 = dz^2 + u^2(z)d\varphi^2,$$

where z stands for the arc length parameter on the rotating curve, $u(z)$ for the distance of the point to the axis of rotation (we assume that u is a trigonometric polynomial), and φ for the angle of rotation. We assume that the field v is directed along the parallels, $v = a(z)\frac{\partial}{\partial\varphi}$, where a is a trigonometric polynomial, in which case, the variables in the spectral equation can be separated and the asymptotic behavior of the spectrum can be calculated by using equations similar to (5). The Riemann surface Λ is given by the equation

$$p^2 + ina(z) = \lambda$$

(n is an integer constant entering the separation of variables), and the spectral graph is defined from Eq. (9) in which $V = na$.

3.2 Sphere

The sphere is obtained by rotating a smooth curve (the graph of a function $f(z)$) around the z axis which intersects the curve at two points at which the tangent to the curve is perpendicular to the axis of rotation (the poles of the surface). We assume that

$$f(z) = \sqrt{(z - z_1)(z - z_2)}k(z),$$

where z_1 and z_2 are the poles of the surface, $k(z)$ is a polynomial, and $k(z) > 0$ for $z \in [z_1, z_2]$. As far as the field v is concerned, it is assumed that

$$v = a(z)\frac{\partial}{\partial \varphi},$$

where $a(z)$ is a polynomial. The Riemann surface is given in C^2 by the equation

$$p^2 f(z)^2 + ina(z) = \lambda;$$

it is punctured not only at the points at infinity but also at the zeros of f (i.e., at the poles of M). The asymptotics of the spectrum is still defined by the Eq. (5); analytical equation (8) are replaced by the equations

$$\int_{z_j}^{z_k} \sqrt{(f_z^2 + 1)(ina(z) + \lambda)}dz = \pi h(m_{ij} + \mu/4),$$

where z_i and z_j are the zeros and poles of the integrand (in particular, the poles of the surface of revolution M can be taken as the limits of integration).

As an example, consider the simplest case of the standard sphere ($f = \sqrt{1 - z^2}$) and take $a(z) = z$. In this case, the Riemann surface is homeomorphic to the torus with three punctures, namely, at the points $z = \pm 1$ and at the point at infinity. The cycles are depicted in Fig. 3.

The cycle γ_1 goes around the points –1 and 1, the cycles γ_2 and γ_3 go around the points $i\lambda/n$, –1 and the points $i\lambda/n$, 1, respectively. Every cycle defines the corresponding quantization conditions, which are of the form

$$\frac{1}{\pi h}\int_{-1}^{1} \sqrt{\frac{inz - \lambda}{1 - z^2}}dz = \frac{1}{2} + m_1$$

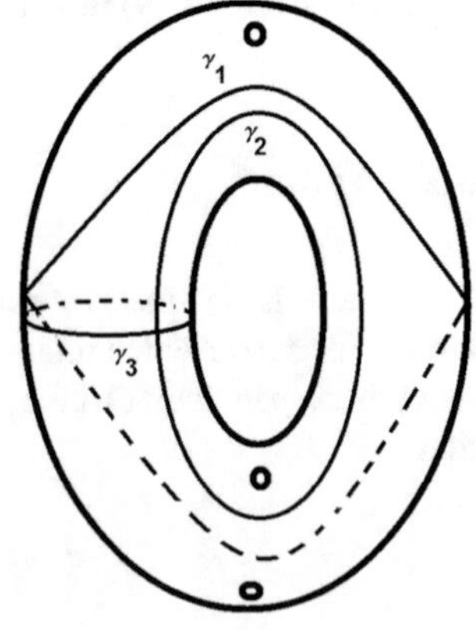

Fig. 3 Cycles on Riemann surface

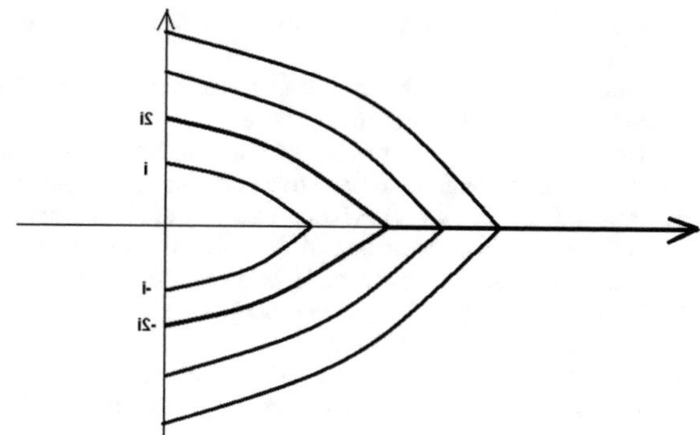

Fig. 4 Spectral graph with countably many edges

for the cycle γ_1,

$$\frac{1}{\pi h}\int_{-1}^{i\lambda/n}\sqrt{\frac{inz-\lambda}{1-z^2}}dz = m_2$$

for the cycle γ_2, and

$$\frac{1}{\pi h}\int_{1}^{i\lambda/n}\sqrt{\frac{inz-\lambda}{1-z^2}}dz = m_3$$

for the cycle γ_3.

To every quantization condition, there corresponds its own sequence of eigenvalues.

Remark 12 In contrast to the preceding section, the quantization conditions corresponding to a surface of revolution involve an integer n (the constant arising in the course of the separation of variables). The asymptotic eigenvalues and the edges of the spectral graph depend on n; thus, the graph consists now of countably many edges. For the standard sphere and for $a = z$, this graph is shown in Fig. 4.

Acknowledgements I thank the referee for very useful comments. The research was supported by the Russian Scientific Foundation (grant 16-11-10069).

References

1. Maslov, V.P.: Asymptotic Methods and Perturbation Theory. MGU (1965)
2. Maslov, V.P., Fedoriuk, M.V.: Quasiclassical Approximation for the Equations of Quantum Mechanics. Nauka (1976)
3. Davies, E.B.: Pseudospectra of differential operators. Oper. Theory **4**(3), 243–262 (2000)
4. Evgrafov, M.A., Fedoriuk, M.V.: Asymptotic behavior of solutions of the equation $w'' - p(z, \lambda)w = 0$ as $\lambda \to \infty$ in the complex z-plane. Uspekhi Mat Nauk **21**(2), 3–50 (1966)

5. Fedoryuk, M.V.: Asymptotic Analysis: Linear Ordinary Differential Equations. Springer, Berlin (1993)
6. Gohberg, I.T., Krein, M.G.: Introduction to the Theory of Linear Nonself-adjoint Operators. American Mathematical Society (1969)
7. Trefethen, L.N.: Pseudospectra of linear operators. In: ISIAM 95: Proceedings of the Third International Congress of Industrial and Applied Mathmatics, pp. 401–434 (1996)
8. Drazin, R.G., Reid, W.H.: Hydrodynamic Stability. Cambridge (1981)
9. Zel'dovich, Y.B., Ruzmaikin, A.A.: The hydromagnetic dynamo as the source of planetary, solar, and galactic magnetism. Uspekhi Fiz Nauk **152**(2), 263–284 (1987)
10. Bender, C.M., Brody, D.C., Jones, H.F., Meister, B.K.: Faster than Hermitian quantum mechanics. Phys. Rev. Lett. **98** (2007)
11. Gulden, T., Janas, M., Koroteev, P., Kamenev, A.: Statistical mechanics of coulomb gases as quantum theory on Riemann surfaces. JETP **144**(9) (2013)
12. Stepin, S.A.: Nonself-adjoint singular perturbations: a model of the passage from a discrete spectrum to a continuous spectrum. Rus. Math. Surv. **50**(6), 1311–1313 (1995)
13. Shkalikov, A.A.: On the limit behavior of the spectrum for large values of the parameter of a model problem. Math. Notes **62**(5), 796–799 (1997)
14. Stepin, S.A.: A model of the transition from a discrete spectrum to a continuous spectrum in singular perturbation theory. (Russian) Fundam. Prikl. Mat. **3**(4), 1199–1227 (1997)
15. Arzhanov, A.A., Stepin, S.A.: Semiclassical spectral asymptotics and the Stokes phenomenon for the weber equation. Dokl. Akad. Nauk **378**(1), 18–21 (2001)
16. Tumanov, S.N., Shkalikov, A.A.: On the limit behaviour of the spectrum of a model problem for the Orr-Sommerfeld equation with poiseuille profile. Izv Math. **66**(4), 829–856 (2002)
17. Shkalikov, A.A.: Spectral portraits of the Orr-Sommerfeld operator with large reynolds numbers. J. Math. Sci. **124**(6), 5417–5441 (2004)
18. D'yachenko, A.V., Shkalikov, A.A.: On a model problem for the Orr–Sommerfeld equation with linear profile. Funktsional Anal i Prilozhen **36**(3), 228–232 (2002)
19. Stepin, S.A., Titov, V.A.: On the concentration of spectrum in the model problem of singular perturbation theory. Dokl. Math. **75**(2), 197–200 (2007)
20. Pokotilo, V.I., Shkalikov, A.A.: Semiclassical approximation for a nonself-adjoint Sturm-Liouville problem with a parabolic potential. Math. Notes **86**(3), 442–446 (2009)
21. Kusainova, L.K., Monashova, AZh, Shkalikov, A.A.: Asymptotics of the eigenvalues of the second-order nonself-adjoint differential operator on the axis. Mat. Zametki **93**(4), 630–633 (2013)
22. Stepin, S.A., Fufaev, V.V.: Phase integral method in the problem of quasiclassical localization of spectra. Dokl. Math. **91**(3), 318–322 (2015)
23. Tumanov, S.N., Shkalikov, A.A.: The limit spectral graph in semiclassical approximation for the SturmLiouville problem with complex polynomial potential. Dokl. Math. **92**(3), 773–777 (2015)
24. Stepin, S.A., Fufaev, V.V.: The phase-integral method in a problem of singular perturbation theory. Izvestiya: Math. **81**(2), 359–390 (2017)
25. Tumanov, S.N., Shkalikov, A.A.: Eigenvalue dynamics of a PT-symmetric Sturm Liouville operator and criteria for similarity to a self-adjoint or a normal operator. Dokl. Math. **96**(3), 607–611 (2017)
26. Galtsev, S.V., Shafarevich, A.I.: Spectrum and pseudospectrum of nonself-adjoint Schrödinger operators with periodic coefficients. Mat Zametki **80**(3), 456–466 (2006)
27. Galtsev, S.V., Shafarevich, A.I.: Quantized Riemann surfaces and semiclassical spectral series for a nonself-adjoint Schrödinger operator with periodic coefficients. Theor. Math. Phys. **148**(2), 206–226 (2006)
28. Esina, A.I., Shafarevich, A.I.: Quantization conditions on Riemannian surfaces and the semiclaical spectrum of the schrödinger operator with complex potential. Mat Zametki **88**(2), 209–227 (2010)
29. Esina, A.I., Shafarevich, A.I.: Asymptotics of the spectrum and the eigenfunctions of the operator of magnetic induction on a two-dimensional compact surface of revolution. Mat Zametki **95**(3), 417–432 (2014)

30. Esina, A.I., Shafarevich, A.I.: Analogs of Bohr – Sommerfeld – Maslov quantization conditions ton Riemann surfaces and spectral series of nonself-adjoint operators. Rus. J. Math. Phys. **20**(2), 172–181 (2013)
31. Esina, A.I.: Asymptotic solutions of the induction equation ("Asimptoticheskie resheniya uravneniya indukcii", in Russian). Ph.D Thesis (kandidatskaya dissertaciya), Moscow State University, Moscow (2014)
32. Roohian, H., Shafarevich, A.I.: Semiclassical asymptotics of the spectrum of a nonself-adjoint operator on the sphere. Rus. J. Math. Phys. **16**(2), 309–315 (2009)
33. Roohian, H., Shafarevich, A.I.: Semiclassical asymptotic behavior of the spectrum of a nonself-adjoint elliptic operator on a two-dimensional surface of revolution. Rus. J. Math. Phys. **17**(3), 328–334 (2010)
34. Nekhaev, D.V., Shafarevich, A.I.: A quasiclassical limit of the spectrum of a Schrodinger operator with complex periodic potential. Sbornik: Math. **208**(10), 1535–1556 (2017)

Semilocal Monodromy of Rigid Local Systems

Toshio Oshima

Abstract The rigid local system on $\mathbb{P}^1 \setminus S$ with a set S of finite points is realized as a rigid Fuchsian differential equation $\mathscr{M}$ of Schlesinger canonical form. Here "rigid" means that the equation is uniquely determined by the equivalence classes of residue matrices of $\mathscr{M}$ at the points in S. The *semilocal monodromy* in this paper is the conjugacy class of the monodromy matrix obtained by the analytic continuation of the solutions of $\mathscr{M}$ along an oriented *simple* closed curve γ on $\mathbb{C} \setminus S$. Since it corresponds to the sum of residue matrices at the singular points surrounded by γ and the equation $\mathscr{M}$ is obtained by applying additions and middle convolutions to the trivial equation, we study the application of the middle convolution to the sums of residue matrices. In this way we give an algorithm calculating this semilocal monodromy, which also gives the local monodromy at the irregular singular point obtained by the confluence of these points.

Keywords Rigid local system · Middle convolution · Monodromy · Fuchsian differential equation

MSC Primary 34M35 · Secondary 34M03

1 Introduction

The global theory of Fuchsian differential equations has been greatly developed after the work of Katz [3] on rigid local systems, which we will shortly explain.

Fuchsian differential equation of Schlesinger canonical form is

$$\mathscr{M} : \frac{du}{dx} = \sum_{i=1}^{p} \frac{A_i}{x - c_i} u \tag{1}$$

T. Oshima (✉)
Josai University, 2-2-20 Hirakawacho, Chiyodaku, Tokyo 102–0093, Japan
e-mail: oshima@ms.u-tokyo.ac.jp

G. Filipuk et al. (eds.), *Formal and Analytic Solutions of Diff. Equations*,
Springer Proceedings in Mathematics & Statistics 256,
https://doi.org/10.1007/978-3-319-99148-1_10

with $A_i \in M(n, \mathbb{C})$. Here n is the rank of the equation, u is a column vector of n unknown functions, $M(n, \mathbb{C})$ denotes the set of square matrices of size n with components in $\mathbb{C}$, A_i is called the *residue matrix* at $x = c_i$ and the residue matrix at $x = \infty$ equals $-(A_1 + \cdots + A_p)$ which we denote by A_{p+1}. The equation $\mathscr{M}$ is called *irreducible* in Schlesinger canonical form if there exists no non-trivial proper subspace $V \subset \mathbb{C}^n$ satisfying $A_i V \subset V$ for $i = 1, \ldots, p$.

Definition 1 For a matrix $A \in M(n, \mathbb{C})$ we put $\{\mu_1, \ldots, \mu_r\} = \{\mu \in \mathbb{C} \mid \operatorname{rank}(A - \mu) < n\}$. Then there exist positive numbers n_j and $m_{j,\nu}$ for $1 \leq j \leq r$ and $1 \leq \nu \leq n_j$ such that

$$\operatorname{corank}(A - \mu_j)^\nu = m_{j,1} + \cdots + m_{j,\nu} \quad (\nu = 1, \ldots, n_j), \tag{2}$$

$$\operatorname{rank}(A - \mu_j)^{n_j} = \operatorname{rank}(A - \mu_j)^{n_j+1}. \tag{3}$$

Here $\sum_{j=1}^r \sum_{\nu=1}^{n_j} m_{j,\nu}=n$. Note that the set $\{[\mu_j]_{m_{j,\nu}} \mid 1 \leq \nu \leq n_j \text{ and } 1 \leq j \leq r\}$, which we call the *eigenvalue class* of A and write by (EC) of A for simplicity, determines the conjugacy class of matrices containing A. The matrix A is semisimple if and only if $n_1 = \cdots = n_r = 1$ and in this case the symbol $[\mu_j]_{m_{j,1}}$ means that μ_j is an eigenvalue of A with multiplicity $m_{j,1}$. We may simply write μ_j in place of $[\mu_j]_1$.

Let $\{[\lambda_{i,\nu}]_{m_{i,\nu}} \mid 1 \leq \nu \leq n_i\}$ be the eigenvalue classes of A_i for $i = 1, \ldots, p+1$, respectively. The *index of rigidity* of $\mathscr{M}$ defined by Katz [3] equals

$$\operatorname{idx} \mathscr{M} := \sum_{i=1}^{p+1} \sum_{\nu=1}^{n_i} m_{i,\nu}^2 - (p-1)n^2.$$

An irreducible equation $\mathscr{M}$ is called *rigid* if the conjugacy classes of A_i for $i = 1, \ldots, p+1$ uniquely determine the simultaneous conjugacy class of $(A_1, \ldots, A_{p+1})$, which means that the local structure of $\mathscr{M}$ at singular points uniquely determines the global structure of $\mathscr{M}$. Katz [3] proved that an irreducible equation $\mathscr{M}$ is rigid if and only if $\operatorname{idx} \mathscr{M} = 2$ by introducing two types of operations of the equations. They are additions and middle convolutions and keep the irreducibility and the index of rigidity. The addition $\operatorname{Ad}\big((x - c_k)^{\lambda_k}\big)$ is defined by the transformation $A_i \mapsto A_i + \lambda_k \delta_{i,k} \quad (i = 1, \ldots, p)$ with $\lambda_k \in \mathbb{C}$ and $1 \leq k \leq p$, which corresponds to the transformation $u \mapsto (x - c_k)^{\lambda_k} u$. The middle convolution mc_μ corresponds to the fractional derivation $u \mapsto \left(\frac{d}{dx}\right)^{-\mu} u$ with $\mu \in \mathbb{C}$, which will be explained in the next section.

Katz [3] proved that any rigid local system is transformed into the trivial equation $u' = 0$ of rank 1 by successive applications of additions and middle convolutions. Since these operations are invertible, any rigid local system is constructed and realized in the form (1) from the trivial equation by successive applications of these operations. The author [4] interpreted the middle convolution for linear ordinary differential equations with polynomial coefficients, reduced various analysis of rigid Fuchsian ordinary differential equations to the study of solutions under the middle convolution and got many general results for solutions of rigid Fuchsian differential equations, such as their integral representations, connection formulas, series expansions, irreducibility of monodromy groups, contiguous relations etc. Note that any rigid local system is uniquely realized by a rigid single Fuchsian differential equation without an apparent singularity (cf. [6, Lemma 2.1]).

Dettweiler–Reiter [1] interpreted the middle convolution mc_μ introduced by Katz into an operation of the tuple of residue matrices $(A_1, \ldots, A_{p+1})$. In fact, they explicitly gave the conjugacy classes of residue matrices $\bar{A}_j$ in terms of those of $A_1, \ldots, A_{p+1}$. Here $(\bar{A}_1, \ldots, \bar{A}_{p+1})$ is the tuple of residue matrices of $\mathrm{mc}_\mu \mathscr{M}$.

Let I be a subset of $\{1, \ldots, p\}$ and put $A_I = \sum_{i \in I} A_i$. We show that the residue class of $\bar{A}_I = \sum_{i \in I} \bar{A}_i$ is explicitly determined by the residue classes of $\bar{A}_1, \ldots, \bar{A}_{p+1}$ and $\bar{A}_I$, which is a generalization of a result in [1] and the main purpose of this paper.

Definition 2 Let γ be an oriented simple closed curve γ in $\mathbb{C} \setminus \{c_1, \ldots, c_p\}$. We may assume

$$\frac{1}{2\pi\sqrt{-1}} \int_\gamma \frac{dz}{x - c_i} = \begin{cases} 1 & (i \in I) \\ 0 & (i \notin I) \end{cases} \tag{4}$$

with a subset $I \subset \{1, \ldots, p\}$. The *semilocal monodromy* of $\mathscr{M}$ for $\{c_i \mid i \in I\}$ is the conjugacy class of the monodromy matrix M of the solutions of $\mathscr{M}$ along the path γ. The semilocal monodromy of $\mathscr{M}$ for $\{c_i \mid i \in \{1, \ldots, p+1\} \setminus I\}$ is the conjugacy class of the matrix M^{-1}.

Suppose $\mathscr{M}$ is rigid. Then the semilocal monodromy does not depend on the positions of c_i if (4) is valid. Hence it is the class containing $e^{2\pi\sqrt{-1}\bar{A}_I}$ if any difference of eigenvalues of A_I is not a non-zero integer. Note that it follows from Corollary 1 that any eigenvalue of the semilocal monodromy of $\mathscr{M}$ is a certain product of integer powers of eigenvalues of local monodromies at singular points. This is not valid when γ is not simple as is given in the first example in Sect. 4.

Suppose the points c_i for $i \in I$ coalesce into one confluent point c_I and the rigid equation $\mathscr{M}$ is changed into an equation $\mathscr{M}'$ with an irregular singular point $c_I \in \mathbb{C}$. We may assume that the semilocal monodromy does not change in the confluence and then we get the local monodromy of $\mathscr{M}'$ at c_I. This is the same for the confluence of the points c_i for $i \in \{1, \ldots, p+1\} \setminus I$.

2 Middle Convolution of a Sum of Residue Matrices

The convolution $\tilde{A}_k$ of the residue matrices A_k of $\mathscr{M}$ is given by

$$\tilde{A}_k = k \begin{pmatrix} 0 & \cdots & 0 & \cdots & 0 \\ \vdots & \cdots & \vdots & \cdots & \vdots \\ A_1 & \cdots & A_k + \mu & \cdots & A_p \\ \vdots & \cdots & \vdots & \cdots & \vdots \\ 0 & \cdots & 0 & \cdots & 0 \end{pmatrix} \in M(pn, \mathbb{C}) \qquad (1 \le k \le p)$$

$$= \Big((A_j + \mu\delta_{i,j})\delta_{i,k} \Big)_{\substack{1\le i\le p \\ 1\le j\le p}} \tag{5}$$

Here $\tilde{A}_k$ are block matrices of size p whose entries are square matrices of size n and $\tilde{A}_{p+1} = -(\tilde{A}_1 + \cdots + \tilde{A}_p)$. Let $\mu \in \mathbb{C}$ with $\mu \neq 0$. Then the subspaces

$$\mathscr{K}_j := j \begin{pmatrix} 0 \\ \vdots \\ \operatorname{Ker} A_j \\ 0 \\ \vdots \end{pmatrix} \simeq \operatorname{Ker} A_j \quad (j = 1, \ldots, p),$$

$$\mathscr{K}_{p+1} := \left\{ \begin{pmatrix} v \\ \vdots \\ v \end{pmatrix} \mid A_{p+1}v = \mu v \right\} \simeq \operatorname{Ker}(A_{p+1} - \mu) \text{ and } \mathscr{K} := \bigoplus_{j=1}^{p+1} \mathscr{K}_j$$

of $\mathbb{C}^{pn}$ are invariant under the linear transformations defined by $\tilde{A}_j$ for $j = 1, \ldots, p$. Then $\tilde{A}_j$ induce linear transformations of $\mathbb{C}^{pn}/\mathscr{K}$ and the corresponding matrices with respect to a base of $\mathbb{C}^{pn}/\mathscr{K}$ are denoted by $\bar{A}_j$, respectively. Then the equation

$$\bar{\mathscr{M}} : \frac{d\bar{u}}{dx} = \sum_{i=1}^{p} \frac{\bar{A}_i}{x - c_i}\bar{u}$$

is the middle convolution $\mathrm{mc}_\mu \mathscr{M}$ of $\mathscr{M}$ and the tuple of matrices $\big(\bar{A}_1, \ldots, \bar{A}_p, \bar{A}_{p+1}\big)$ is the middle convolution of the tuple $(A_1, \ldots, A_{p+1})$. Here $\bar{A}_{p+1} = -(\bar{A}_1 + \cdots + \bar{A}_p)$.

Put $\bar{A}_I := \sum_{i\in I} \bar{A}_i$, $\tilde{A}_I := \sum_{i\in I} \tilde{A}_i$ and

$$\iota_j(v) := j \begin{pmatrix} 0 \\ \vdots \\ v \\ 0 \\ \vdots \end{pmatrix} \qquad (v \in \mathbb{C}^n,\ 1 \le j \le p).$$

For simplicity, we assume $I = \{1, \dots, k\}$ with $1 \le k \le p$. Then

$$\tilde{A}_I = \begin{pmatrix} A_1 + \mu & A_2 & \cdots & A_k & A_{k+1} & \cdots & A_p \\ A_1 & A_2 + \mu & \cdots & A_k & A_{k+1} & \cdots & A_p \\ \vdots & \vdots & \ddots & \vdots & \cdots & \vdots & \vdots \\ A_1 & A_2 & \cdots & A_k + \mu & A_{k+1} & \cdots & A_p \\ 0 & 0 & \cdots & 0 & 0 & \cdots & 0 \\ \vdots & \vdots & \vdots & \vdots & \vdots & \vdots & \vdots \end{pmatrix} \in M(pn, \mathbb{C}). \tag{6}$$

(Here the column containing A_k and the row containing $A_k + \mu$ are labelled k.)

By the linear automorphism on $\mathbb{C}^{pn}$ defined by the matrix

$$P = \begin{pmatrix} I_n & & & & & \\ -I_n & I_n & & & & \\ \vdots & & \ddots & & & \\ -I_n & & & I_n & & \\ & & & & I_n & \\ & & & & & \ddots \end{pmatrix} \in M(pn, \mathbb{C}),$$

(with the k-th row and column labelled k) the linear transformation $\tilde{A}_I$ on $\mathbb{C}^{pn}$ and the subspaces $\mathscr{K}_j$ are changed into

$$\tilde{A}'_I := P\tilde{A}_I P^{-1} = \begin{pmatrix} A_1 + \cdots + A_k + \mu & A_2 & \cdots & A_k & A_{k+1} & \cdots & A_p \\ & \mu & & & & & \\ & & \ddots & & & & \\ & & & \mu & & & \\ 0 & 0 & \cdots & 0 & 0 & \cdots & 0 \\ \vdots & \vdots & \vdots & \vdots & \vdots & \vdots & \vdots \end{pmatrix},$$

$$\mathscr{K}'_1 := \left\{ \begin{pmatrix} v \\ -v \\ \vdots \\ -v \\ 0 \\ \vdots \end{pmatrix} \middle| \, v \in \operatorname{Ker} K_1 \right\}, \quad \mathscr{K}'_{p+1} := \left\{ \begin{pmatrix} v \\ 0 \\ \vdots \\ 0 \\ v \\ \vdots \end{pmatrix} \middle| \, v \in \operatorname{Ker}(K_{p+1} - \mu) \right\},$$

(the k-th entries being $-v$ and 0 respectively)

$$\mathscr{K}'_j := \mathscr{K}_j \quad (2 \le j \le p) \text{ and } \mathscr{K}' := \bigoplus_{j=1}^{p+1} \mathscr{K}'_j. \tag{7}$$

Here we note that

$$\begin{aligned}(\tilde{A}'_I-\lambda)^\nu\iota_1(w)&=\iota_1\big((A_I+\mu-\lambda)^\nu w\big)\quad (w\in\mathbb{C}^n,\ \nu=1,2,\ldots),\\ \operatorname{corank}(\tilde{A}'_I-\lambda)^\nu|_{\mathbb{C}^{pn}/\mathcal{K}'}&=\operatorname{corank}(A_I+\mu-\lambda)^\nu\quad(\lambda\in\mathbb{C}\setminus\{0,\mu\},\ \nu=1,2,\ldots)\\ \operatorname{corank}(\tilde{A}'_I-\mu)^{pn}|_{\mathbb{C}^{pn}/\mathcal{K}'}&=\dim\operatorname{Ker}A_I^n+(k-1)n-\sum_{i=1}^k\dim\mathcal{K}_i,\\ \operatorname{corank}(\tilde{A}'_I-0)^{pn}|_{\mathbb{C}^{pn}/\mathcal{K}'}&=\dim\operatorname{Ker}(A_I+\mu)^n+(p-k)n-\sum_{j=k+1}^p\dim\mathcal{K}_j.\end{aligned}$$

Since (EC) of $(\tilde{A}'_I-\lambda)^\nu|_{\mathbb{C}^{pn}/\mathcal{K}'}$ equals (EC) of $\bar{A}_I$, we have the following theorem by the above expression.

Theorem 1 *Retain the assumption $\mu\neq 0$ and $I\subset\{1,\ldots,p\}$. We have*

$$\begin{cases}\dim\operatorname{Ker}(\bar{A}_I-\lambda)^\nu=\dim\operatorname{Ker}(A_I+\mu-\lambda)^\nu\quad(\forall\lambda\in\mathbb{C}\setminus\{0,\mu\},\ \nu=1,2,\ldots),\\ \dim\operatorname{Ker}(\bar{A}_I-\mu)^{pn}=\dim\operatorname{Ker}A_I^n+(k-1)n-\sum_{i=1}^k\dim\mathcal{K}_i,\\ \dim\operatorname{Ker}(\bar{A}_I-0)^{pn}=\dim\operatorname{Ker}(A_I+\mu)^n+(p-k)n-\sum_{i=k+1}^{p+1}\dim\mathcal{K}_i.\end{cases}$$

Suppose

$$\operatorname{Ker}A_I\subset\operatorname{Ker}A_1\cap\cdots\cap\operatorname{Ker}A_k \tag{8}$$

and

$$\operatorname{Ker}(A_I+\mu)=\{0\}\ \text{or}\ k=p. \tag{9}$$

Then if A_I is semisimple, so is $\bar{A}_I$.

Proof Note that the assumption (8) implies $\iota_1(\operatorname{Ker}A_I)\subset\bigoplus_{i=1}^k\mathcal{K}'_i$. Then the claims in the theorem are clear by the argument just before the theorem. □

Remark 1 (i) If a subset $J\subset\{0,\ldots,p+1\}$ contains $p+1$, we have a similar result for $\bar{A}_J=\sum_{j\in J}A_j$ by the fact $\bar{A}_{\{0,\ldots,p\}\setminus J}+\bar{A}_J=0$.

(ii) The condition (8) in the theorem is valid if

$$\dim\operatorname{Ker}A_I\leq\max\Big\{0,n-\sum_{i=1}^k\operatorname{codim}\operatorname{Ker}A_i\Big\}.$$

(iii) Dettweiler-Reiter [1] obtained (EC) of A_I when $\#I = 1$. Theorem 1 is a generalization of their result. As is given in [1] a multiplicative version of Theorem 1 may be possible.

(iv) Haraoka [2] showed that the rigid equation $\mathscr{M}$ can be extended to a KZ equation

$$\frac{\partial \tilde{u}}{\partial x_i} = \sum_{\substack{0 \le \nu \le p \\ \nu \ne i}} \frac{A_{i,\nu}}{x_i - x_\nu} \tilde{u} \quad (0 \le i \le p)$$

with $x_0 = x, x_j = c_j$ and $A_{0,j} = A_j$ $(j = 1, \dots, p)$. Here $A_{i,j} = A_{j,i}$ and $A_{i,i} = 0$. Put $A_{i,p+1} := -(A_{i,0} + \cdots + A_{i,p})$ and $\tilde{A}_I := \sum_{1 \le \nu < \nu' \le k} A_{i_\nu, i_{\nu'}}$ for $I = \{i_1, \dots, i_k\} \subset \{0, 1, \dots, p+1\}$. Then the author [7] studied the simultaneous conjugacy class of the tuple $(\tilde{A}_I, \tilde{A}_J)$ when $I \cap J = \emptyset$ or $I \subset J$ which assures $[\tilde{A}_I, \tilde{A}_J] = 0$. Since $A_{\{1,\dots,k\}} = \tilde{A}_J - \tilde{A}_I$ with $I = \{1, \dots, k\}$ and $J = \{0, \dots, k\}$, we have (EC) of $A_{\{1,\dots,k\}}$ by this simultaneous conjugacy class. In fact, this is the original idea of this paper.

Corollary 1 *Let $\mathscr{L}$ be the integer lattice spanned by the eigenvalues of the residue matrices $A_1, \dots, A_{p+1}$ of a rigid Fuchsian equation $\mathscr{M}$. Then any eigenvalue of $A_I = \sum_{i \in I} A_i$ is in $\mathscr{L}$ for any $I \subset \{1, \dots, p+1\}$.*

Proof We can reduce $\mathscr{M}$ to the trivial equation and construct it from the trivial equation by applying suitable operations $\mathrm{Ad}\big((x - x_j)^{\lambda_j}\big)$ and mc_μ with $\lambda_j,\ \mu \in \mathscr{L}$, which is explained in the next section. Hence the corollary follows from the theorem.

3 Semilocal Monodromy

Let $\{[\lambda_{j,\nu}]_{n_j} \mid \nu = 1, \dots, m_{j,\nu}\}$ be the eigenvalue classes of the residue matrices A_j of $\mathscr{M}$ given in (1) for $j = 1, \dots, p+1$. Then the *generalized Riemann scheme* of $\mathscr{M}$ is defined by

$$\{\lambda_{\mathbf{m}}\} = \begin{Bmatrix} x = c_1 & \dots & c_p & \infty \\ [\lambda_{1,1}]_{m_{1,1}} & \dots & [\lambda_{p,1}]_{m_{p,1}} & [\lambda_{p+1,1}]_{m_{p+1,1}} \\ \vdots & \vdots & \vdots & \vdots \\ [\lambda_{1,n_1}]_{m_{1,n_1}} & \dots & [\lambda_{p,n_p}]_{m_{p,n_p}} & [\lambda_{p+1,n_{p+1}}]_{m_{p+1,n_{p+1}}} \end{Bmatrix} \tag{10}$$

and

$$\mathbf{m} = m_{1,1} \cdots m_{1,n_1}, m_{2,1} \cdots m_{2,n_2}, \dots, m_{p+1,1} \cdots m_{p+1,n_{p+1}}$$

which express the $(p+1)$ tuples of partitions of n

$$n = m_{j,1} + \cdots + m_{j,n_j} \quad (j = 1, \ldots, p+1) \tag{11}$$

and is called the *spectral type* of $\mathcal{M}$. We define $\operatorname{rank} \mathbf{m} = n$.

The spectral type $\mathbf{m}$ is *ordered* if

$$m_{j,1} \geq m_{j,2} \geq \cdots \geq m_{j,n_j} \quad (j = 1, \ldots, p+1).$$

For a given spectral type $\mathbf{m}$, $s\mathbf{m}$ denotes the corresponding ordered spectral type.

For an ordered spectral type $\mathbf{m}$ we define

$$d(\mathbf{m}) := \sum_{j=1}^{p+1} m_{j,1} - (p-1)\operatorname{rank} \mathbf{m}, \tag{12}$$

$$\partial \mathbf{m} := \mathbf{m}' = \left(m'_{j,\nu}\right)_{\substack{\nu=1,\ldots,n_j \\ j=1,\ldots,p+1}} \text{ with}$$

$$m'_{j,\nu} = m_{j,\nu} - d(\mathbf{m})\delta_{\nu,1} \quad (\nu = 1, \ldots, n_j,\ j = 1, \ldots, p+1). \tag{13}$$

Here some $m'_{j,\nu}$ may be zero. Then such $m'_{j,\nu}$ are omitted and n_j may be decreased.

It is proved by Katz [3] that $\mathcal{M}$ is rigid if there exists a non-negative integer r such that $(\partial s)^\nu \mathbf{m}$ are tuples of partitions of positive integers for $\nu \in \{1, \ldots, r\}$ and moreover

$$\operatorname{rank} \mathbf{m} > \operatorname{rank} \partial s\mathbf{m} > \operatorname{rank} (\partial s)^2 \mathbf{m} > \cdots > \operatorname{rank} (\partial s)^r \mathbf{m} = 1. \tag{14}$$

Here $(p+1)$ tuples of partition $\mathbf{m}$ mean that $m_{j,\nu}$ are non-negative integer in (11).

Suppose (10) is the generalized Riemann scheme of $\mathcal{M}$. Suppose moreover $\operatorname{rank} \mathcal{M} > 1$ and $\mathbf{m}$ is ordered by replacing $\mathbf{m}$ by $s\mathbf{m}$ if necessary. Applying $\prod_{j=1}^{p} \operatorname{Ad}\left((x - c_j)^{-\lambda_{j,1}}\right)$ to $\mathcal{M}$, we may assume $\lambda_{j,1} = \cdots = \lambda_{j,p} = 0$. Then we apply $\operatorname{mc}_{\lambda_{p+1}}$ to the system, we get a rigid Fuchsian equation with the spectral type $\partial \mathbf{m}$. The sequence (14) of spectral types corresponds to this procedure.

Katz [3] moreover showed that if $\mathbf{m}$ are tuples of partitions with this property (14), then for any $\lambda_{j,\nu}$ satisfying Fuchs condition

$$\sum_{j=1}^{p+1} \sum_{\nu=1}^{n_j} m_{j,\nu} \lambda_{j,\nu} = 0,$$

there exists a Fuchsian equation $\mathcal{M}$ with the generalized Riemann scheme (10), which is rigid for a generic $\lambda_{j,\nu}$. This follows from the fact $\operatorname{Ad}\left((x - c_j)^{\lambda}\right) \circ \operatorname{Ad}\left((x - c_j)^{-\lambda}\right) = \operatorname{mc}_{-\mu} \circ \operatorname{mc}_{\mu} = \operatorname{id}$.

A necessary and sufficient condition for the irreducibility of the monodromy group of the solutions of $\mathcal{M}$ is explicitly given (cf. [4, Proposition 10.16] and [6]). Then (EC) of the local monodromy matrix at $x = c_j$ is given by

$$\left\{[e^{2\pi\sqrt{-1}\lambda_{j,\nu}}]_{m_{j,\nu}} \mid \nu = 1, \ldots, n_j\right\}$$

if $\mathscr{M}$ is rigid and irreducible, which is given in [4, Remark 10.11 (iii)]. This is not obvious when there exist $\nu < \nu'$ with $\lambda_{j,\nu} - \lambda_{j,\nu'} \in \mathbb{Z} \setminus \{0\}$ but this is proved as follows.

If $\lambda_{j,\nu} - \lambda_{j,\nu'} \notin \mathbb{Z}$ for any ν and ν' with $1 \le \nu < \nu' \le n_j$, the claim is obvious.

Suppose (EC) of a matrix $A(t)$ with the continuous parameter $t \in [0, 1]$ is given by $\{[\lambda_\nu(t)]_{m_\nu} \mid \nu = 1, \ldots, r\}$ for $t \in (0, 1]$. We may assume $\lambda_\nu(t)$ are continuous functions on $[0, 1]$. Then (EC) of a matrix $A(0)$ weakly equals $\{[\lambda_\nu(0)]_{m_\nu} \mid \nu = 1, \ldots, r\}$ (cf. [5, Proposition 3.3]). Here "weakly" means that the condition (2) is replaced by

$$\operatorname{corank} (A - \mu_j)^\nu \ge m_{j,1} + \cdots + m_{j,\nu} \quad (\nu = 1, \ldots, n_j) \tag{15}$$

in Definition 1. Then the index of rigidity with respect to the local monodromy matrices implies the above statement.

Proposition 1 *Let* $\mathscr{M}$ *in* (1) *be a rigid Fuchsian differential equation and for* $I \subset \{1, \ldots, p\}$, *let* $\left\{[\lambda_\nu]_{m_\nu} \mid \nu = 1, \ldots, r\right\}$ *be (EC) of* $A_I = \sum_{i \in I} A_i$. *Suppose* $\lambda_\nu - \lambda_{\nu'} \notin \mathbb{Z} \setminus \{0\}$ *for* $1 \le \nu < \nu' \le r$. *Then (EC) of the semilocal monodromy of* $\mathscr{M}$ *for* $\{c_i \mid i \in I\}$ *equals* $\left\{[e^{2\pi\sqrt{-1}\lambda_\nu}]_{m_\nu} \mid \nu = 1, \ldots, r\right\}$.

Proof Since the equation is rigid, the semilocal monodromy does not depend on the points c_i and we may choose points c_i $(i \in I)$ as a single point, which implies the proposition. There may be a better understanding of this proof if we consider c_j as variables (cf. Remark 1 (iv)). □

Remark 2 (i) We expect that the semilocal monodromy for a rigid spectral type **m** with a generalized Riemann scheme (10) is semisimple if the exponents $\lambda_{j,\nu}$ are generic under the Fuchs condition. Note that the semisimplicity of local monodromies do not assure that of a semilocal monodromy (cf. (16) with $\lambda_1 + \cdots + \lambda_k + \mu = 0$).

We also expect that by the continuation of parameters $\lambda_{j,\nu}$ with the rigidity, we also determine a semilocal monodromy even if it is not semisimple as in the case of the local monodromy.

(ii) The algorithm calculating (EC) of A_I given in this paper is implemented in a computer algebra, which is contained in [8].

4 Examples

We start with Gauss hypergeometric equation, which is characterized by the spectral type 11, 11, 11. Applying the operation $\mathrm{mc}_\gamma \circ \mathrm{Ad}\big((x-1)^\beta\big) \circ \mathrm{Ad}\big(x^\alpha\big)$ to the trivial equation, we get

$$\frac{du}{dx} = \left(\frac{\begin{pmatrix} \alpha+\gamma & \beta \\ 0 & 0 \end{pmatrix}}{x} + \frac{\begin{pmatrix} 0 & 0 \\ \alpha & \beta+\gamma \end{pmatrix}}{x-1} \right) u$$

with the Riemann scheme

$$\begin{Bmatrix} x = 0 & 1 & \infty \\ 0 & 0 & -\gamma \\ \alpha+\gamma & \beta+\gamma & -\alpha-\beta-\gamma \end{Bmatrix}.$$

Under a suitable base of solutions the local monodromy matrices M_0 at $x = 0$ and M_1 at $x = 1$ are given by

$$M_0 = \begin{pmatrix} ac & (b-1)c \\ 0 & 1 \end{pmatrix}, \quad M_1 = \begin{pmatrix} 1 & 0 \\ a-1 & bc \end{pmatrix}$$

with $a = e^{2\pi\sqrt{-1}\alpha}$, $b = e^{2\pi\sqrt{-1}\beta}$ and $c = e^{2\pi\sqrt{-1}\gamma}$.

The monodromy matrix corresponding to a simple closed curve $|z| = 2$ is given by M_1M_0 and (EC) of M_1M_0 is $\{c,\ abc\}$ if the equation is irreducible.

The monodromy matrix corresponding to a closed curve C with $\frac{1}{2\pi\sqrt{-1}}\int_C \frac{dz}{z} = -1$ and $\frac{1}{2\pi\sqrt{-1}}\int_C \frac{dz}{z-1} = 1$ is given by $M_1M_0^{-1}$. The eigenvalue of $M_1M_0^{-1}$ is not a rational function of a, b and c. For example, if $a = c = -1$, the eigenvalue t satisfies $t^2 + 3(b-1)t - b = 0$.

Applying $\mathrm{mc}_\mu \circ \prod_{j=1}^{p} \mathrm{Ad}\big((x-c_j)^{\lambda_j}\big)$ to the trivial equation, we get Jordan-Pochhammer equation $\mathscr{M}$ with the generalized Riemann scheme

$$\begin{Bmatrix} x = c_1 & \cdots & c_p & \infty \\ [0]_{p-1} & \cdots & [0]_{p-1} & [-\mu]_{p-1} \\ \lambda_1+\mu & \cdots & \lambda_p+\mu & -\lambda_1-\cdots-\lambda_p-\mu \end{Bmatrix}.$$

This equation is characterized by the spectral type $\overbrace{(p-1)1, (p-1)1, \ldots, (p-1)1}^{p+1}$. The monodromy group of this equation is irreducible if and only if any one of the $(p+2)$ numbers $\lambda_1, \ldots, \lambda_p, \mu, \lambda_1 + \cdots + \lambda_p + \mu$ is not an integer (cf. [4, Sect. 13.3]).

Then (EC) of $A_{1,\ldots,k}$ with $1 \le k \le p$ equals

$$\big\{\lambda_1 + \cdots + \lambda_k + \mu,\ [0]_{p-k},\ [\mu]_{k-1}\big\}$$

and (EC) of the semilocal monodromy for $\{c_1, \ldots, c_k\}$ equals

$$\big\{e^{2\pi\sqrt{-1}(\lambda_1+\cdots+\lambda_k+\mu)},\ [1]_{p-k},\ [e^{2\pi\sqrt{-1}\mu}]_{k-1}\big\} \tag{16}$$

if the equation has an irreducible monodromy (cf. (7)). Replacing

$$c_j \text{ by } \frac{1}{\tilde{c}_j} \text{ and } \lambda_j \text{ by } \sum_{i=j}^{p} \frac{\tilde{\lambda}_i}{\tilde{c}_j \prod\limits_{k+1\le\nu\le i,\ \nu\ne j} (\tilde{c}_j - \tilde{c}_\nu)} \text{ for } j = k+1, \dots, p,$$

we get an irregular singularity at $x = \infty$ by the confluence given by $\tilde{c}_j \to 0$ for $j = k+1, \dots, p$ which corresponds to a *versal addition* defined in [4, Sect. 2.3] (cf. [4, Sect. 13.3]). This versal addition depends holomorphically on $\tilde{c}_j$ and equals $\mathrm{Ad}\left(e^{-\tilde{\lambda}_{k+1}x - \frac{\tilde{\lambda}_{k+2}}{2}x^2 - \cdots - \frac{\tilde{\lambda}_p}{p-k}x^{p-k}} \prod_{j=1}^{k}(x - c_j)^{\lambda_j}\right)$ when $\tilde{c}_{k+1} = \cdots = \tilde{c}_p = 0$. Then the conjugacy class of the semilocal monodromy matrix for $\{c_1, \dots, c_k\}$ is kept invariant under the confluence and (EC) of the inverse of the local monodromy matrix at the irregular singular point equals (16).

Acknowledgements This work was supported by the JSPS grant-in-aid for scientific research B, No. 25287017.

References

1. Dettweiler, T., Reiter, S.: An algorithm of Katz and its applications to the inverse Galois problems. J. Symb. Comput. **30**, 761–798 (2000)
2. Haraoka, Y.: Middle convolution for completely integrable systems with logarithmic singularities along hyperplane arrangements. Adv. Stud. Pure Math. **62**, 109–136 (2012)
3. Katz, N.M.: Rigid Local Systems. Annals of Mathematics Studies, vol. 139. Princeton University Press, Princeton (1995)
4. Oshima, T.: Fractional calculus of Weyl algebra and Fuchsian differential equations. In: MSJ Memoirs, vol. 11. Mathematical Society of Japan (2012)
5. Oshima, T.: Classification of Fuchsian systems and their connection problem. RIMS Kôkyûroku Bessatsu B **37**, 163–192 (2013)
6. Oshima, T.: Reducibility of hypergeometric equations. In: Analytic, Algebraic and Geometric Aspects of Differential Equations. Trends in Mathematics, pp. 425–453. Birkhäuser (2017)
7. Oshima, T.: Transformations of KZ type equations. RIMS Kôkyûroku Bessatsu B **61**, 141–162 (2017)
8. Oshima, T.: os_muldif.rr, a library of a computer algebra Risa/Asir (2008–2017). ftp://akagi.ms.u-tokyo.ac.jp/pub/math/muldif/

On the Newton Polygon of a Moser-Irreducible Linear Differential System

Moulay Barkatou

Abstract In this paper we consider the problem of computing the Newton polygon, as well as the associated Newton polynomials, of a linear differential system with meromorphic coefficients $x\frac{dY}{dx} = A(x)Y$, having an irregular singularity at the origin. We give a new estimate of the Katz invariant of such a system and prove a generalization of an old theorem from [2] which states that for a Moser-irreducible system, and under a simply checkable condition, the leading terms of the exponential part of the formal solutions can be directly computed from the leading terms of the coefficients of the characteristic polynomial of the matrix $A(x)$.

Keywords Systems of linear differential equations · Irregular singularities
Newton polygon · Exponential growth · Katz invariant · Formal solutions

MSC Primary 34M35 · Secondary 34M30, 34M25, 33F10

1 Introduction

Throughout this paper we let $\mathbb{K} = \mathbb{C}[[x]][x^{-1}]$ be the field of Laurent power series in x and we denote by ϑ the Euler derivation $\vartheta := x\frac{d}{dx}$. We consider a first-order system of linear differential equations with coefficients in $\mathbb{K}$:

$$[A] \qquad \vartheta Y = A(x)Y, \tag{1}$$

where $A(x)$ is an $n \times n$ matrix with entries in $\mathbb{K}$:

$$A(x) = x^{-q} \sum_{k=0}^{\infty} A_k x^k, \tag{2}$$

M. Barkatou (✉)
XLIM UMR 7252 CNRS ; University of Limoges, 123 Avenue Albert Thomas,
87060 Limoges Cedex, France
e-mail: moulay.barkatou@unilim.fr

G. Filipuk et al. (eds.), *Formal and Analytic Solutions of Diff. Equations*,
Springer Proceedings in Mathematics & Statistics 256,
https://doi.org/10.1007/978-3-319-99148-1_11

here q is a nonnegative integer and the A_k's are constant matrices with $A_0 \neq 0$ when $q > 0$. The number q is called the *Poincaré rank* of the system (1).

In the sequel, we will use the notation $[A]$ to refer to a system of the form (1) whose coefficient matrix is A.

It is well-known (see [19, 20] or [1]) that any system (1) has a formal fundamental matrix solution (FFMS for short) of the form

$$Y(x) = \Phi(x^{1/s})x^R \exp\{Q(x^{-1/s})\} \tag{3}$$

where s is an integer ≥ 1 called the *ramification index*,

$$Q(x^{-1/s}) = diag(Q_1(x^{-1/s}), \ldots, Q_n(x^{-1/s}))$$

is a diagonal matrix containing polynomials in $x^{-1/s}$ without constant terms, R is a constant matrix commuting with Q, and $\Phi(x^{1/s})$ is a formal series in $x^{1/s}$. It is clear that, apart from the ordering of the diagonal elements, the matrix $Q(x^{-1/s})$ is invariant with respect to the class of gauge transformations over $\overline{\mathbb{K}} = \cup_{m\geq 1}\mathbb{C}((x^{1/m}))$. We will refer to $Q(x^{-1/s})$ as the *exponential part* of the system (1).

The system $[A]$ has a regular singularity at $x = 0$ if and only if $Q = 0$. This is, in particular, the case when the Poincaré rank q is zero. For a system with irregular singularity, the "degree" in x^{-1} of $Q(x^{-1/s})$ is a positive rational number called the *Katz invariant* or the *exponential growth* [1] of the system $[A]$ (notation $\kappa(A)$). It determines the growth of $\exp\{Q(x^{-1/s})\}$ as $x \to 0$, which in turn controls the asymptotic behavior of the solutions as $x \to 0$. It is known that the Katz invariant of $[A]$ is equal to the largest slope of the *Newton polygon* $N_d(A)$ of $[A]$ which is defined as the Newton polygon of any scalar differential equation obtained from $[A]$ by the choice of a cyclic vector (see Sect. 2). More generally, the degree in x^{-1} of each nonzero entry $Q_j(x^{-1/s})$ of $Q(x^{-1/s})$ is equal to a positive slope k_j of $N_d(A)$ and its leading coefficient is a root of the Newton polynomial associated to k_j. Therefore, it is important to be able to compute these datas directly from the coefficient matrix A without resorting to the cyclic vector method. In an old paper [2] we designed an algorithm to compute the exponential part of any system of the form (1). This algorithm relies on a method, which we developed in the same paper, for computing the Katz invariant and its corresponding *Newton polynomial* directly from the leading terms of the coefficients of the characteristic polynomial of the coefficient matrix of the given system. In the present paper we generalize these results by showing that for a *Moser-irreducible* system $[A]$ with positive Poincaré rank q (satisfying a simply checkable sufficient condition), the leading terms of the exponential parts Q_j (with $\deg_{1/x}(Q_j) \geq q - 1$) can computed directly from the leading terms of the coefficients of the characteristic polynomial of A. We also give a new estimation $\kappa(A)$ improving existing estimation (proved so far in the literature) [2, 8, 12]. Note that, using a different approach, M. Miyake [16] has obtained results, on the estimation of the exponential growth of a Moser-irreducible, which are similar to some of ours.

[1] In the present paper we will rather use the first denomination: "Katz invariant".

Another important question is to estimate the number of coefficients of A which determine $\kappa(A)$. It has been shown in [14] that the exponential part Q of a system $[A]$ of size n and Poincaré rank q is completely determined by the first nq coefficients $A_0, A_1, \ldots, A_{nq-1}$ in the series expansion (2) of $A(x)$. It follows that nq is equally an upper bound for the number of coefficients which are involved in the determination of the Newton polygon and polynomials of the system $[A]$. However, if one is interested only in the leading term of Q, one can expect that fewer coefficients are needed. Combining results from [14] and our new estimate of $\kappa(A)$, we give an explicit bound on the number of coefficients of A which are required for the computation of the leading term of the exponential part.

Notation

Let $\mathcal{O} = \mathbb{C}[[x]]$ be the ring of formal power series in x with coefficients in the field of complex numbers $\mathbb{C}$. For $a \in \mathbb{K}$ we denote by $\mathrm{v}(a)$ the order in x of a, ($\mathrm{v}(0) = +\infty$). For a nonzero element $a \in \mathbb{K}$, we define the leading coefficient of a and denote by $\mathrm{LC}(a)$ the coefficient of $x^{\mathrm{v}(a)}$ in the expansion of a. We define the leading term of a as $\mathrm{LT}(a) = \mathrm{LC}(a)x^{\mathrm{v}(a)}$.

By $Mat_n(\mathcal{O})$, $Mat_n(\mathbb{K})$ we denote the ring of $n \times n$ matrices whose elements lie in $\mathcal{O}$, $\mathbb{K}$ respectively. We write $GL_n(\mathbb{K})$ for the group of invertible matrices in $Mat_n(\mathbb{K})$. By I_n we denote the identity matrix of order n. By $diag(a, b, c, \ldots)$ we denote the square diagonal (*resp.* block-diagonal) matrix whose diagonal elements are $a, b, c, \ldots$. By $\mathrm{Comp}(c_i)_{0 \le i \le n-1}$ we denote the companion matrix

$$\mathrm{Comp}(c_i)_{0 \le i \le n-1} = \begin{pmatrix} 0 & 1 & 0 & & 0 \\ 0 & 0 & 1 & & 0 \\ \vdots & \vdots & \vdots & \ddots & \vdots \\ 0 & 0 & 0 & \ldots & 1 \\ -c_0 & -c_1 & -c_2 & \ldots & -c_{n-1} \end{pmatrix}$$

If $A = (a_{i,j})$ is a matrix in $\mathbb{K}$, we define its *valuation* by $\mathrm{v}(A) = \min_{i,j} (\mathrm{v}(a_{i,j}))$. The *leading matrix* of a nonzero matrix A (notation $\mathrm{LC}(A)$ or simply A_0) is the coefficient of $x^{\mathrm{v}(A)}$ in the series expansion of A.

We will use the notation χ_A to denote the characteristic polynomial of a square matrix A which is defined here as: $\chi_A(\lambda) = \det(\lambda I_n - A)$.

2 A Few Reminders

In this section we give a few reminders of various notions and known facts which will be used in the sequel.

Gauge Transformation and Equivalent Systems

A *gauge transformation* of (1) is a linear substitution $Y(x) = T(x)Z(x)$ with $T \in GL_n(\mathbb{K})$. It transforms (1) into an *equivalent* system of the same form, whose coefficient matrix is given by $T[A] := T^{-1}AT - T^{-1}\vartheta T$. Two matrices $A, B \in Mat_n(\mathbb{K})$ are called *equivalent* over $\mathbb{K}$ if there exists $T \in GL_n(\mathbb{K})$ such that $B = T[A]$. One can also consider gauge transformations T with entries in the field of formal Puiseux series $\overline{\mathbb{K}} = \cup_{m\geq 1}\mathbb{C}((x^{1/m}))$, in that case the matrix $T[A] \in Mat_n(\overline{\mathbb{K}})$ and the matrices $A, T[A]$ (and the corresponding systems) are called *equivalent* over $\overline{\mathbb{K}}$.

Reduction of a System to a Single Scalar Equation

It is well known that any scalar differential equation of order n with coefficients in $\mathbb{K}$

$$D(y) = \vartheta^n y + c_{n-1}\vartheta^{n-1} y + \cdots + c_1\vartheta y + c_0 y = 0, \tag{4}$$

can be reduced to a first order system of the form (1) whose coefficient matrix is the companion matrix $\mathrm{Comp}(c_i)_{0\leq i\leq n-1}$. It is also true that any linear system (1) is gauge equivalent to a system whose coefficient matrix is the companion matrix $\mathrm{Comp}(c_i)_{0\leq i\leq n-1}$ of a scalar differential operator of the form (4). This fact is known as the *Cyclic vector Lemma*. Many proofs of this result can be found in the literature [10, 11, 18]. In fact, there exist an algorithm [3] which given a system (1) with coefficients in $\mathbb{K}$ produces a nonsingular polynomial transformation T such that $T[A]$ is the companion matrix of a scalar equation.

2.1 *Newton Polygon and Polynomials*

In this section, we recall the definition of *differential* Newton polygon and polynomials of a differential system and their links to the leading terms of the exponential parts in its formal solutions. We also introduce the notion of *algebraic* Newton polygon and polynomials of a differential system. The latter are directly obtained from the coefficient matrix of the system and will be used later to compute (at least partially) the differential Newton polygon and polynomials for a Moser-irreducible system.

We start by defining the Newton polygon and polynomials of a finite sequence of elements of K.

Newton Polygon of a Finite Sequence of Elements of $\mathbb{K}$

For $(u, v) \in \mathbb{R}^2$, put $\mathcal{Q}(u, v) = \{(x, y) \in \mathbb{R}^2 |\ x \leq u,\ y \geq v\}$. For a finite sequence $c = (c_i)_{0\leq i\leq n}$ of elements of $\mathbb{K}$ with $c_n \neq 0$, we denote by $\mathcal{Q}(c)$ the union of

$\mathcal{Q}(i, v(c_i))$ for $0 \leq i \leq n$ and $c_i \neq 0$. The *Newton polygon* of c, denoted by $N(c)$, is the border of the intersection of $(\mathbb{R}^+ \times \mathbb{R})$ with the convex hull of the set $\mathcal{Q}(c)$.

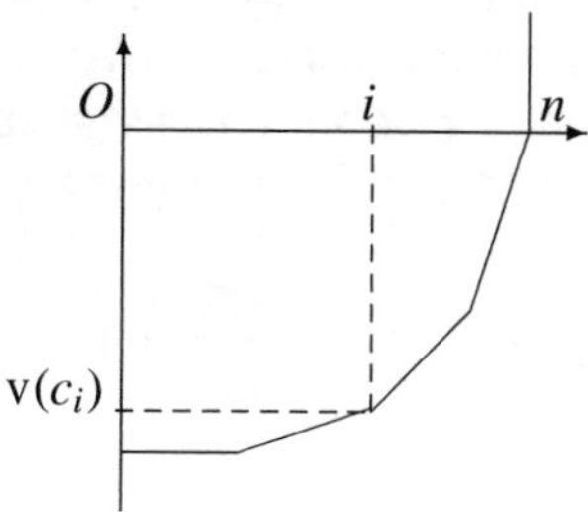

Note that the slopes of the non-vertical sides of $N(c)$ are non-negative rational numbers. The *length* of a side S of $N(c)$ is, by definition, the length of the projection of S on the $x-$axis.

Let k be a slope of $N(c)$ and S_k the corresponding side. We denote by $I(k)$ the set of the points $(i, v(c_i))$ that lie on the side S_k. We put $i_{right}(k) := \max I(k)$ and $i_{left}(k) := \min I(k)$ if $k > 0$ and $i_{left}(k) = 0$ if $k = 0$. Note that for $k > 0$, the set $I(k)$ has at least two elements and in this case $i_{left} < i_{right}$. The *Newton polynomial* associated to a slope k is defined as

$$P_k(\lambda) = \sum_{i \in I(k)} \mathrm{LC}(c_i)\lambda^{(i - i_{left}(k))}$$

This is a polynomial of degree $\ell_k = i_{right}(k) - i_{left}(k)$ the length of the side S_k. Note that for $k > 0$, $P_k(\lambda)$ has a nonzero constant term.

Differential Newton Polygon and Polynomials of a Differential System

The *Newton polygon and polynomials* of a differential operator, of order n, $D = \sum_{i=0}^{n} c_i \vartheta^i \in \mathbb{K}[\vartheta]$ (respectively, a polynomial $P = \sum_{i=0}^{n} c_i \lambda^i \in \mathbb{K}[\lambda]$ of degree n) are defined as the Newton polygon and polynomials of the sequence $(c_i)_{0 \leq i \leq n}$ of its coefficients. We will use the notation $N(D)$ (or $N(P)$).

Recall (see [15]) that for a differential operator $D \in \mathbb{K}[\vartheta]$ the slopes of the sides of $N(D)$ coincide with the "degrees" in $1/x$ of the polynomials Q_j occurring in the exponential part of a FFMS of the companion system of D. Moreover, the roots of the associated Newton polynomials give the leading coefficients of the polynomials Q_j. As all the systems that are equivalent to a given differential system $[A]$ have the same exponential part, one can define the *Newton polygon and polynomials* of $[A]$, notation $N_d(A)$, as the ones of any equivalent scalar operator obtained from $[A]$ using the cyclic vector method.[2]

[2]Direct proofs of the fact that this Newton polygon does not depend of the chosen cyclic vector can be found, for instance, in [13] or [18].

Katz Invariant

By definition the *Katz invariant* of the system (1) is the "degree" in x of the exponential part $Q(x^{-1/s})$ in a FFMS of (1). It coincides with the largest slope of the Newton polygon $N(A)$ of the system (1). We will denote it by $\kappa(A)$.

Thus, if $D = \sum_{i=0}^{n} c_i \vartheta^i$ is a scalar operator obtained from the system (1) via a cyclic vector, then the Katz invariant of system (1) is given by

$$\kappa(A) = \max\left(0, \max_{0 \leq j < n}\left(\frac{-\mathrm{v}(c_j)}{n-j}\right)\right) \tag{5}$$

Algebraic Newton Polygon and Polynomials of a Differential System

Given a system $[A]$, we define its *algebraic Newton polygon and polynomials* as the Newton polygon and polynomials of the characteristic polynomial of A: $\chi_A(\lambda) = \det(\lambda I - A)$. We will use the notation $N_a(A) = N(\chi_A)$ to distinguish it from the *differential* Newton polygon $N_d(A)$. Note that for a companion matrix C one has $N_d(C) = N_a(C)$.

We define the *algebraic Katz invariant* of $[A]$ (notation $\kappa_a(A)$) as the biggest slope of $N_a(A)$.

Remark 1 Note that the algebraic Newton polygon and polynomials are not invariant under gauge transformations, unlike the differential Newton polygon and polynomials.

2.2 *Moser-Irreducible Differential Systems*

In this section, we recall the notion of Moser-irreducible systems and show some of their properties which will be used in the sequel of this paper.

2.2.1 Moser-Irreducible Matrices

To a system of the form (1) whose coefficient matrix A is given by (2), Moser [17] associates the two rational numbers

$$m(A) := \max\left(q + \frac{\mathrm{rank}(A_0)}{n},\ 1\right) \quad \textit{and} \quad \mu(A) = \min\left\{m(T[A]) \mid T \in GL_n(\mathbb{K})\right\}.$$

We will refer to $m(A)$ and $\mu(A)$ as respectively the *Moser rank* and the *Moser invariant* of the system [A]. The system (1) has a regular singularity at $x = 0$ if and only if $\mu(A) = 1$.

The matrix A is called *Moser-irreducible* if $m(A) = \mu(A)$, otherwise it is called *Moser-reducible*. In other words, the matrix A is Moser-reducible if there exists a matrix $T \in GL_n(\mathbb{K})$ such that $m(T[A]) < m(A)$.

In [17] Moser proved the following theorem which gives a simple criterion for a matrix to be Moser-reducible:

Theorem 1 (Moser [17]) *If $m(A) > 1$, the matrix A is Moser-reducible if and only if the polynomial*

$$\mathscr{B}(A, \lambda) := x^{rank(A_0)} \det (\lambda I - x^{q-1}A(x))_{|x=0}$$

vanishes identically in λ.

Moreover, when A is Moser-reducible then the reduction can be carried out with a simple transformation of the form $T = (T_0 + xT_1)diag(1, \ldots, 1, x, \ldots, x)$, where T_0, T_1 are constant matrices with $\det T_0 \neq 0$.

Applying Moser's Theorem several times, if necessary, $\mu(A)$ can be determined. Thus, a polynomial matrix T such that $m(T[A]) = \mu(A)$ can be computed in this way. There are various efficient algorithms to construct such a transformation [5–7].

Proposition 1 *Suppose that $[A]$ is Moser-irreducible with $m(A) > 1$, write $\chi_A(\lambda) = \sum_{i=0}^{n} a_i\lambda^i$ and put $(q, r, d) := (-\mathrm{v}(A), rank(A_0), \deg \mathscr{B}(A, \lambda))$. Then*

$$q = \min \{m \in \mathbb{N} |\, \mathrm{v}(a_i) + (n - i)m \geq 0, \ \textit{for all}\ 0 \leq i \leq n\}, \tag{6}$$

$$r = \max_{0 \leq i \leq n} ((q - 1)(i - n) - \mathrm{v}(a_i)), \tag{7}$$

$$d = \max \{0 \leq i \leq n \mid r = (q - 1)(i - n) - \mathrm{v}(a_i)\}. \tag{8}$$

Proof First, we remark that $\sum_{i=0}^{n} x^{(n-i)q} a_i\lambda^i = \det(\lambda I_n - x^q A) \in \mathscr{O}[\lambda]$ since $x^q A \in \mathscr{O}$. Hence $\mathrm{v}(a_i) + (n - i)q \geq 0,\ \textit{for all}\ 0 \leq i \leq n$. We should prove now that q is the smallest integer with this property, when A is Moser-irreducible. Put $\theta_A(\lambda, x) := x^r \det (\lambda I_n - x^{q-1}A(x)) = \sum_{i=0}^{n} \tilde{a}_i(x)\lambda^i$. We note that

$$\theta_A(\lambda, x) = x^{r+n(q-1)} \det(x^{1-q}\lambda I_n - A(x)) = \sum_{i=0}^{n} x^{r+(n-i)(q-1)} a_i\lambda^i$$

hence $\tilde{a}_i = x^{r+(n-i)(q-1)} a_i$ for all i. Since the matrix $x^{q-1}A(x) = x^{-1}(A_0 + O(x))$ has at most r linearly independent rows with valuation -1, one sees that $\theta_A(\lambda, x) \in \mathscr{O}[\lambda]$. Moreover, if we replace x^r, in the definition of θ_A, by a higher power $x^{r'}$, then we get that $\theta_A(\lambda, x) \in x\mathscr{O}[\lambda]$. In other words, we get that $\mathrm{v}(a_i) + (n - i)(q - 1) + r = \mathrm{v}(\tilde{a}_i) \geq 0$ for all i, and that r is the smallest integer satisfying this property. Thus (7) holds. Now, if A is Moser-irreducible, the polynomial $\mathscr{B}(A, \lambda) = \theta_A(\lambda, 0) = \sum_{i=0}^{n} \tilde{a}_i(0)\lambda^i$ is not identically zero, hence $\mathrm{v}(\tilde{a}_i) = 0$ for some $i < n$. This implies that the set $\{i \,|\, \mathrm{v}(a_i) + (n - i)(q - 1) + r = 0\}$ is not empty and its maximum is equal to d, the degree of $\mathscr{B}(A, \lambda)$, which proves (8). In particular, one has $\mathrm{v}(a_d) +$

$(n-d)(q-1) = -r < 0$, which implies that $v(a_d) + (n-d)m < 0$ for all integer $0 \le m \le (q-1)$, which in turn implies (6). □

Remark 2 1. The relations (6), (7) and (8) can be, geometrically, interpreted as follows:

a. The number q is the smallest nonnegative integer q such that the straight line $L_1 : y = (x-n)q$ stays below $N_a(A)$, the algebraic Newton polygon of A.
b. The number r is the smallest positive integer such that the straight line $L2 : y = (x-n)(q-1) - r$ stays below $N_a(A)$.
c. The point $(d, (d-n)(q-1) - r)$ belongs to $N_a(A) \cap L_2$ and when $d > 0$ the intersection $N_a(A) \cap L_2$ is a side of $N_a(A)$.

2. It follows that the intersection of $N_a(A)$ with the half-plane $x \ge d$ lie below the line $L_3 : y = (q - 1 + \frac{r}{n-d})(x-n)$. In other words, if $(i, v(a_i)) \in N_a(A)$ and $i \ge d$ then

$$v(a_i) \le (q - 1 + \frac{r}{n-d})(i-n) \tag{9}$$

In the following proposition, we gather a few facts which will be useful in the sequel. They can be, almost straightforwardly, deduced from some results proven in [9] (see Theorem 3.3) and [8] (see Lemma 4.1).

Proposition 2 *Suppose that the system* $[A]$ *is Moser-irreducible with a positive Poincaré rank* $q > 0$ *and let* $d := \deg \mathcal{B}(A, \lambda)$. *Then*

1. *the degree* d *is equal to the sum of the lengths of the slopes of the differential Newton polygon* $N_d(A)$ *which are less than or equal to* $(q-1)$.
2. *if* $d > 0$ *and* $q > 1$ *then* $N_d(A)$ *has a side of slope* $(q-1)$ *if and only if* $\mathcal{B}(A, \lambda)$ *contains at least two monomials and in that case the corresponding Newton polynomial is* $\lambda^{-\nu}\mathcal{B}(A, \lambda)$ *where* ν *is the largest integer* ν *such that* λ^{ν} *divides* $\mathcal{B}(A, \lambda)$,
3. *if* $d > 0$ *and* $q = 1$ *then the Newton polygon* $N_d(A)$ *has a side with slope* 0 *with an associated Newton polynomial* $P_0(\lambda)$ *of degree* d *and having roots that differ by integers from the roots of* $\mathcal{B}(A, \lambda)$.

2.2.2 Moser-Reduction of a Scalar Equation

Consider a scalar differential equation of the form (4) and let $C = \mathrm{Comp}(c_i)_{0 \le i \le n-1}$ be its companion matrix.

It has been shown in [17] that $\mu(C)$ can be computed explicitly from the valuations of the coefficients c_i. This is expressed by the following theorem.

Theorem 2 (Moser [17]) *Put* $c_n = 1$ *and define the integers* q^* *and* r^* *by*

$$q^* = \min \{m \in \mathbb{N} | \mathrm{v}(c_i) + (n-i)m \geq 0 \text{ for all } 0 \leq i \leq n\} \tag{10}$$

$$r^* = \max_{0 \leq i \leq n} ((q^* - 1)(i - n) - \mathrm{v}(c_i)) \tag{11}$$

Then $\mu(C) = q^* + \frac{r^*}{n}$.

Proof We won't repeat Moser's proof here but, for sake of clarity and for later reference, we give very briefly the main idea of it:

Consider the matrix $B := T[C]$ where $T := x^{\alpha} = \mathrm{diag}(x^{\alpha_0}, \ldots, x^{\alpha_{n-1}})$, with

$$\alpha_i = \min\{(n-i)q^*, r^* + (n-i)(q^* - 1)\} \text{ for } 0 \leq i \leq n-1.$$

Using the special structure of B, one can check that $\mathrm{v}(B) = -q^*$, $\mathrm{rank}(B_0) = r^*$ and show that $\mathscr{B}(B, \lambda) \neq 0$. It follows that B is Moser-irreducible and hence $\mu(B) = q^* + \frac{r^*}{n}$. Let us notice that from Moser's proof, it appears that the degree d^* of $\mathscr{B}(B, \lambda)$ is given by

$$d^* = \max \{0 \leq i \leq n \mid r^* = (q^* - 1)(i - n) - \mathrm{v}(c_i)\}. \tag{12}$$

□

Remark 3 Consider a differential system $[A]$ and let C be a companion matrix equivalent to A. Let (q^*, r^*, d^*) denote the triplet associated with C. Since q^*, r^*, d^* are defined by equations similar to (6), (7) and (8), it is clear that the properties listed in Remark 2 are equally true, mutatis mutandis, when we replace $q, r, d, N_a(A)$ by $q^*, r^*, d^*, N_d(A)$. These properties can be summarized in the following way:

1. The points $(n, 0)$ and $(d^*, (d^* - n)(q^* - 1) - r^*)$ are vertices of $N_d(A)$.
2. $N_d(A) \cap \{(x, y) \in \mathbb{R}^2 \,|\, x \geq d\}$ is a subset of the compact region delimited by the triangle defined by the three straight lines: $L_1^* : y = (x - n)q^*$, $L_2^* : y = (x - n)(q^* - 1) - r^*$ and $L_3^* : y = (q^* - 1 + \frac{r^*}{n - d^*})(x - n)$.
   In particular, one has

$$\mathrm{v}(c_i) \leq (q^* - 1 + \frac{r^*}{n - d^*})(i - n) \tag{13}$$

   for all $i \geq d^*$ such that the point $(i, v(c_i))$ lies on a side of $N_d(A)$.

Remark 4 We have equally, the following properties which are easy to establish

1. The integer d^* is equal to the sum of the lengths of all the slopes of $N_d(A)$ which are less than or equal to $(q^* - 1)$.
2. When $\kappa := \kappa(A) < q^*$, one has $d^* \leq n - r^* - 1$ and $d^* \leq i_{left}(\kappa) \leq n - r^* - 1$ (recall that $i_{left}(\kappa)$ is the smallest integer i for which $\kappa = -\frac{\mathrm{v}(c_i)}{(n-i)}$).

3. If the Newton polygon $N_d(A)$ has a single non-vertical side with a slope greater than $(q^* - 1)$ (this is the case if and only if $i_{left}(\kappa) = d^*$) then

$$\kappa = -\frac{\mathrm{v}(c_{d^*})}{(n - d^*)} = q^* - 1 + \frac{r^*}{n - d^*}$$

4. The point $(d^*, (d^* - n)(q^* - 1) - r^*)$ is a vertex of $N_d(A)$ which belongs to $S_{q^*-1} := N_d(A) \cap L_2^*$. More precisely, if $d^* \geq 1$ and $q^* = 1$ then S_{q^*-1} is a side of $N_d(A)$ with slope 0 and $\mathscr{B}(B, \lambda)$ is the corresponding Newton polynomial [3]; and when $d^* \geq 1$ and $\mathscr{B}(B, \lambda)$ contains at least two monomials then S_{q^*-1} is a side of $N_d(A)$ with slope $(q^* - 1)$ whose Newton polynomial is $\lambda^{-\nu}\mathscr{B}(B, \lambda)$ where ν is the largest integer ν such that λ^ν divides $\mathscr{B}(B, \lambda)$.

3 An Estimate of the Katz Invariant

In this section, we prove a new estimate for the Katz invariant of a Moser-irreducible system $[A]$ and give a new bound on the number of the coefficients, in the expansion of $A(x)$, which determine $\kappa(A)$.

3.1 Estimation of the Katz Invariant of a Moser-Irreducible System

Consider a scalar differential equation of order n, $D(y) = 0$, of the form (4) and keep the notation of the previous sections. We write κ for $\kappa(D)$.

Lemma 1 *Let q^*, r^*, d^* be defined as above and assume that $q^* > \kappa$. Then we have*

$$q^* - 1 + \frac{r^*}{n - d^*} \leq \kappa \leq q^* - \frac{1}{r^* + 1} \tag{14}$$

Proof The first inequality follows from the equation $r^* = (q^* - 1)(d - n) - \mathrm{v}(c_d)$ and the inequality $\frac{-\mathrm{v}(c_d)}{n-d} \leq \kappa$. Let $i_0 := i_{left}(\kappa)$. Using the equality $\frac{-\mathrm{v}(c_{i_0})}{n-i_0} = \kappa$ and the inequality $(q^* - 1)(i_0 - n) - \mathrm{v}(c_{i_0}) \leq r^*$ we obtain

$$\kappa \leq q^* - 1 + \frac{r^*}{n - i_0}$$

As $n - i_0 \geq r^* + 1$ we get the second inequality in (14). □

We are able now to give an estimate for the Katz invariant of a Moser-irreducible system (1) with a positive Poincaré rank q.

[3] Here B is the matrix defined in the proof of Theorem 2.

Lemma 2 *Consider a Moser-irreducible system (1) with positive Poincaré q. Consider any companion matrix C obtained from* $[A]$ *via the cyclic vector method and calculate* q^*, r^* *and* d^* *using formulas (10), (11) and (12). Then we have* $q^* = q$, $r^* = rank(A_0)$ *and* $d^* = \deg \mathscr{B}(A, \lambda)$. *This implies, in particular, that* q^*, r^* *and* d^* *are independent of the choice of cyclic vector.*

Proof Since $[A]$ is Moser-irreducible and $q > 0$ then the system $\vartheta Y = AY$ has an irregular singularity at the origin. Let $C = \text{Comp}(-c_i)_{0 \le i \le n-1}$ be a companion matrix equivalent to A and let the triplet (q^*, r^*, d^*) be defined by formulas (10), (11) and (12). Then $\mu(C) = q^* + r^*/n$. The matrices C and A are equivalent and A is Moser-irreducible, so $q^* + r^*/n = \mu(C) = \mu(A) = m(A) = q + rank(A_0)/n$. This implies $q^* = q$ and $r^* = rank(A_0)$. The equality $d^* = \deg \mathscr{B}(A, \lambda)$ follows from the fact that $N_d(A) = N_d(C)$, and item 1 in Proposition 2 and item 1 of Remark 4. □

Proposition 3 *Consider a Moser-irreducible system* $[A]$ *with positive Poincaré rank* q *and let* $r = rank(A_0)$ *and* $d = \deg \mathscr{B}(A, \lambda)$. *We also assume that its leading matrix* A_0 *is nilpotent, otherwise we have* $\kappa(A) = q$. *Then we have*

$$q - 1 + \frac{r}{n-d} \le \kappa(A) \le q - \frac{1}{r+1} \tag{15}$$

Proof Let $C = \text{Comp}(-c_i)_{0 \le i \le n-1}$ be any companion matrix equivalent to A. Since $[A]$ is Moser-irreducible and $q > 0$, the system $[A]$ has an irregular singularity and its Katz invariant $\kappa(A) = \kappa(C)$ is given by (5). By Lemma 2 the numbers q, r and d coincide respectively with the integers q^*, r^*, d^* given by the formulas (10), (11) and (12). Since A_0 is nilpotent we have $q > \kappa(A)$ and we can apply (14) to get the estimations (15). □

Remark 5 Using the fact that $r \ge 1$ in one hand and $n - d \ge r + 1$ on the other hand, one sees that the (15) implies the following weaker estimate

$$q - 1 + \frac{1}{n-d} \le \kappa(A) \le q - \frac{1}{n-d}$$

which has been shown in [8] (see Theorem 5.1).

Corollary 1 *With the notation and assumptions of Proposition 3, if* $d = n - r - 1$ *then* $\kappa(A) = q - 1 + \frac{r}{n-d} = q - \frac{1}{r+1}$.

Remark 6 With the notation and assumptions of Proposition 3, the following properties hold:

1. if $r = n - 1$ then necessarily $d = n - r - 1 = 0$ (since $d \le n - r - 1$ does always hold) and in this case the differential Newton polygon of $[A]$ consists of a single side of length n and slope $\kappa(A) = q - \frac{1}{n}$;
2. if we suppose that the differential Newton polygon of $[A]$ consists of a single side of length n and slope κ, then $i_{left}(\kappa) = 0 = d$ and $\kappa = q - 1 + \frac{r}{n}$.

3.2 Number of Coefficients Involved in the Computation of $\kappa(A)$

Another important question is to estimate the number of coefficients of A which determine the $\kappa(A)$. Let's recall that it has been proven, in [14](see Remark 1.3, page 7), that for any system $[A]$ of Poincaré rank q, if N is any integer satisfying $N > n(q - \kappa)$, the coefficients $A_0, A_1, \ldots, A_{N-1}$ determine $\kappa(A)$ as well as the corresponding Newton polynomial. Applying this result to the case where $[A]$ is Moser-irreducible, we get, thanks to the estimate (15), that we can take $N > n(1 - \frac{r}{n-d})$. In particular, if $d = 0$ only the first $n - r + 1$ coefficients are needed. It turns out that even when $d > 0$ we still need only the first $n - r - d + 1$ coefficients[4] of A to compute $\kappa(A)$ as shown by the following proposition[5]:

Proposition 4 *Consider a system* $[A]$ *with the notation and assumptions of Proposition 3. Then* $\kappa(A)$ *as well as the corresponding Newton polynomial are determined by the coefficients* $A_0, A_1, \ldots, A_{n-d-r}$.

Proof Since $[A]$ is Moser irreducible, according to a result in [8] (Lemma 4.1 p. 259), there exists a transformation $T \in GL_n(\mathcal{O})$ such that $T[A]$ is a block-diagonal matrix $T[A] = \mathrm{diag}(x^{-q}D^{11}, x^{-q+1}D^{22})$ where the D^{11}, D^{22} have their coefficients in $\mathcal{O}$, D^{11} is of dimension $n - d$ and $[x^{-q}D^{11}]$ is Moser-irreducible with $\mathrm{rank} D_0^{11} = r$. Moreover, $\deg \mathcal{B}(x^{-q}D^{11}, \lambda) = 0$. It follows that $\kappa(x^{-q}D^{11})$, as well as the corresponding Newton polynomial, are determined by the first the first $n - d - r + 1$ coefficients in the series D^{11}. We conclude by noticing that in one hand these coefficients depend only on $A_0, A_1, \ldots, A_{n-d-r}$, since $T \in GL_n(\mathcal{O})$; and in the other hand $[A]$ and $[x^{-q}D^{11}]$ have the same Katz invariant as well as the same corresponding Newton polynomial. □

4 Computing the Katz Invariant and the Corresponding Newton Polynomial

In this section, we address the problem of determining the Katz invariant $\kappa(A)$ and the associated Newton polynomial $P_{\kappa(A)}$, directly from the characteristic polynomial of the matrix A without computing first, any equivalent companion matrix.

We generalize and refine results from our old paper [2] (see Theorem 1, page 11) where it has been shown that the Katz invariant and its corresponding Newton polynomial of a Moser-irreducible system $[A]$, of size n and Poincaré rank q, can be explicitly computed from the leading terms of the coefficients of the characteristic polynomial of $[A]$ provided that $q > n - \mathrm{rank}(A_0)$. In fact we will prove a more general result (see Theorem 3 below), where we state that for a Moser-irreducible system

[4]Note that $n(1 - \frac{r}{n-d}) \geq n - r - d$.

[5]Note that this proposition is similar to Theorem 5.2 in [8].

$[A]$, of size n and Poincaré rank $q > 0$, all the slopes greater than $(q-1)$ as well as the corresponding Newton polynomials can be explicitly computed from the leading terms of the coefficients of the characteristic polynomial of $[A]$ under the (sufficient) condition $q > (r+1)(1-\frac{r}{n-d})$ where $(r, d) = (\text{rank}(A_0), \deg \mathscr{B}(A, \lambda))$.

4.1 Three Preliminary Lemmas

We start by proving the following three lemmas which are refinements of Lemma 3, Lemma 4 and Proposition 1 from [2].

Lemma 3 *Let A be a matrix in $Mat_n(\mathbb{K})$ with $q = -\text{v}(A) > 0$ and H a matrix in $Mat_n(\mathscr{O})$. Write*

$$\det(\lambda I - A + H) - \det(\lambda I - A) = t_{n-1}\lambda^{n-1} + t_{n-2}\lambda^{n-2} \cdots + t_0$$

with $t_{n-1}, t_{n-2}, \ldots, t_0$ in $\mathbb{K}$. Then

$$\text{v}(t_{n-i}) \geq (1-i)q + \max(0, i-r-1) \qquad \textit{for all } i \in \{1, \ldots, n\}.$$

where $r = \text{rank}(A_0)$, the rank of the leading matrix A_0 of A.

Proof Note first that we can assume without any loss of generality that A has exactly r rows with valuation equal to $-q$ and the other rows have valuation greater or equal to $-q+1$. Now, let $i \in \{1, \ldots, n\}$, then t_{n-i} consists in a sum of terms of the form $h.\prod_{j=1}^{i-1} u_j$ where h is an entry of H and the u_j's are entries in different rows of A or H; so among $u_1, \ldots, u_{i-1}$ at most r the elements are of valuation $-q$, the other (if any) have their valuation greater or equal to $-q+1$. This implies that

$$\text{v}(h.\prod_{j=1}^{i-1} u_j) = \text{v}(h) + \sum_{j=1}^{i-1} \text{v}(u_j) \geq \begin{cases} -q(i-1) & \text{for } \ i-1 \leq r \\ -qr + (-q+1)(i-r-1) & \text{for } \ i-1 \geq r. \end{cases}$$

□

Lemma 4 *Let $[A]$ and $[B]$ be two systems over $\mathbb{K}$ that are equivalent and both Moser-irreducible with $m(A) = m(B) = q + \frac{r}{n} > 1$. Write*

$$\det(\lambda I - A) = \sum_{i=0}^{n} a_i \lambda^i \quad \textit{and} \quad \det(\lambda I - B) = \sum_{i=0}^{n} b_i \lambda^i.$$

Then

$$\text{v}(a_{n-i} - b_{n-i}) \geq (1-i)q + \max(0, i-r-1) \qquad \textit{for all } i \in \{1, \ldots, n\}.$$

Proof Let $T \in GL(n, \mathbb{K})$ such that $B = T[A]$. Write T in Smith normal form, i.e., $T(x) = P(x)x^{\gamma}Q(x)$ where P and Q are matrices in $Mat_n(\mathcal{O})$ with $\det P(0) \neq 0$, $\det Q(0) \neq 0$ and $\gamma = diag(\gamma_1, \ldots, \gamma_n)$ with $\gamma_1, \ldots, \gamma_n$ in $\mathbb{Z}$ and $\gamma_1 \leq \gamma_2 \cdots \leq \gamma_n$. Put $\tilde{A} = P[A]$ and $\tilde{B} = Q^{-1}[B]$. Then one has $\tilde{B} = x^{-\gamma}\tilde{A}x^{\gamma} - \gamma$ and $\mathrm{v}(\tilde{A}) = \mathrm{v}(A)$, $\mathrm{v}(B) = \mathrm{v}(\tilde{B})$. Moreover, the matrices $\tilde{A}_0$ and A_0 (respectively, $\tilde{B}_0$ and B_0) are similar, so these four matrices have the same rank r.

One has

$$\begin{aligned}
\det(\lambda I - \tilde{A}) &= \det(\lambda I - P^{-1}AP + P^{-1}\vartheta P) = \det(\lambda I - A + (\vartheta P)P^{-1}),\\
\det(\lambda I - \tilde{B}) &= \det(\lambda I - QBQ^{-1} + Q\vartheta Q^{-1}) = \det(\lambda I - B - Q^{-1}\vartheta Q),\\
\det(\lambda I - \tilde{B}) &= \det(\lambda I - x^{-\gamma}\tilde{A}x^{\gamma} + \gamma) = \det(\lambda I - \tilde{A} + \gamma).
\end{aligned}$$

Notice that, since P, Q, P^{-1}, Q^{-1} are in $Mat_n(\mathcal{O})$, the matrices $(\vartheta P)P^{-1}$, $Q^{-1}\vartheta Q$ are also $Mat_n(\mathcal{O})$. If we denote by $\tilde{a}_i$ (respectively, $\tilde{b}_i$) the coefficient of λ^i in the characteristic polynomial of $\tilde{A}$ (respectively, $\tilde{B}$) then, by applying lemma 3, we get that the three quantities $\mathrm{v}(a_{n-i} - \tilde{a}_{n-i})$, $\mathrm{v}(b_{n-i} - \tilde{b}_{n-i})$ and $\mathrm{v}(\tilde{b}_{n-i} - \tilde{a}_{n-i})$ are greater or equal to $(1-i)q + \max(0, i-r-1)$.

Now as

$$\mathrm{v}(a_{n-i} - b_{n-i}) \geq \min(\mathrm{v}(a_{n-i} - \tilde{a}_{n-i}), \mathrm{v}(\tilde{a}_{n-i} - \tilde{b}_{n-i}), \mathrm{v}(\tilde{b}_{n-i} - b_{n-i}))$$

we get that

$$\mathrm{v}(a_{n-i} - b_{n-i}) \geq (1-i)q + \max(0, i-r-1)$$

for $i \in \{1, \ldots, n\}$. □

Lemma 5 *Let $[A]$ be Moser-irreducible with $m(A) = q + \frac{r}{n} > 1$. Write $\det(\lambda I - A) = \sum_{i=0}^{n} a_i\lambda^i$. Then for any equivalent companion matrix $C = \mathrm{Comp}(c_i)_{0 \leq i \leq n-1}$, one has*

$$\mathrm{v}(a_i - c_i) \geq (i - n + 1)q + \max(0, n-i-r-1), \tag{16}$$

for $i = 0 \cdots n-1$.

Proof As $[A]$ is Moser-irreducible we know that q and $r := rank(A_0)$ coincide respectively with the numbers q^* and r^* defined by (10) and (11). For $i \in \{0, \ldots, n-1\}$, define

$$\alpha_i = \min\{(n-i)q, r + (n-i)(p-1)\} \text{ for } 0 \leq i \leq n-1$$

as in the proof of Theorem 2 and consider the matrix $B := x^{\alpha}[C]$. We know that $[B]$ is Moser-irreducible. We can then apply Lemma 4 and get that

$$\mathrm{v}(a_i - b_i) \geq (1 - (n-i))q + \max(0, n-i-r-1), \qquad \text{for } i = 0 \cdots n-1$$

where the b_i's are the coefficients of characteristic polynomial of B.

Now, since

$$\det(\lambda I - C) = \det(\lambda I - x^{-\alpha} C x^{\alpha}) = \det(\lambda I - B - \alpha).$$

we can apply Lemma 3 to the matrices B and α to get

$$\mathrm{v}(b_i - c_i) \geq (1 - (n - i))q + \max(0, n - i - r - 1).$$

Hence

$$\mathrm{v}(a_i - c_i) \geq \min(\mathrm{v}(a_i - b_i), \mathrm{v}(b_i - c_i)) \geq (1 - n + i)q + \max(0, n - i - r - 1).$$

□

4.2 Main Results

We are now able to state and prove our main results.

Theorem 3 *Let $[A]$ be a Moser-irreducible system and (q, r, d) the associated Moser-triplet. Put $\delta := 1 - \frac{r}{n-d}$. Then*

(i) If $q \geq (r+1)\delta$ then $N_d(A)$ and $N_a(A)$ coincide on the right half-plane $x \geq d$.
(ii) If $q > (r+1)\delta$ then one equally has that for each slope k (common to $N_d(A)$ and $N_a(A)$) the corresponding differential and algebraic Newton polynomials are equal.

Proof Let $C = \mathrm{Comp}(c_i)_{0 \leq i \leq n-1}$ be a companion matrix equivalent to $[A]$ under $GL(n, \mathbb{K})$. Let $J := \{i \geq d \text{ s. t. } (i, \mathrm{v}(a_i)) \in N_a(A) \text{ or } (i, \mathrm{v}(c_i)) \in N_d(A)\}$. We will prove first that under condition $q > (r+1)\delta$, the Laurent series a_i and c_i have the same leading term for each $i \in J$. For this it is sufficient to prove that $\mathrm{v}(a_i - c_i) > \min(\mathrm{v}(a_i), \mathrm{v}(c_i))$ for all $i \in J$.

Since $[A]$ is Moser-irreducible, we know that all the points belonging to $N_a(A) \cup N_d(A)$ lie in the half-plane $y \leq (q - \delta)(x - n)$; hence for $i \in J$ at least one of the two inequalities (13) or (9) holds. We also know, by Lemma 5, that inequality (16) holds for all i. So, it is sufficient to prove that condition $q > (r+1)\delta$ implies that the r.h.s in (16) is greater than the r.h.s in (13) or (9). In other words we should prove that the following quantities

$$\varepsilon_i := (i - n + 1)q + \max(0, n - i - r - 1) - (i - n)(q - \delta)$$

are positive when $q > (r+1)\delta$. This follows from the fact that $\min_i \varepsilon_i = q - (r+1)\delta$. Indeed, for $i > n - r - 1$, one has $\varepsilon_i = q - \delta(n - i) > q - \delta(r+1)$; for $i < n - r - 1$, one has

$$\varepsilon_i = q - r - 1 + (1 - \delta)(n - i) > q - r - 1 + (1 - \delta)(r + 1) = q - (r + 1)\delta,$$

and finally for $i = n - r - 1$ we have $\varepsilon_i = q - \delta(r + 1)$. This shows that in the case $q - \delta(r + 1) > 0$ the two properties (i) and (ii) hold.

Now we will see that when $q = (r + 1)\delta$, (i) is still true but there are cases (see item 4 below) where the second point (ii) of the theorem is not guaranteed.

1. For all $i \geq d$ such that $i \neq f := n - r - 1$ and at least one of the inequalities (13) or (9) hold, the coefficients a_i and c_i have the same leading term, even if $q = (r + 1)\delta$.
2. If one of the two points $(f, \mathrm{v}(a_f))$ and $(f, \mathrm{v}(c_f))$ lies strictly below the line $L_3 : y = (q - \delta)(x - n)$ then the coefficients a_f and c_f will have the same leading term, even if $q = (r + 1)\delta$. Indeed, if $\mathrm{v}(a_f) < (q - \delta)(f - n)$, then $\mathrm{v}(c_f - a_f) - \mathrm{v}(a_f) > \varepsilon_f = q - (r + 1)\delta \geq 0$.
3. Thus, if the two points $(f, \mathrm{v}(a_f))$ and $(f, \mathrm{v}(c_f)$ lies strictly below the line $L_3 : y = (q - \delta)(x - n)$, or if $(f, \mathrm{v}(a_f)) \notin N_a(A)$ and $(f, \mathrm{v}(c_f) \notin N_d(A)$ or if $f = d$ then the conclusion of the theorem holds under the weaker condition $q \geq (r + 1)\delta$.
4. Suppose that $q = (r + 1)\delta$. If $f > d$ and $(f, \mathrm{v}(a_f)) \in N_a(A) \cap L_3$, then, on one hand, the intersection of $N_a(A)$ with the half-plane $x \geq d$ consists in a single side of slope $q - \delta$ and, on the other hand the point $(f, \mathrm{v}(c_f))$ lies on or above the line L_3 (due to item 2). This, combined with the item 1 above, imply that $N_d(A)$ coincide with $N_d(A)$ on the right of the line $x = d$ but we cannot draw any conclusion about the corresponding Newton polynomials.

□

Corollary 2 *Consider a Moser-irreducible system (1) with Poincaré rank $q > 0$. Let $r = rank(A_0)$ and $d = \deg \mathscr{B}(A, \lambda)$. Let the a_i's denote the coefficients of the characteristic polynomial of $A(x)$. If $q > (r + 1)(1 - \frac{r}{n-d})$ then $\kappa(A)$ and the corresponding Newton polynomial are given by*

$$\kappa(A) = \max_{0 \leq i < n} \Big(\frac{-\mathrm{v}(a_i)}{n - i}\Big), \tag{17}$$

$$P_\kappa(\lambda) = \sum_{h=0}^{\eta} \mathrm{LC}(a_{i_h})\lambda^{(i_h - i_0)} \tag{18}$$

where $0 \leq i_0 < i_1 < \cdots < i_\eta = n$ denote the integers i for which $\kappa(n - i) = -\mathrm{v}(a_i)$.

Remark 7 In [12] (Th. 2, p. 76), it has been shown that the Katz invariant of a Moser-irreducible system $[A]$, with $m(A) = q + \frac{r}{n} > 1$, can be obtained from the characteristic polynomial of A under the condition $q > r + 1$. The authors of [12] did not state anything about the Newton polynomial. As $q > r + 1$ implies our condition $q > (r + 1)\delta$ in Theorem 3, one sees that our results generalize and improve Theorem 2 in [12].

It also improves Theorem 1 in [2] which requires the condition $q > n - r$ which equally implies $q > (r + 1)\delta$.

Example 1 Consider the system $\vartheta Y = A(x)Y$ where the matrix A is given by

$$A = x^{-2}\begin{pmatrix} -5x & 5x^2 & -2x^2 & -9x \\ 5 & 3x & 2x & -4x \\ 4x^2 & -6x^2 & -5x & 2x^3 \\ 2-2x & -5x^2 & 3x & -6x \end{pmatrix}$$

The size of the system is $n = 4$, its Poincaré rank is $q = 2$. The matrix A_0 is nilpotent and has rank $r = 1$. The system $[A]$ is Moser-irreducible and its corresponding Moser polynomial is given by

$$\mathscr{B}(A,\lambda) := x^r \det(\lambda I - A_0/x - A_1)_{|x=0} = 18(\lambda-3)(\lambda+5).$$

In particular it has degree $d = 2$. Hence $\delta := 1 - \frac{r}{n-d} = \frac{1}{2}$. Here the condition $q > n - r$, required in Theorem 1 from [2], is not satisfied. The condition $q > r + 1$ in Theorem 2 of [12] is not satisfied neither. Hence none of the mentioned theorems can be applied. However, our weaker condition $q > (r + 1)\delta$ holds here and hence we can affirm that $N_d(A)$ and $N_a(A)$ coincide on the right half-plane $x \geq d = 2$ and that, for each slope k (common to $N_d(A)$ and $N_a(A)$) the corresponding differential and algebraic Newton polynomials are equal.

The coefficients $(a_i)_{i=0\ldots4}$ of $\chi_A := \det(\lambda I - A)$ are given by

$$\begin{aligned} a_4(x) &= 1, \quad a_3(x) = 13x^{-1} \\ a_2(x) &= 18x^{-3} - 6x^{-2} - 8x^{-1} + 2 \\ a_1(x) &= 36x^{-4} - 601x^{-3} - 132x^{-2} - 20x^{-1} + 12 \\ a_0(x) &= -270x^{-5} - 2449x^{-4} - 1504x^{-3} - 424x^{-2} - 96x^{-1} \end{aligned}$$

The algebraic Newton polygon $N_a(A)$ is the convex hull with non-negative slopes of the set

$$\{(0,-5), (1,-4), (2,-3), (3,-1), (4,0)\}.$$

It has two sides with respective slopes 1 and $\frac{3}{2}$. The Katz-invariant of $[A]$ is then equal to $\frac{3}{2}$ and the corresponding Newton polynomial is $\lambda^2 + 18$.

Note that, from Proposition 2 and Remark 2, we know that $N_d(A)$ and $N_a(A)$ have another common side of slope $q - 1$ whose corresponding Newton polynomial is $\mathscr{B}(A,\lambda)$. So in this example, $N_d(A)$ and $N_a(A)$ do coincide everywhere (not only on the half-plane $x \geq d$).

Remark 8 Our Theorem 3 cannot be directly applied to a Moser-irreducible system $[A]$ for which $q < (r + 1)\delta$. However, it has been shown in [2] (see Lemma 5 and Remark 5 pp. 12–13) that after a suitable ramification $x = t^m$ one can construct a Moser-irreducible system $[\tilde{A}]$ which is equivalent to $[A]$ over $\mathbb{C}((t))$ and such that

$\mu(\tilde{A}) = \tilde{q} + \tilde{r}/n$ with $\tilde{q} > n - \tilde{r}$. Thus, one can apply Theorem 3 to $[\tilde{A}]$ and then get the slopes of $N_d(A)$ that are greater $\tilde{q} - 1$ as well as the corresponding Newton polynomials (see Lemma 1 in [2] for the relationship between $N_d(\tilde{A})$ and $N_d(A)$).

References

1. Balser, W., Jurkat, W.B., Lutz, D.A.: A general theory of invariants for meromorphic differential equations; Part I, formal invariants. Funkcial. Ekvac. **22**, 197–227 (1979)
2. Barkatou, M.A.: An algorithm to compute the exponential part of a formal fundamental matrix solution of a linear differential system. J. Appl. Alg. Eng. Commun. Comput. **8**(1), 1–23 (1997)
3. Barkatou, M.A.: An algorithm for computing a companion block diagonal form for a system of linear differential equations. AAECC **4**, 185–195 (1993)
4. Barkatou, M.A.: Rational Newton Algorithm for computing formal solutions of linear differential equations. In: Proceedings of the International Symposium on Algebraic Computation 1988. Lecture Notes in Computer Science, vol. 358. Springer, Berlin (1989)
5. Barkatou, M.A.: A rational version of Moser's algorithm. In Proceedings of the International Symposium on Symbolic and Algebraic Computation, pp. 297–302. ACM Press, New York (1995)
6. Barkatou, M.A., Pflügel, E.: Computing super-irreducible forms of systems of linear differential equations via Moser-reduction: a new approach. In: Proceedings of the International Symposium on Symbolic and Algebraic Computation, pp. 1-8. ACM Press, New York (2007)
7. Barkatou, M.A., Pflügel, E.: On the Moser- and super-reduction algorithms of systems of linear differential equations and their complexity. J. Symb. Comput. **44**(8), 1017–1036 (2009)
8. Pflügel, E.: An improved estimate for the maximal growth order of solutions of linear differential systems. Arch. Math **83**, 256–263 (2004)
9. Pflügel, E.: Effective formal reduction of linear differential systems. Appl. Algebra Eng. Commun. Comput. **10**, 153–187 (2000)
10. Cope, F.: Formal solutions of irregular linear differential equations. Am. J. Math. **58**, 130–140 (1936)
11. Deligne, P.: Equations différentielles à points singuliers réguliers. Lectures Notes in Mathematics, vol. 163. Springer, Berlin (1970)
12. Hilali, A., Wazner, A.: Un algorithme de calcul de l'invariant de Katz d'un système différentiel linéaire. Annales de l'Institut Fourier, Tome XXXVI-Fasicule **3**, 67–83 (1986)
13. Levelt, A.H.M. : Stabilizing differential operators: a method for computing invariants at irregular singularities. In: Singer, M. (ed.), Differential Equations and Computer Algebra, Computational Mathematics and Applications, pp. 181–228. Academic Press Ltd., Massachusetts (1991)
14. Lutz, D.A., Schäfke, R.: On the identification and stability of formal invariants for singular differential equations. Linear Algebra Appl. **72**, 1–46 (1985)
15. Malgrange, B.: Sur le réduction formelle des équations différentielles à singularités irrégulières, Preprint Grenoble (1979)
16. Miyake, M.: Exponential growth order of Moser irreducible system and a counterexample for Barkatou's conjucture. RIMS Kokyuroku, No. 2020 (2017)
17. Moser, J.: The order of a singularity in Fuchs' theory. Math. Z. **72**, 379–398 (1960)
18. Ramis, J.P.: Théorèmes d'indices Gevrey pour les équations différentielles ordinaires, pp. 57–60, Pub. IRMA, Strasbourg, France (1981)
19. Turritin, H.L.: Convergent solutions of ordinary linear homogeneous differential equations in the neighborhood of an irregular singular point. Acta Matm. **93**, 27–66 (1955)
20. Wasow, W.: Asymptotic Expansions For Ordinary Differential Equations. Interscience Publishers, New York (1965)

Part IV
Related Topics

Uniqueness Property for ρ-Analytic Functions

Grzegorz Łysik

Abstract We give necessary and sufficient geometric conditions in order that the triangle equality $\|x + y\| = \|x\| + \|y\|$ hold in a real normed vector space. Using these conditions we derive a uniqueness property for ρ-analytic functions on $\mathbb{R}^n$.

Keywords Analytic functions · Triangle equality · Uniqueness property

MSC Primary 26E05 · Secondary 46B20, 32A05

1 Introduction and Statement of the Main Result

In [1] we have given a characterization of analyticity of a function u in terms of integral means of u over Euclidean balls $B(x, R)$. Namely we have proved that a complex valued function u continuous on an open set $\Omega \subset \mathbb{R}^n$ is real analytic on Ω if and only if there exist functions $u_l \in C^0(\Omega, \mathbb{C})$ for $l \in \mathbb{N}_0$ and $\varepsilon \in C^0(\Omega, \mathbb{R}_+)$ such that

$$\frac{1}{|B(x, R)|} \int_{B(x,R)} u(y)\, dy = \sum_{l=0}^{\infty} u_l(x) R^l$$

locally uniformly in $\{(x, R) : x \in \Omega, 0 \le R < \varepsilon(x)\}$. This characterization justifies introduction of a definition of analytic functions on metric measure spaces, see [1, Definition 2]. On the other hand it is well known that real analytic functions possess the uniqueness property: *If a function vanishes on a nonempty open subset of a connected set Ω, then it vanishes on Ω.* The main aim of the paper is to prove that this property holds for ρ-analytic functions where ρ is a metric generated by an arbitrary norm on $\mathbb{R}^n$, see Theorem 1 below.

G. Łysik (✉)
Faculty of Mathematics and Natural Science, Jan Kochanowski University, Świętokrzyska 15, 25-406 Kielce, Poland
e-mail: glysik@ujk.edu.pl

G. Filipuk et al. (eds.), *Formal and Analytic Solutions of Diff. Equations*, Springer Proceedings in Mathematics & Statistics 256,
https://doi.org/10.1007/978-3-319-99148-1_12

Let Ω be an open subset of $\mathbb{R}^n$, $\|\cdot\|$ a norm on $\mathbb{R}^n$ and ρ the associated metric, i.e., $\rho(x, y) = \|x - y\|$ for $x, y \in \mathbb{R}^n$. For a function $u \in C^0(\Omega)$ define the *solid mean value function* by

$$M_\rho(u;\ x, R) = \frac{1}{|B_\rho(x, R)|} \int_{B_\rho(x,R)} u(y)\, dy, \quad x \in \Omega,\ \ 0 < R < \mathrm{dist}_\rho(x, \partial\Omega),$$

where $B_\rho(x, R)$ is a ball with respect to the metric ρ with center at x and radius R. For $R = 0$ set $M_\rho(u;\ x, 0) = u(x)$.

Definition 1 We say that a function $u \in C^0(\Omega)$ is *ρ-analytic on Ω* if there exist functions $u_l \in C^0(\Omega)$ for $l \in \mathbb{N}_0$ and $\varepsilon \in C^0(\Omega, \mathbb{R}_+)$ such that

$$M_\rho(u;\ x, R) = \sum_{l=0}^{\infty} u_l(x) R^l$$

locally uniformly in $\{(x, R) : x \in \Omega, 0 \le R < \varepsilon(x)\}$.

Theorem 1 *Let Ω be a connected open subset of $\mathbb{R}^n$ and let $u \in C^0(\Omega)$ be a ρ-analytic function on Ω. If u vanishes on a nonempty open set $U \subset \Omega$, then $u \equiv 0$ on Ω.*

2 The Triangle Equality

Let X be a normed vector space over $\mathbb{R}$. In the proof of Theorem 1 we need to know when the *triangle equality* holds, i.e., when for given elements $x, y \in X$ one has

$$\|x + y\| = \|x\| + \|y\|. \tag{1}$$

The problem of a characterization of vectors x and y for which the triangle equality (1) holds has been studied for different types of vector spaces, see [2–4] and references within. For instance, it is well known that if X is strictly convex, i.e., $\frac{x+y}{2}$ belongs to the open unit ball $B(1)$ for any x, y from the unit sphere $S(1)$, $x \neq y$, then (1) holds for nonzero vectors $x, y \in X$ if and only if $\frac{x}{\|x\|} = \frac{y}{\|y\|}$. In Banach space setting Nakamoto and Takashi proved in [3, Theorem 2] that (1) holds for nonzero vectors $x, y \in X$ if and only if there exists an extremal point f in the closed unit ball $\overline{B}^*(1)$ of the dual space X^* such that $f(x) = \|x\|$ and $f(y) = \|y\|$. However we were not able to find in the existing literature necessary and sufficient geometric conditions for the triangle equality (1) in an arbitrary normed vector space.

In this Section we give a necessary and sufficient geometric conditions for the triangle equality (1). Namely we have the following

Theorem 2 *Let $(X, \|\cdot\|)$ be a real normed vector space and $x, y \in X \setminus \{0\}$. Then the following conditions are equivalent:*

(a) *the triangle equality (1) holds;*
(b) *the unit sphere* $S(1)$ *contains the segment* $[\frac{x}{\|x\|}, \frac{y}{\|y\|}]$ *(possibly reduced to a point);*
(c) $x + y \neq 0$ *and* $\frac{x+y}{\|x+y\|} = \alpha \frac{x}{\|x\|} + (1-\alpha)\frac{y}{\|y\|}$ *with some* $0 < \alpha < 1$.

Proof $(b) \Rightarrow (a)$. By homogeneity of the norm we can assume that $\|x\| + \|y\| = 1$. Indeed if $\|x\| + \|y\| = \lambda > 0$, then putting $\widetilde{x} = x/\lambda$, $\widetilde{y} = y/\lambda$ we have $\|\widetilde{x}\| + \|\widetilde{y}\| = 1$, $x/\|x\| = \widetilde{x}/\|\widetilde{x}\|$ and $y/\|y\| = \widetilde{y}/\|\widetilde{y}\|$. So $\|x\| = \alpha$ and $\|y\| = 1-\alpha$ with some $0 < \alpha < 1$. Since by assumption for any $0 < \beta < 1$, $\|\beta \cdot x/\|x\| + (1-\beta) \cdot y/\|y\|\| = 1$ we get $\|x\| + \|y\| = 1 = \|\alpha \cdot x/\|x\| + (1-\alpha) \cdot y/\|y\|\| = \|x+y\|$.
$(a) \Rightarrow (b)$. Assume that $x/\|x\| \neq y/\|y\|$ and $S(1)$ does not contain the segment $[x/\|x\|, y/\|y\|]$. By the convexity of the closed unit ball $\overline{B}(1)$ there exists $0 < \alpha < 1$ such that $\|z_\alpha\| < 1$ where $z_\alpha = \alpha \cdot x/\|x\| + (1-\alpha) \cdot y/\|y\|$. Take a point $\zeta \in (x/\|x\|, z_\alpha]$. Then $\zeta = \beta \cdot x/\|x\| + (1-\beta) \cdot z_\alpha$ with some $0 \leq \beta < 1$. By the triangle inequality we get $\|\zeta\| \leq \beta + (1-\beta)\|z_\alpha\| < 1$. Analogously $\|\zeta\| < 1$ for any point $\zeta \in [z_\alpha, y/\|y\|)$. Hence for any $0 < \beta < 1$ we get

$$\left\|\beta \frac{x}{\|x\|} + (1-\beta)\frac{y}{\|y\|}\right\| < 1.$$

Putting $\beta = \frac{\|x\|}{\|x\|+\|y\|}$ in the above inequality we get

$$\left\|\frac{x}{\|x\|+\|y\|} + \frac{y}{\|x\|+\|y\|}\right\| < 1.$$

Hence $\|x+y\| < \|x\| + \|y\|$ and we get a contradiction.
$(a) \Rightarrow (c)$. The implication is clear since

$$\frac{x+y}{\|x+y\|} = \frac{\|x\|}{\|x+y\|}\frac{x}{\|x\|} + \frac{\|y\|}{\|x+y\|}\frac{y}{\|y\|} = \alpha\frac{x}{\|x\|} + (1-\alpha)\frac{y}{\|y\|} \tag{2}$$

with $0 < \alpha = \frac{\|x\|}{\|x+y\|} < 1$, and $\|x+y\| > 0$ if (a) holds.
$(c) \Rightarrow (a)$. If vectors x and y are linearly dependent, it can be assumed that $y = \lambda x$ with $\lambda \neq 0$ and $\lambda \neq -1$. Then we get $\frac{y}{\|y\|} = \mathrm{sgn}\lambda \frac{x}{\|x\|}$ and $\frac{x+y}{\|x+y\|} = \mathrm{sgn}(1+\lambda)\frac{x}{\|x\|}$. Hence (c) implies that

$$\mathrm{sgn}\lambda = \alpha + (1-\alpha)\mathrm{sgn}(1+\lambda)$$

with some $0 < \alpha < 1$ and this is possible only if $\lambda > 0$ in which case (1) clearly holds. On the other hand if x and y are linearly independent, then by (2) we get $\alpha = \frac{\|x\|}{\|x+y\|}$ and $1-\alpha = \frac{\|y\|}{\|x+y\|}$. So $1 = \frac{\|x\|}{\|x+y\|} + \frac{\|y\|}{\|x+y\|}$ and (a) follows. □

Recall that a point $z \in \overline{B}_\rho(0,r)$ is called an *extremal point* of $\overline{B}_\rho(0,r)$ if for any $x, y \in \overline{B}_\rho(0,r)$, $x \neq y$, and $\alpha \in [0,1]$ if $\alpha x + (1-\alpha)y = z$, then $\alpha = 0$ or $\alpha = 1$. Clearly extremal points belong to $S_\rho(0,r)$. Equivalently $z \in S_\rho(0,r)$ is an extremal point of $S_\rho(0, r)$ if $z = \alpha x + (1-\alpha)y$ with $x, y \in S_\rho(0, r)$ and $0 < \alpha < 1$, then $x = y = z$.

As a direct consequence of the definition of extremal points and Theorem 2 we get

Corollary 1 *Let $x, y \in X \setminus \{0\}$ be such that $x + y$ be an extremal point of $S_\rho(0, r)$ with some $r > 0$. Then $\|x + y\| = \|x\| + \|y\|$ if and only if $y = \lambda x$ with some $\lambda > 0$.*

3 Auxiliary Lemmas

In the proof of Theorem 1 we also need the following lemmas.

Lemma 1 *Let ρ be a metric on $\mathbb{R}^n$ associated with a norm $\|\cdot\|$, $0 < \delta < \sigma/2$ and let $z \in S_\rho(0, \sigma)$ be an extremal point of $S_\rho(0, \sigma)$. Let $W \subset S_\rho(0, 1)$ be a non-empty, open neighborhood of the point z/σ and let $\{A^i\}_{i=1}^\infty$ be a dense set of points in $S_\rho(0, \delta) \setminus \delta W$. Set*

$$T = \bigcup_{i=1}^{\infty} B(A^i, \sigma - \delta).$$

Then $\mathrm{dist}_\rho(z, T) = \inf_{y \in T} \|z - y\| > 0$ *and so $B_\rho(0, \sigma) \setminus T$ contains an open ball.*

Proof It is sufficient to show that there exists $c > 0$ such that $\|z - A^i\| > \sigma - \delta + c$ for any $i \in \mathbb{N}$. To the contrary assume that for any $j \in \mathbb{N}$ there exists $i \in \mathbb{N}$ such that $\sigma - \delta \le \|z - A^i\| \le \sigma - \delta + 1/j$. Due to the compactness of the set $S_\rho(0, \delta) \setminus W$ one can choose a subsequence $\{A^{i_j}\}_{j=1}^\infty$ of $\{A^i\}_{i=1}^\infty$ convergent to a point $A^\infty \in S_\rho(0, \delta) \setminus W$ such that $\|z - A^\infty\| = \sigma - \delta$. The triangle inequality gives $\sigma = \|z\| = \|z - A^\infty + A^\infty\| \le \|z - A^\infty\| + \|A^\infty\| = \sigma - \delta + \delta$. So we get the triangle equality $\|z - A^\infty + A^\infty\| = \|z - A^\infty\| + \|A^\infty\| = \sigma$, which by Theorem 2 implies that $z/\|z\|$ lies in the segment $[\frac{A^\infty}{\|A^\infty\|}, \frac{z - A^\infty}{\|z - A^\infty\|}]$. Since $z/\|z\|$ is an extremal point of $S_\rho(0, 1)$ this segment reduces to the point $z/\|z\|$, which implies that $A^\infty = \delta z/\sigma$. So the sequence $\{A^{i_j}\}_{j=1}^\infty \subset S_\rho(0, \delta) \setminus W$ converges to $\delta z/\sigma \in W$, which gives a contradiction. The last statement is clear. □

Lemma 2 *Let $z \in B_\rho(0, r + \sigma) \setminus B_\rho(0, r)$ with $0 < \sigma < r/2$. Then there exists $A \in B_\rho(0, r)$ such that z is an extremal point of $S_\rho(A, \sigma)$.*

Proof If z is an extremal point of $S_\rho(0, \|z\|)$ take $A = \left(1 - \frac{\sigma}{\|z\|}\right)z$. Then $A \in B_\rho(0, r)$ and z is an extremal point of $S_\rho(A, \sigma)$ since $z - A = \frac{\sigma}{\|z\|}z$.
Otherwise there exist extremal points $x_1, x_2 \in S_\rho(0, \|z\|)$, $x_1 \ne x_2$, such that $z = \alpha x_1 + (1-\alpha)x_2$ with some $0 < \alpha \le 1/2$. Since $x_2 - z = \alpha(x_2 - x_1)$ we have $\alpha = \frac{\|x_2 - z\|}{\|x_2 - x_1\|}$ and $1 - \alpha = \frac{\|z - x_1\|}{\|x_2 - x_1\|}$. Set

$$A = \alpha x_1 + \beta x_2 \quad \text{with } \beta = 1 - \alpha - \frac{\sigma}{\|x_2\|} > 0.$$

Then by Theorem 2, $\|A\| = \|\alpha x_1 + \left(1 - \alpha - \frac{\sigma}{\|x_2\|}\right)x_2\| = \alpha\|x_1\| + \left(1 - \alpha - \frac{\sigma}{\|x_2\|}\right)$ $\|x_2\| = \|z\| - \sigma < r$ since $S_\rho(0, \|z\|)$ contains the segment $[x_1, x_2]$. Hence $A \in B_\rho(0, r)$ and $z = A + \frac{\sigma}{\|x_2\|}x_2$ is an extremal point of $S_\rho(A, \sigma)$. □

4 Proof of Theorem 1

Proof It is sufficient to show that if u vanishes on a ball $B_\rho(x, r) \Subset \Omega$ of radius $r > 0$ and a center $x \in \Omega$, then it vanishes on a ball $B_\rho(x, r + \sigma)$ with some $\sigma > 0$. Clearly we can assume that $x = 0$, i.e., u vanishes on a ball $B_\rho(0, r) \Subset \Omega$ with some $r > 0$. Since u is ρ-analytic on Ω there exist functions $u_l \in C^0(\Omega)$ for $l \in \mathbb{N}_0$ and $\varepsilon \in C^0(\Omega; \mathbb{R}_+)$ such that for any $x \in \overline{B}_\rho(0, r)$ and $0 < R < R_0 := \inf_{x \in \overline{B}_\rho(0,r)} \varepsilon(x)$,

$$\frac{1}{|B_\rho(x, R)|} \int_{B_\rho(x,R)} u(y)dy = \sum_{l=0}^{\infty} u_l(x)R^l.$$

Note that if $x \in B_\rho(0, r)$, then $\int_{B_\rho(x,R)} u(y)dy = 0$ for small $R > 0$. By the analyticity of the function $(0, R_0) \ni R \mapsto \sum_{l=0}^{\infty} u_l(x)R^l$ we conclude that for any $x \in B_\rho(0, r)$ and $0 < R < R_0$,

$$\int_{B_\rho(x,R)} \operatorname{Re} u(y)dy = 0.$$

Take a point $z \in B_\rho(0, r + \sigma) \setminus \overline{B}_\rho(0, r)$ where $0 < \sigma < \min(\frac{r}{2}, R_0)$ and assume that $u(z) \neq 0$. Multiplying eventually by a constant one can assume that $\operatorname{Re} u(z) > 0$. By the continuity of u one can find $0 < \delta < \min(\sigma, \|z\|/2, r + \sigma - \|z\|, \|z\| - r)$ such that $\operatorname{Re} u(x) > 0$ for $x \in B_\rho(z, \delta)$. By Lemma 2 we can find a point $A^0 \in B_\rho(0, r)$ such that z is an extremal point of $\overline{B}_\rho(A^0, \sigma)$. By the homogeneity of the norm $\|\cdot\|$ and translation invariance of the metric ρ it follows that for every $N \in \mathbb{N}$ there exists an open subset W_N of $S_\rho(0, 1)$ such that

$$S_\rho(A^0, \sigma) \cap B_\rho(z, \delta/N) = \{x \in \mathbb{R}^n : x = A^0 + \sigma w \text{ with } w \in W_N\}.$$

Clearly W_N is nonempty since $(z - A^0)/\sigma \in W_N$. For $0 < d < \min(\delta, r - \|A^0\|)$ let $\{A^i\}_{i \in \mathbb{N}}$ be a dense countable subset of $S_\rho(A^0, d) \setminus (A^0 + dW_N)$. Then $\|A^i\| \leq \|A^0\| + d < r$ for $i \in \mathbb{N}$. So $A^i \in B_\rho(0, r)$ for $i \in \mathbb{N}$. Consider the set

$$T_N = \bigcup_{i=1}^{\infty} B_\rho(A^i, \sigma - d).$$

If N is sufficiently big, then $B_\rho(A^0, \sigma) \setminus \overline{B}_\rho(z, \delta) \subset T_N \subset B(A^0, \sigma)$. Since the integrals of $\operatorname{Re} u$ over $B_\rho(A^0, \sigma)$ and over $B_\rho(A^i, \sigma - d)$, $i \in \mathbb{N}$, vanish we conclude that the integral of Re u over $B_\rho(A^0, \sigma) \setminus \overline{T}_N \subset B_\rho(z, \delta)$ vanishes. On the other hand, by Lemma 1, the set $B_\rho(A^0, \sigma) \setminus \overline{T}_N$ has a positive Lebesgue measure, which gives a contradiction. Hence u vanishes on $B_\rho(0, r + \sigma)$ and the proof is finished. □

References

1. Łysik, G.: A characterization of real analytic functions. Ann. Acad. Sci. Fenn. Math. **43**, 475–482 (2018)
2. Barraa, M., Boumazgour, M.: Inner derivations and norm equality. Proc. Am. Math. Soc. **130**, 471–476 (2002)
3. Nakamoto, R., Takahashi, S.: Norm equality condition in triangular inequality. Sci. Math. Jpn. **55**, 463–466 (2002)
4. Rajić, R.: Characterization of the norm triangle equality in pre-Hilbert C*-modules and applications. J. Math. Inequalities **3**, 347–355 (2009)

On the Algebraic Study of Asymptotics

Naofumi Honda and Luca Prelli

Abstract The aim of this paper is to review a new functorial interpretation of asymptotics, which allows the construction of new sheaves of multi-asymptotically developable functions closely related with asymptotics along a subvariety with a simple singularity such as a cusp. This requires some new geometrical and combinatorial notions underlying the multi-normal deformation of a real analytic manifold and the construction of the multi-specialization functor along a family of submanifolds.

Keywords Algebraic analysis · Asymptotic expansions · Specialization · Normal deformation · Subanalytic sheaves

MSC Primary 32C38 · Secondary 35A27, 41A60, 34M30

1 Introduction

Asymptotically developable expansions of holomorphic functions on a sector are an important tool to study ordinary differential equations with irregular singularities.

In higher dimension H. Majima introduced in [1] the notion of strongly asymptotically developable functions along a normal crossing divisor. These functions are related with Whitney holomorphic functions on a multi-sector, as proven in [2].

The aim of this paper is to report on the generalization and the functorial construction of asymptotics, with the notion of multi-asymptotic expansions. Locally we can construct new sheaves of multi-asymptotically developable functions closely

N. Honda
Department of Mathematics, Faculty of Science, Hokkaido University,
Sapporo 060-0810, Japan
e-mail: honda@math.sci.hokudai.ac.jp

L. Prelli (✉)
Dipartimento di Matematica, Università degli studi di Padova,
Via Trieste 63, 35121 Padova, Italy
e-mail: lprelli@math.unipd.it

G. Filipuk et al. (eds.), *Formal and Analytic Solutions of Diff. Equations*,
Springer Proceedings in Mathematics & Statistics 256,
https://doi.org/10.1007/978-3-319-99148-1_13

related with asymptotics along a subvariety with a simple singularity such as a cusp. The results are extracted from [3].

2 Sheaves on a Subanalytic Site

In order to perform a functorial costruction of the asymptotics, we need a sheaf theoretical interpretation of Whithey holomorphic functions. Whitney holomorphic functions do not define a sheaf in the usual sense (i.e. with the usual topology), since they do not satisfy gluing conditions.

Example 1 Let us consider z^{-1} on $\mathbb{C} \setminus \{0\}$. Then it defines a Whithey holomorphic function on each set $B_n = \{n^{-1} < |z| < 1\}$, but it does not define a Whithey function on $\bigcup_n B_n = \{|z| < 1\} \setminus \{0\}$.

In order to overcome these kind of problems Kashiwara and Schapira introduced in [4] the subanalytic site: they equipped an analytic manifold with a Grothendieck topology. Denote by $\mathrm{Op}(X_{sa})$ the category of open subanalytic subsets of X. One endows $\mathrm{Op}(X_{sa})$ with the following topology: $S \subset \mathrm{Op}(X_{sa})$ is a covering of $U \in \mathrm{Op}(X_{sa})$ if for any compact K of X there exists a finite subset $S_0 \subset S$ such that $K \cap \bigcup_{V \in S_0} V = K \cap U$. We will call X_{sa} the subanalytic site.

In this new setting Whitney holomorphic functions form a sheaf in the (derived category of) subanalytic site associated to an analytic manifold.

Let $\mathrm{Mod}(\mathbb{C}_{X_{sa}})$ denote the category of sheaves on X_{sa} and let $\mathrm{Mod}_{\mathbb{R}\text{-c}}(\mathbb{C}_X)$ be the abelian category of $\mathbb{R}$-constructible sheaves on X. We denote by $\rho : X \to X_{sa}$ the natural morphism of sites. The inverse image ρ^{-1} sends subanalytic sheaves to the category of sheaves on X with its usual topology.

Let X, Y be two real analytic manifolds, and let $f : X \to Y$ be a real analytic map. The six Grothendieck operations $\mathcal{H}om$, $\otimes$, f^{-1}, f_*, $f_{!!}$ and $f^!$ are well defined for subanalytic sheaves (in the derived category). Refer to [4] or [5] for more details.

3 Multi-Normal Deformation

We refer to [6] for the definition of the classical normal deformation. For simplicity, we assume $X = \mathbb{C}^n$, with coordinates $z = (z_1, \dots, z_n)$. Let $\chi = \{M_1, \dots, M_\ell\}$ be a family of submanifolds, $M_j = \{z_i = 0,\ i \in I_j\}$, $I_j \subseteq \{1, \dots, n\}$. We associate to χ an action $\mu_j(z, \lambda) = (\lambda^{a_{j1}} z_1, \dots, \lambda^{a_{jn}} z_n)$ with $a_{ji} \in \mathbb{N}_0$ ($\mathbb{N}_0 := \mathbb{N} \cup \{0\}$), $a_{ji} \neq 0$ if $i \in I_j$, $a_{ji} = 0$ otherwise. We call A_χ the matrix (a_{ji}) associated to the action.

Let $A_\chi = (a_{ji})$ be an $\ell \times n$ matrix with $a_{ji} \in \mathbb{N}_0$, $a_{ji} \neq 0$ if $i \in I_j$, $a_{ji} = 0$ otherwise. We can define a general normal deformation $\widetilde{X} = \mathbb{C}^n \times \mathbb{C}^\ell$ with the map $p : \widetilde{X} \to X$ defined by

$$p(x, t) = (\varphi_1(t)x_1, \dots, \varphi_n(t)x_n)$$

with

$$\varphi_i(t) = \prod_{j=1}^{\ell} t_j^{a_{ji}} \qquad (i = 1, 2, \ldots, n). \tag{1}$$

Comparing with the matrix A_χ, when $t \in (\mathbb{R}^+)^\ell$ we have

$$(\log \varphi_1, \ldots, \log \varphi_n) = (\log t_1, \ldots, \log t_\ell) A_\chi.$$

Set $S_\chi = \{t_1 = \cdots = t_\ell = 0\}$. Let $s : S_\chi \hookrightarrow \widetilde{X}$ be the inclusion, $\Omega = \{t_1, \ldots, t_\ell > 0\}$, $M = \bigcap_{i=1}^{\ell} M_i$. We get a commutative diagram

$$\begin{array}{ccccc} S_\chi & \xrightarrow{s} & \widetilde{X} & \xleftarrow{i_\Omega} & \Omega \\ \downarrow \tau & & \downarrow p & \swarrow \tilde{p} & \\ M & \xrightarrow{i} & X. & & \end{array}$$

For simplicity we assume that $\ell \leq n$ and the $\ell \times \ell$ submatrix $A_{\chi\ell}$ made from the first ℓ-columns and the first ℓ-rows in A_χ is invertible. We are interested in the zero section S_χ of $\widetilde{X}$ defined by $\{t_i = 0,\ i = 1, \ldots, \ell\}$. In particular (for simplicity) points $\xi = (\xi_1, \ldots, \xi_n)$, $\xi_i \neq 0$, $i = 1, \ldots, \ell$, in S_χ.

Example 2 Let us consider some examples in $\mathbb{C}^2$.

(Majima) Let $X = \mathbb{C}^2 \ni (z_1, z_2)$, $M_i = \{z_i = 0\}$, $i = 1, 2$. Consider the matrix

$$A_\chi = \begin{pmatrix} 1 & 0 \\ 0 & 1 \end{pmatrix}, \quad \varphi_1 = t_1, \quad \varphi_2 = t_2.$$

We have $\widetilde{X} = (z_1, z_2, t_1, t_2)$, $p : (z_1, z_2, t_1, t_2) \to (z_1 t_1, z_2 t_2)$.
(Takeuchi) Let $X = \mathbb{C}^2 \ni (z_1, z_2)$, $M_1 = \{0\}$, $M_2 = \{z_2 = 0\}$. Consider the matrix

$$A_\chi = \begin{pmatrix} 1 & 1 \\ 0 & 1 \end{pmatrix}, \quad \varphi_1 = t_1, \quad \varphi_2 = t_1 t_2.$$

We have $\widetilde{X} = (z_1, z_2, t_1, t_2)$, $p : (z_1, z_2, t_1, t_2) \to (z_1 t_1, z_2 t_1 t_2)$. This is the binormal deformation of [7].
(Cusp) Let $X = \mathbb{C}^2 \ni (z_1, z_2)$, $M_1 = M_2 = \{0\}$. Consider the matrix

$$A_\chi = \begin{pmatrix} 3 & 2 \\ 1 & 1 \end{pmatrix}, \quad \varphi_1 = t_1^3 t_2, \quad \varphi_2 = t_1^2 t_2.$$

We have $\widetilde{X} = (z_1, z_2, t_1, t_2)$, $p : (z_1, z_2, t_1, t_2) \to (z_1 t_1^3 t_2, z_2 t_1^2 t_2)$.
(Generalized cusp) Let $X = \mathbb{C}^2 \ni (z_1, z_2)$, $M_1 = M_2 = \{0\}$. Let $k \in \mathbb{N}$ and consider the matrix

$$A_\chi = \begin{pmatrix} 2k+1 & 2 \\ k & 1 \end{pmatrix}, \quad \varphi_1 = t_1^{2k+1} t_2^k, \quad \varphi_2 = t_1^2 t_2.$$

We have $\widetilde{X} = (z_1, z_2, t_1, t_2)$, $p : (z_1, z_2, t_1, t_2) \to (z_1 t_1^{2k+1} t_2^k, z_2 t_1^2 t_2)$.

4 Multi-Sectors

Let $\xi = (\xi_1, \dots, \xi_n) \in S_\chi$ with $\xi_i \neq 0$, $i = 1, \dots, \ell$. Let $\epsilon > 0$, and let $W = W_1 \times \cdots \times W_n$, W_i open conic cone in $\mathbb{C}$ containing the direction ξ_i. Set $|z|_\ell = (|z_1|, \dots, |z_\ell|)$. A multi-sector $S(W, \epsilon)$ is an element of the family $C(\xi)$ defined as follows:

$$S(W,\epsilon) = \left\{ (z_1, \dots, z_n); \begin{array}{l} z_i \in W_i \quad (i = 1, \dots, n), \\ \varphi_i^{-1}(|z|_\ell) < \epsilon \quad (i \leq \ell), \\ |\xi_i| - \epsilon < \dfrac{|z_i|}{\varphi_i(\varphi^{-1}(|z|_\ell))} < |\xi_i| + \epsilon \quad (i > \ell) \end{array} \right\},$$

where $\epsilon > 0$, and W_i are cones in $\mathbb{C}$ containing the direction ξ_i and φ_i^{-1} is a rational polynomial in ℓ variables such that $\varphi_i(\varphi^{-1}(z)) = z_i$, $i = 1, \dots, \ell$. Comparing with the matrix $A_{\chi\ell}$, when $t \in (\mathbb{R}^+)^\ell$ we have

$$(\log \varphi_1^{-1}, \dots, \log \varphi_\ell^{-1}) = (\log t_1, \dots, \log t_\ell) A_{\chi\ell}^{-1}.$$

We say that $S(W', \epsilon') < S(W, \epsilon)$ (properly contained) if $\overline{W'} \setminus \{0\} \subset W$ and $\epsilon' < \epsilon$. The main geometrical properties of a multi-sector S are the following:

- S is locally cohomologically trivial. That is, $\mathrm{RHom}(\mathbb{C}_S; \mathbb{C}_X) = \mathbb{C}_{\overline{S}}$,
- S is 1-regular, that is, there exists a constant $C > 0$ satisfying that, for any point p and q in S, there exists a rectifiable curve in S which joins p and q and whose length is $\leq C|p - q|$.

Example 3 Let us consider some examples in $\mathbb{C}^2$.

(Majima) Let $X = \mathbb{C}^2 \ni (z_1, z_2)$, $M_i = \{z_i = 0\}$, $i = 1, 2$. Then

$$A_\chi^{-1} = \begin{pmatrix} 1 & 0 \\ 0 & 1 \end{pmatrix}, \quad \varphi_1^{-1} = t_1, \quad \varphi_2^{-1} = t_2.$$

We have

$$C(\xi) \ni S(W, \epsilon) = \left\{ z \in X; \begin{array}{l} z_i \in W_i \quad (i = 1, 2), \\ |z_i| < \epsilon \quad (i = 1, 2) \end{array} \right\},$$

where $\epsilon > 0$ and W_i conic open subset containing ξ_i.

(Takeuchi) Let $X = \mathbb{C}^2 \ni (z_1, z_2)$, $M_1 = \{0\}$, $M_2 = \{z_2 = 0\}$. Then

$$A_\chi^{-1} = \begin{pmatrix} 1 & -1 \\ 0 & 1 \end{pmatrix}, \quad \varphi_1^{-1} = t_1, \quad \varphi_2^{-1} = \frac{t_2}{t_1}.$$

We have

$$C(\xi) \ni S(W, \epsilon) = \left\{ z \in X;\ \begin{array}{c} z_i \in W_i \quad (i = 1, 2), \\ |z_1| < \epsilon, \\ |z_2| < \epsilon |z_1| \end{array} \right\},$$

where $\epsilon > 0$ and W_i conic open subset containing ξ_i. These are the multi-sectors of [7].
(Cusp) Let $X = \mathbb{C}^2 \ni (z_1, z_2)$, $M_1 = M_2 = \{0\}$. Then

$$A_\chi^{-1} = \begin{pmatrix} 1 & -2 \\ -1 & 3 \end{pmatrix}, \quad \varphi_1^{-1} = \frac{t_1}{t_2}, \quad \varphi_2^{-1} = \frac{t_2^3}{t_1^2}.$$

We have

$$C(\xi) \ni S(W, \epsilon) = \left\{ z \in X;\ \begin{array}{c} z_i \in W_i \quad (i = 1, 2), \\ |z_1| < \epsilon |z_2|, \\ |z_2|^3 < \epsilon |z_1|^2 \end{array} \right\},$$

where $\epsilon > 0$ and W_i conic open subset containing ξ_i.
(Generalized cusp) Let $X = \mathbb{C}^2 \ni (z_1, z_2)$, $M_1 = M_2 = \{0\}$. Then

$$A_\chi^{-1} = \begin{pmatrix} 1 & -2 \\ -k & 2k+1 \end{pmatrix}, \quad \varphi_1^{-1} = \frac{t_1}{t_2^k}, \quad \varphi_2^{-1} = \frac{t_2^{2k+1}}{t_1^2}.$$

We have

$$C(\xi) \ni S(W, \epsilon) = \left\{ z \in X;\ \begin{array}{c} z_i \in W_i \quad (i = 1, 2), \\ |z_1| < \epsilon |z_2|^k, \\ |z_2|^{2k+1} < \epsilon |z_1|^2 \end{array} \right\},$$

where $\epsilon > 0$ and W_i conic open subset containing ξ_i.

5 Multi-Specialization

Let $\rho : S_{\chi sa} \to S_\chi$ denote the natural functor of sites. The multi-specialization along χ is the functor

$$\begin{aligned} \nu_\chi : D^b(\mathbb{C}_{X_{sa}}) &\to D^b(\mathbb{C}_{S_\chi}) \\ F &\mapsto \rho^{-1} s^{-1} R\Gamma_\Omega p^{-1} F. \end{aligned}$$

(Here D^b denotes the bounded derived category.) Thanks to the functor $\rho^{-1} : D^b(\mathbb{C}_{S_{sa}}) \to D^b(\mathbb{C}_S)$ we can calculate the fibers at $\xi \in S_\chi$ which are given by

$$(H^j \nu_\chi F)_\xi \simeq \varinjlim_{S(W,\epsilon)} H^j(S(W,\epsilon); F),$$

where $S(W,\epsilon)$ ranges through the family $C(\xi)$.

Let $\mathcal{O}_X^{\mathrm{w}} \in D^b(\mathbb{C}_{X_{sa}})$ denote the subanalytic sheaf of Whitney holomorphic functions. The sheaf of multi-asymptotically developable holomorphic functions is the multi-specialization $\nu_\chi \mathcal{O}_X^{\mathrm{w}}$ of Whitney holomorphic functions.

6 Multi-Asymptotics

Let $\mathcal{P}_\ell$ be the set of nonempty subsets of $\{1, \ldots, \ell\}$. Let $J \in \mathcal{P}_\ell$. We use the following notations:

- $I_J = \bigcup_{j \in J} I_j$,
- $M_J = \bigcap_{j \in J} M_j$,
- $z_J = (z_i)_{i \in I_J}$, $z_J^C = (z_i)_{i \notin I_J}$,
- $\mathbb{N}_0^J = \left\{(\alpha_1, \ldots, \alpha_n) \in \mathbb{N}_0^n,\ \alpha_i = 0,\ i \notin I_J\right\}$,
- $\pi_J : X \to M_J$ the projection,
- given $S \subset X$, $S_J = \pi_J(S)$.

Let $S := S(W,\epsilon)$ be a multi-sector. We say that $F = \{F_J\}_{J \in \mathcal{P}_\ell}$ is a total family of coefficients of multi-asymptotic expansion along χ on S if each F_J consists of a family $\left\{f_{J,\alpha}\right\}_{\alpha \in \mathbb{N}_0^J}$ of holomorphic functions on S_J.

Given a total family of coefficients $F = \{F_J\}_{J \in \mathcal{P}_\ell}$ and $N = (n_1, \ldots, n_\ell) \in \mathbb{N}_0^\ell$, the approximate function of degree N of F is

$$\mathrm{App}^{<N}(F; z) = \sum_{J \in \mathcal{P}_\ell} (-1)^{\sharp J + 1} \sum_{\alpha \in A_J(N)} \frac{f_{J,\alpha}(z_J^C)}{\alpha!} z^\alpha$$

where

$$A_J(N) = \left\{ \alpha \in \mathbb{N}_0^J;\ \sum_{i \in I_j} a_{ji}\alpha_i < n_j \ \text{ for any } j \in J \right\}.$$

We say that f is multi-asymptotically developable to $F = \{F_J\}$ along χ on $S = S(W, \epsilon)$ if and only if for any cone $S' = S(W', \epsilon')$ properly contained in S and for any $N = (n_1, \ldots, n_\ell) \in \mathbb{N}_0^\ell$, there exists a constant C such that

$$\left| f(z) - \mathrm{App}^{<N}(F; z) \right| \leq C \prod_{1 \leq j \leq \ell} \varphi_j^{-1}(|z|_\ell)^{n_j} \qquad (z \in S').$$

Example 4 Let us consider some examples in $\mathbb{C}^2$.

(Majima) Let $M_{\{1\}} = \{z_1 = 0\}$, $M_{\{2\}} = \{z_2 = 0\}$, $M_{\{1,2\}} = \{0\}$,

$$S(W, \epsilon) = \left\{ z \in X;\ \begin{array}{ll} z_i \in W_i & (i = 1, 2), \\ |z_i| < \epsilon & (i = 1, 2) \end{array} \right\}.$$

We have $S_{\{1,2\}} = \{\mathrm{pt}\}$ and

$$S_{\{1\}} = \left\{ z \in M_1;\ \begin{array}{l} z_2 \in W_2, \\ |z_2| < \epsilon \end{array} \right\}, \quad S_{\{2\}} = \left\{ z \in M_2;\ \begin{array}{l} z_1 \in W_1, \\ |z_1| < \epsilon \end{array} \right\}.$$

A total family of coefficients is

$$F = \left\{ \{f_{\{1\},\alpha_1}(z_2)\}_{\alpha_1 \in \mathbb{N}_0}, \{f_{\{2\},\alpha_2}(z_1)\}_{\alpha_2 \in \mathbb{N}_0}, \{f_{\{1,2\},\alpha}\}_{\alpha \in \mathbb{N}_0^2} \right\},$$

where $f_{\{1\},\alpha_1}(z_2)$ (resp. $f_{\{2\},\alpha_2}(z_1)$) is holomorphic in $S_{\{1\}}$ (resp. $S_{\{2\}}$) and $f_{\{1,2\},\alpha} \in \mathbb{C}$. Let $N = (n_1, n_2) \in \mathbb{N}_0^2$. We have

$$\begin{aligned} \mathrm{App}^{<N}(F; z) = &\sum_{\alpha_1 < n_1} f_{\{1\},\alpha_1}(z_2) \frac{z_1^{\alpha_1}}{\alpha_1!} + \sum_{\alpha_2 < n_2} f_{\{2\},\alpha_2}(z_1) \frac{z_2^{\alpha_2}}{\alpha_2!} \\ &- \sum_{\substack{\alpha_1 < n_1 \\ \alpha_2 < n_2}} f_{\{1,2\},\alpha} \frac{z_1^{\alpha_1} z_2^{\alpha_2}}{\alpha_1! \alpha_2!}. \end{aligned}$$

A holomorphic function f is strongly asymptotically developable if, for any multi-sector S' properly contained in $S(W, \epsilon)$ and for any $N = (n_1, n_2) \in \mathbb{N}_0^2$, there exists a positive constant $C_{S',N}$ such that

$$\left| f(z) - \mathrm{App}^{<N}(F; z) \right| \leq C_{S',N} |z_1|^{n_1} |z_2|^{n_2}$$

with $z \in S'$. This corresponds to Majima's asymptotics of [1].

(Takeuchi) Let $M_{\{1\}} = \{0\}$, $M_{\{2\}} = \{z_2 = 0\}$, $M_{\{1,2\}} = \{0\}$,

$$S(W,\epsilon) = \left\{ z \in X;\ \begin{array}{l} z_i \in W_i \quad (i = 1,2), \\ |z_1| < \epsilon, \\ |z_2| < \epsilon |z_1| \end{array} \right\}.$$

We have $S_{\{1\}} = S_{\{1,2\}} = \{\mathrm{pt}\}$ and

$$S_{\{2\}} = \left\{ z \in M_2;\ \begin{array}{l} z_1 \in W_1, \\ |z_1| < \epsilon \end{array} \right\}.$$

A total family of coefficients is

$$F = \left\{ \{f_{\{1\},\alpha}\}_{\alpha \in \mathbb{N}_0^2}, \{f_{\{2\},\alpha_2}(z_1)\}_{\alpha_2 \in \mathbb{N}_0}, \{f_{\{1,2\},\alpha}\}_{\alpha \in \mathbb{N}_0^2} \right\},$$

where $f_{\{2\},\alpha_2}(z_1)$ is holomorphic in $S_{\{2\}}$ and $f_{\{1\},\alpha}, f_{\{1,2\},\alpha} \in \mathbb{C}$. Let $N = (n_1, n_2) \in \mathbb{N}_0^2$. We have

$$\begin{aligned} \mathrm{App}^{<N}(F; z) = & \sum_{\alpha_1 + \alpha_2 < n_1} f_{\{1\},\alpha} \frac{z_1^{\alpha_1} z_2^{\alpha_2}}{\alpha_1! \alpha_2!} + \sum_{\alpha_2 < n_2} f_{\{2\},\alpha_2}(z_1) \frac{z_2^{\alpha_2}}{\alpha_2!} \\ & - \sum_{\substack{\alpha_1 + \alpha_2 < n_1 \\ \alpha_2 < n_2}} f_{\{1,2\},\alpha} \frac{z_1^{\alpha_1} z_2^{\alpha_2}}{\alpha_1! \alpha_2!}. \end{aligned}$$

A holomorphic function f is strongly asymptotically developable if, for any multi-sector S' properly contained in $S(W,\epsilon)$ and for any $N = (n_1, n_2) \in \mathbb{N}_0^2$, there exists a positive constant $C_{S',N}$ such that

$$\left| f(z) - \mathrm{App}^{<N}(F; z) \right| \le C_{S',N} |z_1|^{n_1 - n_2} |z_2|^{n_2}$$

with $z \in S'$. This corresponds to Takeuchi's asymptotics of [8].

(Cusp) Let $M_{\{1\}} = M_{\{2\}} = M_{\{1,2\}} = \{0\}$,

$$S(W,\epsilon) = \left\{ z \in X;\ \begin{array}{l} z_i \in W_i \quad (i = 1,2), \\ |z_1| < \epsilon |z_2|, \\ |z_2|^3 < \epsilon |z_1|^2 \end{array} \right\}.$$

We have $S_{\{1\}} = S_{\{2\}} = S_{\{1,2\}} = \{\mathrm{pt}\}$. A total family of coefficients is given by

$$F = \left\{ \{f_{\{1\},\alpha}\}_{\alpha \in \mathbb{N}_0^2}, \{f_{\{2\},\alpha}\}_{\alpha \in \mathbb{N}_0^2}, \{f_{\{1,2\},\alpha}\}_{\alpha \in \mathbb{N}_0^2} \right\},$$

where $f_{\{1\},\alpha}, f_{\{2\},\alpha}, f_{\{1,2\},\alpha} \in \mathbb{C}$. Let $N = (n_1, n_2) \in \mathbb{N}_0^2$. We have

$$\begin{aligned}\mathrm{App}^{<N}(F; z) = &\sum_{3\alpha_1+2\alpha_2<n_1} f_{\{1\},\alpha}\frac{z_1^{\alpha_1} z_2^{\alpha_2}}{\alpha_1!\alpha_2!} + \sum_{\alpha_1+\alpha_2<n_2} f_{\{2\},\alpha}\frac{z_1^{\alpha_1} z_2^{\alpha_2}}{\alpha_1!\alpha_2!}\\ &- \sum_{\substack{3\alpha_1+2\alpha_2<n_1\\ \alpha_1+\alpha_2<n_2}} f_{\{1,2\},\alpha}\frac{z_1^{\alpha_1} z_2^{\alpha_2}}{\alpha_1!\alpha_2!}\end{aligned}$$

A holomorphic function f is strongly asymptotically developable if, for any multi-sector S' properly contained in $S(W, \epsilon)$ and for any $N = (n_1, n_2) \in \mathbb{N}_0^2$, there exists a positive constant $C_{S',N}$ such that

$$\left|f(z) - \mathrm{App}^{<N}(F; z)\right| \le C_{S',N}|z_1|^{n_1-2n_2}|z_2|^{3n_2-n_1}$$

with $z \in S'$.

(Generalized cusp) Let $M_{\{1\}} = M_{\{2\}} = M_{\{1,2\}} = \{0\}$,

$$S(W, \epsilon) = \left\{ z \in X;\ \begin{array}{c} z_i \in W_i \quad (i = 1, 2),\\ |z_1| < \epsilon|z_2|^k,\\ |z_2|^{2k+1} < \epsilon|z_1|^2 \end{array} \right\}.$$

We have $S_{\{1\}} = S_{\{2\}} = S_{\{1,2\}} = \{\mathrm{pt}\}$. A total family of coefficients is given by

$$F = \left\{\{f_{\{1\},\alpha}\}_{\alpha\in\mathbb{N}_0^2}, \{f_{\{2\},\alpha}\}_{\alpha\in\mathbb{N}_0^2}, \{f_{\{1,2\},\alpha}\}_{\alpha\in\mathbb{N}_0^2}\right\},$$

where $f_{\{1\},\alpha}, f_{\{2\},\alpha}, f_{\{1,2\},\alpha} \in \mathbb{C}$. Let $N = (n_1, n_2) \in \mathbb{N}_0^2$. We have

$$\begin{aligned}\mathrm{App}^{<N}(F; z) = &\sum_{(2k+1)\alpha_1+2\alpha_2<n_1} f_{\{1\},\alpha}\frac{z_1^{\alpha_1} z_2^{\alpha_2}}{\alpha_1!\alpha_2!} + \sum_{k\alpha_1+\alpha_2<n_2} f_{\{2\},\alpha}\frac{z_1^{\alpha_1} z_2^{\alpha_2}}{\alpha_1!\alpha_2!}\\ &- \sum_{\substack{(2k+1)\alpha_1+2\alpha_2<n_1\\ k\alpha_1+\alpha_2<n_2}} f_{\{1,2\},\alpha}\frac{z_1^{\alpha_1} z_2^{\alpha_2}}{\alpha_1!\alpha_2!}\end{aligned}$$

A holomorphic function f is strongly asymptotically developable if, for any multi-sector S' properly contained in $S(W, \epsilon)$ and for any $N = (n_1, n_2) \in \mathbb{N}_0^2$, there exists a positive constant $C_{S',N}$ such that

$$\left|f(z) - \mathrm{App}^{<N}(F; z)\right| \le C_{S',N}|z_1|^{n_1-2n_2}|z_2|^{(2k+1)n_2-kn_1}$$

with $z \in S'$.

7 Multi-Specialization and Multi-Asymptotics

One can check that multi-asymptotics on a multi-sector S are Whitney holomorphic functions on $S' < S$. Moreover the geometrical properties of a multi-sector imply vanishing of the cohomology of multi-specialization. Combining these two results we have that the sheaf $\nu_\chi \mathcal{O}_X^{\mathrm{w}}$ is concentrated in degree zero and we have

$$\nu_\chi \mathcal{O}_X^{\mathrm{w}}(S) = \{f \text{ holomorphic and multi-asymptotically developable on } S\}.$$

Set $Z := \bigcup_{j=1}^{\ell} M_j$. Let $\mathcal{O}_X^{\mathrm{w}}$, $\mathcal{O}_{X|X\setminus Z}^{\mathrm{w}}$, $\mathcal{O}_{X|Z}^{\mathrm{w}}$ denote the sheaves on the subanalytic site X_{sa} of Whitney holomorphic functions, flat Whitney holomorphic functions and Whitney holomorphic functions on Z respectively. See [8] for more details.

We can prove functorially the exactness of the sequence

$$0 \to \nu_\chi \mathcal{O}_{X|X\setminus Z}^{\mathrm{w}} \to \nu_\chi \mathcal{O}_X^{\mathrm{w}} \to \nu_\chi \mathcal{O}_{X|Z}^{\mathrm{w}} \to 0. \tag{2}$$

In the case of Majima's asymptotics, $\nu_\chi \mathcal{O}_X^{\mathrm{w}}, \nu_\chi \mathcal{O}_{X|X\setminus Z}^{\mathrm{w}}, \nu_\chi \mathcal{O}_{X|Z}^{\mathrm{w}}$ are isomorphic to the sheaves of strongly asymptotically developable functions, flat asymptotics and consistent families of coefficients respectively. In this case (2) is the Borel-Ritt exact sequence for Majima's asymptotics.

So we have obtained a general Borel-Ritt exact sequence for multi-asymptotically developable functions.

Example 5 We end this paper with some examples of consistent families in $\mathbb{C}^2$. For the general definition we refer to [3].
(Majima) The family

$$F = \{\{f_{\{1\},\alpha_1}(z_2)\}, \{f_{\{2\},\alpha_2}(z_1)\}, \{f_{\{1,2\},(\alpha_1,\alpha_2)}\}\}$$

is consistent if

- $f_{\{1\},\alpha_1}(z_2)$ is strongly asymptotically developable to
$$\{f_{\{1,2\},(\alpha_1,\alpha_2)}\}_{\alpha_2 \in \mathbb{N}_0}$$
on $S_{\{1\}}$ for each $\alpha_1 \in \mathbb{N}_0$,
- $f_{\{2\},\alpha_2}(z_1)$ is strongly asymptotically developable to
$$\{f_{\{1,2\},(\alpha_1,\alpha_2)}\}_{\alpha_1 \in \mathbb{N}_0}$$
on $S_{\{2\}}$ for each $\alpha_2 \in \mathbb{N}_0$.

(Takeuchi) The family

$$F = \{\{f_{\{1\},(\alpha_1,\alpha_2)}\}, \{f_{\{2\},\alpha_2}(z_1)\}, \{f_{\{1,2\},(\alpha_1,\alpha_2)}\}\}$$

is consistent if

- $f_{\{1\},(\alpha_1,\alpha_2)} = f_{\{1,2\},(\alpha_1,\alpha_2)}$ for each $(\alpha_1, \alpha_2) \in \mathbb{N}_0^2$,
- $f_{\{2\},\alpha_2}(z_1)$ is strongly asymptotically developable to

$$\{f_{\{1,2\},(\alpha_1,\alpha_2)}\}_{\alpha_1 \in \mathbb{N}_0}$$

on $S_{\{2\}}$ for each $\alpha_2 \in \mathbb{N}_0$.

(Cusp) and (Generalized cusp) The family

$$F = \{\{f_{\{1\},(\alpha_1,\alpha_2)}\}, \{f_{\{2\},(\alpha_1,\alpha_2)}\}, \{f_{\{1,2\},(\alpha_1,\alpha_2)}\}\}$$

is consistent if

- $f_{\{1\},(\alpha_1,\alpha_2)} = f_{\{2\},(\alpha_1,\alpha_2)} = f_{\{1,2\},(\alpha_1,\alpha_2)}$ for each $(\alpha_1, \alpha_2) \in \mathbb{N}_0^2$.

References

1. Majima, H.: Asymptotic Analysis for Integrable Connections with Irregular Singular Points. Lecture notes in math, vol. 1075. Springer, Berlin (1984)
2. Galindo, F., Sanz, J.: On strongly asymptotically developable functions and the Borel-Ritt theorem. Studia Math. **133**, 231–248 (1999)
3. Honda, N., Prelli, L.: Generalization of multi-specializations and multi-asymptotics (2017). arXiv:1507.04572
4. Kashiwara, M., Schapira, P.: Ind-sheaves. Astérisque **271** (2001)
5. Prelli, L.: Sheaves on subanalytic sites. Rend. Sem. Mat. Univ. Padova **120**, 167–216 (2008)
6. Kashiwara, M., Schapira, P.: Sheaves on manifolds. Grundlehren der Math, vol. 292. Springer, Berlin (1990)
7. Schapira, P., Takeuchi, K.: Déformation normale et bispécialisation. C. R. Acad. Sci. Paris Math. **319**, 707–712 (1994)
8. Honda, N., Prelli, L.: Multi-specialization and multi-asymptotic expansions. Adv. Math. **232**, 432–498 (2013)

Deformations with a Resonant Irregular Singularity

Davide Guzzetti

Abstract I review topics of my talk in Alcalá, inspired by the paper [1]. An isomonodromic system with irregular singularity at $z = \infty$ (and Fuchsian at $z = 0$) is considered, such that $z = \infty$ becomes resonant for some values of the deformation parameters. Namely, the eigenvalues of the leading matrix at $z = \infty$ coalesce along a locus in the space of deformation parameters. I give a complete extension of the isomonodromy deformation theory in this case.

Keywords Isomonodromy deformation · Stokes matrices · Coalescing Eigenvalues · Painlevé equations · Frobenius manifolds

MSC Primary 34M56 · Secondary 34M40, 34M35

1 Introduction

In these proceedings, I extract some of the main results of [1], which I have presented at the workshop in Alcalá, September 4–8, 2017. In [1] we have studied deformations of a class of linear differential systems when the eigenvalues of the leading matrix at $z = \infty$ coalesce along a locus in the space of deformation parameters. The above class contains, in particular, the $n \times n$ ($n \in \mathbb{N}$) system

$$\frac{dY}{dz} = A(z, t)Y, \qquad A(z, t) = \Lambda(t) + \frac{A_1(t)}{z}. \tag{1}$$

with singularity of Poincaré rank 1 at $z = \infty$. The matrices $\Lambda(t)$ and $A_1(t)$ are holomorphic functions of $t = (t_1, \ldots, t_n)$ in a polydisc

$$\mathscr{U}_\varepsilon(0) := \{t \in \mathbb{C}^n \mid |t| \le \varepsilon\}, \qquad |t| := \max_{1 \le i \le m} |t_i|, \tag{2}$$

D. Guzzetti (✉)
SISSA, Via Bonomea 265, Trieste, Italy
e-mail: guzzetti@sissa.it

G. Filipuk et al. (eds.), *Formal and Analytic Solutions of Diff. Equations*, Springer Proceedings in Mathematics & Statistics 256,
https://doi.org/10.1007/978-3-319-99148-1_14

in $\mathbb{C}^n$, centered at $t = 0$. Here, $\Lambda(t)$ is diagonal

$$\Lambda(t) := \mathrm{diag}(u_1(t), \ldots, u_n(t)). \tag{3}$$

In these notes, I will consider only the case when the deformation is isomonodromic, and I refer to [1] for a more general discussion including the non-isomonodromic case.

In some important cases for applications to Frobenius manifolds (like quantum cohomology) and Painlevé equations, it may happen that the eigenvalues coalesce along a certain locus Δ in the t-domain, called the *coalescence locus*, where the matrix $\Lambda(t)$ *remains diagonal*. This means that $u_a(t) = u_b(t)$ for some indices $a \neq b \in \{1, \ldots, n\}$ whenever t belongs to Δ, while $u_1(t)$, $u_2(t)$,..., $u_n(t)$ are pairwise distinct for $t \notin \Delta$. So, the point $z = \infty$ for $t \in \Delta$ is a *resonant irregular singularity*. I will assume that $A_1(t)$ is holomorphic at Δ, at least up to Theorem 1.

An isomonodromic system as above appears in the analytic approach to semisimple Frobenius manifolds [2–4], because its monodromy data allow to locally reconstruct the manifold structure (see also [5]). Coalescing eigenvalues arise in Frobenius manifolds *remaining semisimple at the locus of coalescent canonical coordinates*. An important example is the quantum cohomology of Grassmannians (see [6, 7]). For $n = 3$, a special case of system (1) gives an isomonodromic description of the general sixth Painlevé equation, according to [8], and also to [2, 4] for special values of coefficients. Coalescence occurs at critical points of the Painlevé equation, and $A_1(t)$ is holomorphic when the sixth Painlevé transcendents *remain holomorphic at a fixed singularity* of the Painlevé equation (see Sect. 2.1 below).

Unfortunately, the deformation with coalescence is "non-admissible", because it does not satisfy some of the assumptions of the isomonodromy deformation theory of Jimbo-Miwa-Ueno [9, 10]. Indeed, when t varies in a neighbourhood of the coalescence locus, several problems arise with the behaviour of fundamental matrix solutions and monodromy data. A theory when $\Lambda(t)$ remains diagonal at Δ seems to be missing from the literature (see [1] for a thorough review of the literature, while in these proceedings I have reduced the bibliography to a minimum, for lack of space). Therefore, for the sake of the applications mentioned above, in [1] we have developed a complete deformation theory in this case.

One of the main reasons for this extension of the theory is that we became interested in proving a conjecture formulated by Boris Dubrovin at the ICM 1998 in Berlin (see [3]). The qualitative part of the conjecture says that the quantum cohomology of a smooth projective variety (which is a Frobenius manifold) is semisimple if and only if there exists a full exceptional collection in derived categories of coherent sheaves on the variety. The quantitative part establishes an explicit relation between the monodromy data of the system (1) associated with the quantum cohomology, and certain quantities associated with objects of the exceptional collection. We started our investigation with the quantum cohomology of Grassmannians (for projective spaces, most of the work was done in [11], where there are no coalescences). The problem we had to face is that almost all Grassmannians are coalescent (the meaning of "almost all" is well explained in [6]), and the Frobenius structure, and thus the system (1), are known *only* at coalescence points. So, we can compute monodromy data only at a coalescence point. The question is if these data coincide with the locally constant data (the system must be isomonodromic) in a whole neighbourhood of the coalescence point, so with the data of the Frobenius manifold, as defined in [2, 4]. The answer is positive, thanks to the main theorems of [1], which I expose in Sect. 2 below. As a result, we could prove the conjecture for Grassmannians, in [7] and [12].

The simplest differential system, illustrating our problem with non admissible deformations, is the following *Whittaker Isomonodromic System* (all details of the example are worked out in [13])

$$\frac{dY}{dz}=\left[\begin{pmatrix} u_1 & 0 \\ 0 & u_2\end{pmatrix}+\frac{A_1(u)}{z}\right]Y. \tag{4}$$

Away from $\Delta=\{u_1=u_2\}$, the system is isomonodromic if and only if

$$A_1(u)=\begin{pmatrix} a & c(u_1-u_2)^{-b} \\ d(u_1-u_2)^{b} & a-b\end{pmatrix},\qquad a,b,c,d\in\mathbb{C}. \tag{5}$$

So, we see that for b, c and $d\neq 0$, the points of Δ are *branch points* ($b\notin\mathbb{Z}$) or *poles* ($b\in\mathbb{Z}$). This is what we must expect, following [14].

We leave it as an exercise to solve the system by a standard reduction to the Whittaker equation:

$$\frac{d^2w}{dx^2}+\left(-\frac{1}{4}+\frac{\kappa}{x}+\frac{\frac{1}{4}-\mu^2}{x^2}\right)w=0,\quad \mu^2:=\frac{b^2+4cd}{4},\quad \kappa:=-\frac{1+b}{2}.$$

Notice that the eigenvalues $a+1/2+\kappa\pm\mu$ of A_1 are independent of u, as it must be in the isomonodromic case. The elements $Y_{11}(z)$ and $Y_{12}(z)$ of the first row of a fundamental matrix solution are obtained by taking two independent solutions $w_1(x)$ and $w_2(x)$ of the Whittaker equation, through the change of variables

$$Y_{1k}(z)=e^{\frac{1}{2}(u_1+u_2)z}z^{a-\frac{b+1}{2}}w_k(x),\quad x=z(u_1-u_2),\qquad k=1,2.$$

If we use the asymptotic properties of Whittaker functions $W_{\kappa,\mu}(x)$, we can explicitly construct three fundamental solutions $Y_{-1}(z,u)$, $Y_0(z,u)$, $Y_1(z,u)$, which are asymptotic to the following formal solution

$$Y_F(z,u)=\left(I+\frac{F_1}{z}+\frac{F_2}{z^2}+\cdots\right)z^{\mathrm{diag}(A_1)}\begin{pmatrix} e^{u_1z} & 0 \\ 0 & e^{u_2z}\end{pmatrix},$$

for $z(u_1-u_2)\to\infty$ in the successive overlapping sectors

$$\mathcal{S}_{-1}:=S\Big(-\frac{5\pi}{2},-\frac{\pi}{2}\Big),\quad \mathcal{S}_0:=S\Big(-\frac{3\pi}{2},\frac{\pi}{2}\Big),\quad \mathcal{S}_1:=S\Big(-\frac{\pi}{2},\frac{3\pi}{2}\Big).$$

The matrix coefficients F_k are uniquely determined by the equation and depend holomorphically on $u \notin \Delta$. Moreover, one can do a further exercise and compute the Stokes matrices defined by the connection relations

$$Y_0(z) = Y_{-1}(z)\mathbb{S}_{-1}, \quad Y_1(z) = Y_0(z)\mathbb{S}_0, \qquad \mathbb{S}_{-1} = \begin{pmatrix} 1 & s_{-1} \\ 0 & 1 \end{pmatrix}, \quad \mathbb{S}_0 = \begin{pmatrix} 1 & 0 \\ s_0 & 1 \end{pmatrix},$$

The result is $s_0 =$

$$\frac{2\pi i}{c\,\Gamma\left(\frac{1}{2}+\kappa+\mu\right)\Gamma\left(\frac{1}{2}+\kappa-\mu\right)} = \frac{2\pi i}{c\,\Gamma\left(\frac{\sqrt{b^2+4cd}}{2}-\frac{b}{2}\right)\Gamma\left(-\frac{\sqrt{b^2+4cd}}{2}-\frac{b}{2}\right)},$$

and $s_{-1} =$

$$\frac{2\pi i\; c\, e^{-2\pi i\kappa}}{\Gamma\left(\frac{1}{2}+\mu-\kappa\right)\Gamma\left(\frac{1}{2}-\mu-\kappa\right)} = \frac{-2\pi i\; c\, e^{i\pi b}}{\Gamma\left(\frac{\sqrt{b^2+4cd}}{2}+1+\frac{b}{2}\right)\Gamma\left(-\frac{\sqrt{b^2+4cd}}{2}+1+\frac{b}{2}\right)}.$$

We notice that the sectors $\mathscr{S}_r$, $r=-1,0,1$ in the x-plane determine sectors in the z-plane which depend on $\arg(u_1-u_2)$. For example, $\mathscr{S}_1 = S(-\frac{\pi}{2}, \frac{3\pi}{2})$ gives

$$-\frac{\pi}{2} - \arg(u_1-u_2) < \arg z < -\arg(u_1-u_2) + \frac{3\pi}{2}$$

The boundaries of these z-sectors rotate with varying u. Also, the Stokes rays $\Re(z(u_1-u_2)) = 0$ rotate. Therefore, for u in some small open domain $\mathscr{V}$ of the (u_1, u_2)-plane, we can *fix* a z-sector of central opening angle greater than π, which is *independent of* $u \in \mathscr{V}$, and where the asymptotic behaviour holds. But if u varies too much outside $\mathscr{V}$, then the asymptotc behaviour will no longer hold in the previously fixed sector of the z-plane.

We do not have to worry about this problem only if the Stokes matrices are trivial. Triviality, namely $s_0 = s_{-1} = 0$, occurs if and only if one of the following conditions is satisfied

(1) $c = d = 0$ and $b \in \mathbb{C}$,
(2) $cd = mn, \quad b = n - m$,
(3) either $d = 0$ and $b = -m$, or $c = 0$ and $b = n$,

for $n \geq 1$ and $m \geq 1$ be integers. Now, we look back at the expression (5) for A_1 and immediately conclude that the first part of the following proposition holds.

Proposition 1 *If $s_{-1} = s_0 = 0$, then the points of Δ are not branch points, namely:*

(1) $A_1(u)$ is single-valued for a loop $(u_1-u_2) \mapsto (u_1-u_2)e^{2\pi i}$ around the coalescence locus $u_1 = u_2$;

(2) *the fundamental matrix solutions* $Y_r(z, u)$, $r = -1, 0, 1$ *are also single-valued.*

The second part of the proposition requires some additional work with the monodromy properties of Whittaker functions at $x = 0$, and we refer to [13]. Moreover, the following stronger converse statement holds (see [13] for the proof):

Proposition 2 *If* $A_1(u)$ *is holomorphic at* Δ *and both* $(A_1)_{12}$ *and* $(A_1)_{21}$ *vanish as* $u_1 - u_2 \to 0$, *then the* $Y_r(z, u)$*'s are single-valued in* $u_1 - u_2$ *and holomorphic at* Δ. *Moreover, the Stokes matrices have entries* $s_{-1} = s_0 = 0$.

In conclusion, the example teaches us three things

- the locus Δ is in general of *branch points* for both A_1 and the fundamental matrix solutions. Also the coefficients F_k of the formal solution may have poles or branch points at Δ. Moreover, for a fundamental matrix solution, the canonical asymptotic behaviour for $z \to \infty$ holds in a *fixed* (big) sector of central opening angle greater than π in the z-plane provided that u varies in a sufficiently small domain $\mathcal{V}$ of the u-space. The asymptotics in the fixed sector is lost otherwise (precisely, when u goes around Δ).
- if the entries of the Stokes matrices, with indices corresponding to those of the coalescing eigenvalues, vanish a Δ, then the points of Δ are not branch points. This exemplifies one main result of [1], which is Theorem 2 below.
- if $A_1(t)$ is holomorphic in a domain containing Δ, and if its entries, with indices corresponding to those of the coalescing eigenvalues, vanish at Δ, then the fundamental matrix solutions are holomorphic also at Δ and the entries (as above) of the Stokes matrices vanish. This exemplifies another main result of [1], namely Theorem 1 below.

2 Main Results

No loss of generality occurs if we assume that $t = 0$ is a coalescence point in the polydisc (2). In the isomonodromic case, it is known [10] that we can take the eigenvalues of $\Lambda(t)$ to be the deformation parameters, as I did in the previous example. Hence, we can assume that the eigenvalues $u_1(t), \ldots, u_n(t)$ are linear in t:

$$u_a(t) = u_a(0) + t_a, \quad 1 \leq a \leq n. \tag{6}$$

Therefore,

$$\Lambda(t) = \Lambda(0) + \mathrm{diag}(t_1, \ldots, t_n),$$

where $\Lambda(0)$ has $s < n$ distinct eigenvalues of multiplicities $p_1, \ldots, p_s$ respectively, so that $p_1 + \cdots + p_s = n$. In this case, Δ is a union of hyperplanes.

Assume that $A_1(t)$ is holomorphic in $\mathcal{U}_\varepsilon(0)$, so that for $t \notin \Delta$ there is a unique formal solution

$$Y_F(z,t) := \Big(I + \sum_{k=1}^{\infty} F_k(t) z^{-k}\Big) z^{\mathrm{diag}(A_1(t))} e^{\Lambda(t) z}, \tag{7}$$

where the matrices $F_k(t)$ are uniquely determined by the equation and are holomorphic on $\mathscr{U}_\varepsilon(0)\backslash\Delta$. The well-known result of [15] states that, if t varies in a sufficiently small domain of $\mathscr{U}_\varepsilon(0)\backslash\Delta$ (actually, very small in [15]), there exists a t-independent sector and a fundamental matrix solution whose asymptotic representation is Y_F. But if t varies too much in $\mathscr{U}_\varepsilon(0)\backslash\Delta$, namely goes around Δ, this is no longer true, and the reason is that the Stokes rays

$$\{z \in \mathscr{R} \mid \Re e[(u_a(t) - u_b(t))z] = 0\}, \qquad \mathscr{R} := \text{ universal covering of } \mathbb{C}\backslash\{0\},$$

associated with $\Lambda(t)$ rotate with t varying.

To be more precise, suppose we have fixed $t \notin \Delta$, so Stokes rays are frozen. We can consider a half plane $\Pi_1 := \{z \in \mathscr{R} \mid \tau - \pi < \arg z < \tau\}$, having chosen τ so that no Stokes rays associated with $\Lambda(t)$ have the direction τ (and so $\tau + k\pi$, $k \in \mathbb{Z}$). Then, we consider a big sector $\mathscr{S}_1(t) :=$ the open sector containing Π_1 and extending up to the closest Stokes rays of $\Lambda(t)$ outside Π_1. Then, there is a unique fundamental solution $Y_1(z,t) \sim Y_F(z,t)$ for $z \to \infty$ in $\mathscr{S}_1(t)$ (this follows from [16]). Just to fix τ once and for all, we choose it so that, in particular, no Stokes rays associated with $\Lambda(0)$ lie on the ray $\arg z = \tau$.

Now, let t vary, so that Stokes rays start to rotate. First, we let t vary only inside a domain $\mathscr{V}$, or better in its closure $\overline{\mathscr{V}}$, sufficiently small that no Stokes rays cross the direction τ. This is what an *admissible deformation* is. We can take $\mathscr{S}_1(\mathscr{V}) := \bigcap_{t\in\overline{\mathscr{V}}} \mathscr{S}_1(t)$. This sector has central opening angle greater than π, by construction. We conclude (and we prove in [1]) that the unique fundamental solution $Y_1(z,t)$ has asymptotic expansion $Y_F(z,t)$ for $z \to \infty$ in $\mathscr{S}_1(\mathscr{V})$ and $t \in \overline{\mathscr{V}}$.

We can repeat the construction of a family of actual solutions $Y_r(z,t)$, $r \in \mathbb{Z}$, having the asymptotic representation $Y_F(z,t)$ in big sectors $\mathscr{S}_r(\mathscr{V})$ constructed as above, starting from the half planes $\Pi_r := \{z \in \mathscr{R} \mid \tau + (r-2)\pi < \arg z < \tau + (r-1)\pi\}$. Notice that $\mathscr{S}_r(\mathscr{V}) \cap \mathscr{S}_{r+1}(\mathscr{V}) \neq \emptyset$. This allows to define for $t \in \mathscr{V}$ the Stokes matrices $\mathbb{S}_r(t)$ by the following relations

$$Y_{r+1}(z,t) = Y_r(z,t)\, \mathbb{S}_r(t). \tag{8}$$

On the other hand, if t varies too much, leaving $\mathscr{V}$, Stokes rays may cross $\arg z = \tau$. Since the dominance relations, which determine the change of asymptotics, depend on the behaviour of the exponents $\exp\{z(u_a - u_b)\}$, we see that when a ray $\Re e[(u_a(t) - u_b(t))z] = 0$ has crossed $\arg z = \tau$, the asymptotic relation $Y_r(z,t) \sim Y_F(z,t)$ for $z \to \infty$ in $\mathscr{S}_r(\mathscr{V})$ generally must fail. Maybe, it may not fail if $(\mathbb{S}_r)_{ab} = (\mathbb{S}_r)_{ba} = 0$. This is what actually happens.

The Stokes rays cross $\arg z = \tau$ for t along a certain locus in the polydisc, that we call $X(\tau)$. From the above discussion, it is clear that everything is nice in $\mathscr{U}_\varepsilon(0)\backslash(\Delta \cup X(\tau))$. In [1]) we have proved that $\Delta \cup X(\tau)$ is a union of real hyperplanes, which disconnect $\mathscr{U}_\varepsilon(0)$. Every connected component is simply connected and homeomorphic to a ball in $\mathbb{R}^{2n}$. Thus, it is a cell in the topological sense, so we call it *a* τ*-cell.* Summarising, we have the following general facts:

(i) The deformation is called *admissible in* $\mathscr{V}$ if t varies in a domain $\mathscr{V} \subset \mathscr{U}_\varepsilon(0)$, such that its closure $\overline{\mathscr{V}}$ is properly contained in a τ-cell.
(ii) If $t \in \mathscr{V}$, there is a family of actual fundamental matrix solutions $Y_r(z,t)$, $r \in \mathbb{Z}$, uniquely determined by the canonical asymptotic representation $Y_r(z,t) \sim Y_F(z,t)$, for $z \to \infty$ in sectors $\mathscr{S}_r(\mathscr{V})$. Each $Y_r(z,t)$ is holomorphic within $\mathscr{R}$ for large $|z|$, and in $t \in \mathscr{V}$. The asymptotic series $I + \sum_{k=1}^{\infty} F_k(t) z^{-k}$ is uniform in $\overline{\mathscr{V}}$.

Moreover, we have the following problems, to be solved in Theorem 1 below:

(iii) When t crosses $X(\tau)$ and leaves the cell of $\mathscr{V}$, which means that some Stokes ray $\Re e\,[(u_a(t) - u_b(t))z] = 0$, associated with $\Lambda(t)$, cross the admissible direction τ, then the asymptotic representation $Y_r(z,t) \sim Y_F(z,t)$ for $z \to \infty$ in $\mathscr{S}_r(\mathscr{V})$ does no longer hold.
(iv) The locus Δ is expected to be a locus of poles or branch points for the coefficients $F_k(t)$ and for the $Y_r(z,t)$'s.
(v) The *Stokes matrices* $\mathbb{S}_r(t)$ in (8) are expected to diverge as t approaches Δ.

Notice that in order to completely describe the Stokes phenomenon, it suffices to consider only three fundamental matrix solutions, for example $Y_r(z,t)$ for $r = 1, 2, 3$, and $\mathbb{S}_1(t)$, $\mathbb{S}_2(t)$ (this has been done in example, with $r = -1, 0, 1$ and $\mathbb{S}_{-1}$ and $\mathbb{S}_0$).

To complete the general picture, we will define the monodromy data. For given t, a matrix $G(t)$ puts $A_1(t)$ in Jordan form $J(t) := G^{-1}(t)\, A_1(t)\, G(t)$. For a given t, the system (1) has a fundamental solution represented in Levelt form

$$Y^{(0)}(z,t) = G(t)\Big(I + \sum_{l=1}^{\infty} \Psi_l(t) z^l\Big) z^{D(t)} z^{S(t)+R(t)}. \tag{9}$$

in a neighbourhood of $z = 0$. The matrix coefficients $\Psi_l(t)$ of the convergent expansion are constructed by a recursive procedure. $J(t) = D(t) + S(t)$, where $D(t) = \text{diag}(d_1(t), \ldots, d_n(t))$ is a matrix of integers, piecewise constant in t, $S(t)$ is a Jordan matrix whose eigenvalues have real part in $[0, 1[$. The nilpotent matrix $R(t)$ has non-vanishing entries only if some eigenvalues of $A_1(t)$ differ by non-zero integers. It is proved in Theorem 1 that the solution (9) turns out to be holomorphic in $t \in \mathscr{U}_\varepsilon(0)$. Chosen $Y^{(0)}(z,t)$, a *central connection matrix* $C^{(0)}$ is defined by the relation

$$Y_1(z,t) = Y^{(0)}(z,t)\, C^{(0)}(t), \qquad z \in \mathscr{S}_1(\mathscr{V}). \tag{10}$$

The *essential* monodromy data (the name is inspired by a similar definition in [10]) are then

$$\mathbb{S}_1(t),\quad \mathbb{S}_2(t),\quad \text{diag}(A_1(t)),\quad C^{(0)}(t),\quad J(t),\quad R(t). \tag{11}$$

Now, when t tends to a point $t_\Delta \in \Delta$, the limits of the above data may not exist. If the limits exist, they *do not in general give the monodromy data of the system with matrix* $A(z, t_\Delta)$ (see [1, 17]).

Definition 1 An admissible deformation in a small domain $\mathcal{V}$ is **isomonodromic in** $\mathcal{V}$ if the essential monodromy data (11) do not depend on $t \in \mathcal{V}$.

For admissible isomonodromic deformations as defined in Definition 1, the classical theory of Jimbo-Miwa-Ueno [10] applies in $\mathcal{V}$. In Theorem 1 and Corollary 1 below we have extended the theory to the whole $\mathcal{U}_\varepsilon(0)$, including the coalescence locus Δ. In the statement of the theorem, we will not explain how small ε is, since this is a little bit technical point (see [1] and [13]). We also skip the construction of new sectors $\widehat{\mathcal{S}}_r(t)$ and $\widehat{\mathcal{S}}_r = \bigcap_{t\in\mathcal{U}_\varepsilon(0)} \widehat{\mathcal{S}}_r(t)$, which appear in the theorem. They are bigger than $\mathcal{S}_r(\mathcal{V})$, namely $\mathcal{S}_r(\mathcal{V}) \subset \widehat{\mathcal{S}}_r \subset \widehat{\mathcal{S}}_r(t)$, with $t \in \mathcal{U}_\varepsilon(0)$.

Theorem 1 *Consider the system (1), with eigenvalues of $\Lambda(t)$ linear in t as in (6), and with $A_1(t)$ holomorphic on a closed polydisc $\mathcal{U}_\varepsilon(0)$ centred at $t = 0$, with sufficiently small radius ε (as specified in [1]). Let Δ be the coalescence locus in $\mathcal{U}_\varepsilon(0)$, passing through $t = 0$. Let the deformation be isomonodromic in $\mathcal{V} \subset \mathcal{U}_\varepsilon(0)$ as in Definition 1. If the matrix entries of $A_1(t)$ satisfy in $\mathcal{U}_\varepsilon(0)$ the vanishing conditions*

$$\Big(A_1(t)\Big)_{ab} = O(u_a(t) - u_b(t)), \qquad 1 \le a \ne b \le n, \tag{12}$$

whenever $u_a(t)$ and $u_b(t)$ coalesce as t tends to a point of Δ, then the following results hold:

- *The coefficients $F_k(t)$ of the formal solution $Y_F(z, t)$ in (7) are holomorphic on the whole $\mathcal{U}_\varepsilon(0)$.*
- *The three fundamental matrix solutions $Y_r(z, t)$, $r = 1, 2, 3$, initially defined on $\mathcal{V}$, with asymptotic representation $Y_F(z, t)$ for $z \to \infty$ in the sectors $\mathcal{S}_r(\mathcal{V})$ introduced above, can be t-analytically continued as single-valued holomorphic functions on $\mathcal{U}_\varepsilon(0)$, with asymptotic representation*

  $$Y_r(z, t) \sim Y_F(z, t) \quad \text{for } z \to \infty \text{ in wider sectors } \widehat{\mathcal{S}}_r,$$

  for any $t \in \mathcal{U}_{\varepsilon_1}(0)$ and any $0 < \varepsilon_1 < \epsilon$. In particular, they are defined at any $t_\Delta \in \Delta$ with asymptotic representation $Y_F(z, t_\Delta)$. The fundamental matrix solution $Y^{(0)}(z, t)$ is also holomorphic on $\mathcal{U}_\varepsilon(0)$
- *The constant matrices $\mathbb{S}_1$, $\mathbb{S}_2$, and $C^{(0)}$, initially defined for $t \in \mathcal{V}$, are actually* globally defined on $\mathcal{U}_\varepsilon(0)$. *They coincide with the Stokes and connection matrices of the fundamental solutions $Y_r(z, 0)$ and $Y^{(0)}(z, 0)$ of the system*

  $$\frac{dY}{dz} = A(z, 0)Y, \qquad A(z, 0) = \Lambda(0) + \frac{A_1(0)}{z}. \tag{13}$$

Also the remaining t-independent monodromy data in (11) coincide with those of (13).

- *The entries (a, b) of the Stokes matrices are characterised by the following vanishing property whenever $u_a(0) = u_b(0)$, $1 \le a \ne b \le n$:*

$$(\mathbb{S}_1)_{ab} = (\mathbb{S}_1)_{ba} = (\mathbb{S}_2)_{ab} = (\mathbb{S}_2)_{ba} = 0. \tag{14}$$

Thus, under the only condition (12), we have no more problems with rotating Stokes rays, with loss of asymptotic representation in fixed big sectors, and with the appearance of branch points at Δ. I think that this is a remarkable fact.

It is now time to explain when and how system (13) suffices to compute the monodromy data of (1). Let the assumptions of Theorem 1 hold. Then, (13) has a formal solution

$$\mathring{Y}_F(z) = \Big(I + \sum_{k=1}^{\infty} \mathring{F}_k z^{-k}\Big) z^{\mathrm{diag}(A_1(0))} e^{\Lambda(0) z}. \tag{15}$$

Actually, there is a family of formal solutions (15): the coefficients $\mathring{F}_k$ can be recursively constructed from the differential system, but *there is not a unique choice for them.* To each element of the family there correspond unique actual solutions $\mathring{Y}_1(z)$, $\mathring{Y}_2(z)$, $\mathring{Y}_3(z)$ such that $\mathring{Y}_r(z) \sim \mathring{Y}_F(z)$ for $z \to \infty$ in a sector $\mathcal{S}_r \supset \mathcal{S}_r(\mathcal{V})$, $r = 1, 2, 3$, with Stokes matrices defined by

$$\mathring{Y}_{r+1}(z) = \mathring{Y}(z)\, \mathring{\mathbb{S}}_r, \qquad r = 1, 2.$$

Notice that only one element of the family of formal solutions (15) satisfies the condition $\mathring{F}_k = F_k(0)$ for any $k \ge 1$, so that $\mathbb{S}_r = \mathring{\mathbb{S}}_r$.

To complete the picture, let us also choose a fundamental matrix solution $\mathring{Y}^{(0)}(z)$ of (13), in Levelt form in a neighbourhood of $z = 0$, and define the corresponding central connection matrix $\mathring{C}^{(0)}$ such that $\mathring{Y}_1(z) = \mathring{Y}^{(0)}(z)\, \mathring{C}^{(0)}$. The following holds

Corollary 1 *Let the assumptions of Theorem 1 hold. If the diagonal entries of $A_1(0)$ do not differ by non-zero integers, then there is a unique formal solution (15) of the system (13), whose coefficients necessarily satisfy the condition*

$$\mathring{F}_k \equiv F_k(0).$$

Hence, the corresponding fundamental matrix solutions $\mathring{Y}_1(z)$, $\mathring{Y}_2(z)$, $\mathring{Y}_3(z)$ of are such that

$$Y_1(z, 0) = \mathring{Y}_1(z), \quad Y_2(z, 0) = \mathring{Y}_2(z), \quad Y_3(z, 0) = \mathring{Y}_3(z).$$

Moreover, for any $\mathring{Y}^{(0)}(z)$ there exists $Y^{(0)}(z, t)$ such that $Y^{(0)}(z, 0) = \mathring{Y}^{(0)}(z)$. The following equalities hold:

$$\mathbb{S}_1 = \mathring{\mathbb{S}}_1, \quad \mathbb{S}_2 = \mathring{\mathbb{S}}_2, \quad C^{(0)} = \mathring{C}^{(0)}.$$

Corollary 1 has a practical computational importance: the *constant monodromy data* (11) of the system (1) on the whole $\mathcal{U}_\varepsilon(0)$ are computable just by considering the system (13) at the coalescence point $t = 0$. This is useful for applications. For example, it allows to compute the monodromy data of a semisimple Frobenius manifold, such as the quantum cohomology of Grassmannians [6, 7] mentioned in the Introduction, just by considering the Frobenius structure at a coalescence point.

In [1], we also prove the (weaker) converse of Theorem 1. Assume that the deformation is admissible and isomonodromic on a simply connected domain $\mathcal{V} \subset \mathcal{U}_\varepsilon(0)$, and that $A_1(t)$ is holomorphic (only) in $\mathcal{V}$. As a result of [14], the fundamental matrix solutions $Y_r(z, t)$, $r = 1, 2, 3$, and $A_1(t)$ can be analytically continued as *multi-valued functions* on $\mathcal{U}_\varepsilon(0)\backslash\Delta$, with movable poles. Nevertheless, if the vanishing condition (14) holds, then Δ *does not contain branch points* and the *asymptotic behaviour is preserved on big sectors*, according to the following

Theorem 2 *Let $A_1(t)$ be holomorphic on an open simply connected domain $\mathcal{V} \subset \mathcal{U}_\varepsilon(0)$, where the deformation is admissible and isomonodromic as in Definition 1. Let ε be sufficiently small (as specified in [1]). If*

$$(\mathbb{S}_1)_{ab} = (\mathbb{S}_1)_{ba} = (\mathbb{S}_2)_{ab} = (\mathbb{S}_2)_{ba} = 0 \quad \text{whenever } u_a(0) = u_b(0), \quad 1 \le a \ne b \le n,$$

then, the fundamental matrix solutions $Y_r(z, t)$ and $A_1(t)$ admit single-valued analytic continuation on *$\mathcal{U}_\varepsilon(0)\backslash\Delta$ as meromorphic functions of t. Moreover, for any $t \in \mathcal{U}_\varepsilon(0)\backslash\Delta$ which is not a pole of $Y_r(z, t)$ we have*

$$Y_r(z, t) \sim Y_F(z, t) \ \text{for } z \to \infty \text{ in } \widehat{\mathcal{S}}_r(t), \quad r = 1, 2, 3,$$

and $Y_{r+1}(z, t) = Y_r(z, t)\,\mathbb{S}_r$, $r = 1, 2$. The sectors $\widehat{\mathcal{S}}_r(t)$ are described in [1].

2.1 *Applications*

We have no space to explain the applications of Theorem 1 and Corollary 1 to Frobenius manifolds (see [7, 18]). As for Painlevé equations, they provide an alternative to Jimbo's approach for the computation of the monodromy data associated with Painlevé VI transcendents holomorphic at a critical point. As an example, we consider the A_3-algebraic solution of Dubrovin-Mazzocco [19]

$$y(s) = \frac{(1-s)^2\,(1+3s)\,(9s^2-5)^2}{(1+s)\,(243s^6+1539s^4-207s^2+25)}, \qquad t(s) = \frac{(1-s)^3\,(1+3s)}{(1+s)^3\,(1-3s)}, \tag{16}$$

with $s \in \mathbb{C}$, which solves the Painlevé VI equation

$$\frac{d^2y}{dt^2} = \frac{1}{2}\left[\frac{1}{y} + \frac{1}{y-1} + \frac{1}{y-t}\right]\left(\frac{dy}{dt}\right)^2 - \left[\frac{1}{t} + \frac{1}{t-1} + \frac{1}{y-t}\right]\frac{dy}{dt}$$
$$+\frac{1}{2}\frac{y(y-1)(y-t)}{t^2(t-1)^2}\left[\frac{9}{4} + \frac{t(t-1)}{(y-t)^2}\right].$$

The above equation is the isomonodromicity condition for a 3×3 system

$$\frac{dY}{dz} = \left[\begin{pmatrix} 0 & 0 & 0 \\ 0 & t & 0 \\ 0 & 0 & 1 \end{pmatrix} + \frac{A_1(t)}{z}\right] Y, \qquad A_1(t) =: \begin{pmatrix} 0 & \Omega_2 & -\Omega_3 \\ -\Omega_2 & 0 & \Omega_1 \\ \Omega_3 & -\Omega_1 & 0 \end{pmatrix};$$
$$\Omega_1 = i\frac{\sqrt{y-1}\sqrt{y-t}}{\sqrt{t}}\left[\frac{A}{(y-1)(y-t)} + \mu\right], \quad \Omega_2 = i\frac{\sqrt{y}\sqrt{y-t}}{\sqrt{1-t}}\left[\frac{A}{y(y-t)} + \mu\right],$$
$$\Omega_3 = -\frac{\sqrt{y}\sqrt{y-1}}{\sqrt{t}\sqrt{1-t}}\left[\frac{A}{y(y-1)} + \mu\right], \quad A := \frac{1}{2}\left[\frac{dy}{dt}t(t-1) - y(y-1)\right]. \tag{17}$$

The above formulae are in [5]. A holomorphic branch is obtained by letting $s \to -\frac{1}{3}$ in (16), which gives convergent Taylor expansions

$$\Omega_1(t) = \frac{i\sqrt{2}}{8} - \frac{i\sqrt{2}t}{256} + O(t^2), \quad \Omega_3(t) = \frac{i\sqrt{2}}{8} + \frac{i\sqrt{2}t}{256} + O(t^2),$$
$$\Omega_2(t) = -\frac{t}{32} + O(t^2).$$

Since $\lim_{t\to 0}\Omega_2(t) = 0$, Theorem 1 holds. Since $\text{diag}(A_1) = (0,0,0)$, also Corollary 1 holds. Accordingly, the Stokes matrices can be computed using (17) at fixed $t = 0$, which is integrable by reduction to a Bessel equation. Thus, its Stokes matrices can be computed:

$$\mathbb{S}_1 = \begin{pmatrix} 1 & 0 & 1 \\ 0 & 1 & -1 \\ 0 & 0 & 1 \end{pmatrix}, \qquad \mathbb{S}_2 = \mathbb{S}_1^{-T} = \begin{pmatrix} 1 & 0 & 0 \\ 0 & 1 & 0 \\ -1 & 1 & 1 \end{pmatrix}.$$

This result is in accordance with [19]. Notice that $(\mathbb{S}_r)_{12} = (\mathbb{S}_r)_{21} = 0$, as Theorem 1 predicts.

References

1. Cotti, G., Dubrovin, B.A., Guzzetti, D.: Isomonodromy deformations at an irregular singularity with coalescing eigenvalues (2017). arXiv:1706.04808
2. Dubrovin, B.: Geometry of 2D topological field theories. Lect. Notes Math. **1620**, 120–348 (1996)
3. Dubrovin, B.: Geometry and analytic theory of Frobenius manifolds. In: Proceedings of ICM98, vol. 2, pp. 315–326 (1998)

4. Dubrovin, B.: Painlevé trascendents in two-dimensional topological field theory. In: Conte, R. (ed.) The Painlevé Property, One Century later. Springer, Berlin (1999)
5. Guzzetti, D.: Inverse problem and monodromy data for three-dimensional Frobenius manifolds. Math. Phys. Anal. Geom. **4**, 245–291 (2001)
6. Cotti, G.: Coalescence phenomenon of quantum cohomology of grassmannians and the distribution of prime numbers (2016). arXiv:1608.06868
7. Cotti, G., Dubrovin, B.A., Guzzetti, D.: Local moduli of semisimple Frobenius coalescent structures (2017). arXiv:1712.08575
8. Mazzocco, M.: Painlevé sixth equation as isomonodromic deformations equation of an irregular system. CRM Proc. Lect. Notes, Am. Math. Soc. **32**, 219–238 (2002)
9. Fokas, A., Its, A., Kapaev, A., Novokshenov, V.: Painlevé Transcendents: The Riemann-Hilbert Approach. AMS (2006)
10. Jimbo, M., Miwa, T., Ueno, K.: Monodromy preserving deformations of linear ordinary differential equations with rational coefficients (I). Physics **D2**, 306–352 (1981)
11. Guzzetti, D.: Stokes matrices and monodromy of the quantum cohomology of projective spaces. Commun. Math. Phys. **207**(2), 341–383 (1999)
12. Cotti, G., Dubrovin, B.A., Guzzetti, D.: Helix structures in quantum cohomology of Fano manifolds. Ready to appear
13. Cotti, G., Guzzetti, D.: Results on the extension of isomonodromy deformations to the case of a resonant irregular singularity. In: Proceedings for the Workshop at CRM. Pisa 13–17 February 2017. Random Matrices: Theory and Applications (2018). https://doi.org/10.1142/S2010326318400038
14. Miwa, T.: Painlevé Property of monodromy preserving deformation equations and the analyticity of τ functions. Publ. RIMS, Kyoto Univ. **17**, 703–721 (1981)
15. Hsieh, P.-F., Sibuya, Y.: Note on regular perturbation of linear ordinary differential equations at irregular singular points. Funkcial. Ekvac **8**, 99–108 (1966)
16. Balser, W., Jurkat, W.B., Lutz, D.A.: Birkhoff invariants and stokes' multipliers for meromorphic linear differential equations. J. Math. Anal. Appl. **71**, 48–94 (1979)
17. Balser, W., Jurkat, W.B., Lutz, D.A.: A general theory of invariants for meromorphic differential equations; Part I. Form. Invariants. Funkcialaj Evacioj **22**, 197–221 (1979)
18. Cotti, G., Guzzetti, D.: Analytic geometry of semisimple coalescent structures. Random Matrices Theory Appl. **6**(4), 1740004, 36, 53 (2017)
19. Dubrovin, B., Mazzocco, M.: Monodromy of certain Painlevé transcendents and reflection groups. Invent. Math. **141**, 55–147 (2000)

Symmetric Semi-classical Orthogonal Polynomials of Class One on q-Quadratic Lattices

Galina Filipuk and Maria das Neves Rebocho

Abstract In this paper we study discrete semi-classical orthogonal polynomials on non-uniform lattices. In the symmetric class one case we give a closed form expression for the recurrence coefficients of orthogonal polynomials.

Keywords Orthogonal polynomials · Divided-difference operator · Non-uniform lattices · Askey–Wilson operator

MSC Primary 33C45 · Secondary 33C47, 42C05

1 Motivation

Orthogonal polynomials on q-quadratic lattices are part of the discrete families of orthogonal polynomials. These families are widely spread in the literature of special functions and applications. Some works we refer to the interested reader include [1–3], where a comprehensive approach to orthogonal polynomials is given, gathering, amongst many other topics, the analysis of divided difference operators, related problems on classification, connections with the Sturm–Louville theory, etc.

In the present paper we consider the general divided-difference operator $\mathbb{D}$ [4, Eq. (1.1)] having the basic property of leaving a polynomial of degree $n-1$

G. Filipuk
Faculty of Mathematics, Informatics and Mechanics, University of Warsaw, Banacha 2, 02-097 Warsaw, Poland
e-mail: filipuk@mimuw.edu.pl

M. das Neves Rebocho (✉)
Departamento de Matemática, Universidade da Beira Interior, 6201-001 Covilhã, Portugal
e-mail: mneves@ubi.pt

M. das Neves Rebocho
Department of Mathematics, CMUC, University of Coimbra, 3001-501 Coimbra, Portugal

G. Filipuk et al. (eds.), *Formal and Analytic Solutions of Diff. Equations*, Springer Proceedings in Mathematics & Statistics 256,
https://doi.org/10.1007/978-3-319-99148-1_15

when applied to a polynomial of degree n. It is well-know that $\mathbb{D}$ yields the Askey–Wilson operator [5] under appropriate specifications. The related lattice, commonly known as a non-uniform lattice, is obtained from a conic defined by (2) [6, Sec. 2] and involves the q-quadratic lattice due to its parametric representations (more details are given in Sect. 2.1). The main problem to be analysed in the present paper lies within the so-called direct problem (see, e.g., [7]): to extract information on recurrence coefficients of orthogonal polynomials, given some data on the corresponding Stieltjes function or the orthogonality measure. We take the difference equation satisfied by the Stieltjes functions, say $A\mathbb{D}S = C\mathbb{M}S + D$, where $A,\ C,\ D$ are polynomials, subject to restrictions $\deg(A) \leq 3,\ \deg(C) \leq 2$. According to the classification from [8], this is the so-called class one (see Sect. 2.2). Our main goal is to give a closed form expression for the recurrence coefficients of orthogonal polynomials in the symmetric case, that is, when one of the recurrence coefficients is zero (cf. Sect. 3). Such a closed formula is given in terms of the lattice as well as in terms of the polynomials $A,\ C,\ D$. To the best of authors' knowledge, these results are knew in the literature. Furthermore, as the calculus on non-uniform lattices generalizes the calculus on lattices of lower complexity (see [2], [9, Sec. 2]), our results may be regarded as a generalization of some of the results on semi-classical orthogonal polynomials of class one, for instance: [10], on symmetric orthogonal polynomials when $\mathbb{D}$ is Hahn's difference operator; [11], on orthogonal polynomials when $\mathbb{D}$ is the derivative operator; [12], on orthogonal polynomials when $\mathbb{D}$ is the forward difference operator.

The paper is organized as follows. In Sect. 2 we give the definitions and state the basic results which will be used in the forthcoming sections. In Sect. 3 we present the main results of the paper, namely, a closed form expression for the recurrence coefficients of the symmetric orthogonal polynomials (see Theorem 1).

2 Preliminary Results

2.1 The General Divided Difference Operator, q-Quadratic Lattices, and Orthogonal Polynomials

We consider the divided difference operator $\mathbb{D}$ given in [4, Eq. (1.1)], with the property that $\mathbb{D}$ leaves a polynomial of degree $n-1$ when applied to a polynomial of degree n. The operator $\mathbb{D}$, defined on the space of arbitrary functions, is given in terms of two functions, y_1, y_2,

$$(\mathbb{D}f)(x) = \frac{f(y_2(x)) - f(y_1(x))}{y_2(x) - y_1(x)}\,. \tag{1}$$

The functions y_1, y_2 may be defined as the two y-roots of the quadratic equation

$$\hat{a}y^2 + 2\hat{b}xy + \hat{c}x^2 + 2\hat{d}y + 2\hat{e}x + \hat{f} = 0\,, \quad \hat{a} \neq 0\,. \tag{2}$$

In the present paper we shall consider the q-quadratic case, related to the so-called non-uniform lattices, which appears whenever $\lambda\tau \neq 0$, $\lambda = \hat{b}^2 - \hat{a}\hat{c}$, $\tau = \left((\hat{b}^2 - \hat{a}\hat{c})(\hat{d}^2 - \hat{a}\hat{f}) - (\hat{b}\hat{d} - \hat{a}\hat{e})^2\right)/\hat{a}$. In such a case, y_1, y_2 are given by

$$y_1(x) = p(x) - \sqrt{r(x)}\,, \quad y_2(x) = p(x) + \sqrt{r(x)} \tag{3}$$

with p, r polynomials given by

$$p(x) = -\frac{\hat{b}x + \hat{d}}{\hat{a}}\,, \quad r(x) = \frac{\lambda}{\hat{a}^2}\left(x + \frac{\hat{b}\hat{d} - \hat{a}\hat{e}}{\lambda}\right)^2 + \frac{\tau}{\hat{a}\lambda}\,. \tag{4}$$

We shall use the notation $\Delta_y = y_2 - y_1$. From (3), it follows that

$$\Delta_y = 2\sqrt{r}\,. \tag{5}$$

For the q−quadratic case there is a parametric representation of the conic (2), $x = x(s)$, $y = y(s)$, such that [6, pp. 254–255]

$$x(s) = x_c + \xi\sqrt{\hat{a}}\,(q^s + q^{-s})\,, \quad y(s) = y_c + \xi\sqrt{\hat{c}}\,(q^{s-1/2} + q^{-s+1/2})\,, \tag{6}$$

$$x_c = (\hat{a}\hat{e} - \hat{b}\hat{d})/\lambda\,,\; y_c = (\hat{c}\hat{d} - \hat{b}\hat{e})/\lambda\,,\; \xi^2 = \tilde{f}/(4\lambda)\,,\; \tilde{f} = \hat{f} - \hat{a}y_c^2 - 2\hat{b}x_c y_c - \hat{c}x_c^2\,,$$

and q defined by

$$q + q^{-1} = \frac{4\hat{b}^2}{\hat{a}\hat{c}} - 2\,. \tag{7}$$

We have $y_1(x(s)) = y(s)$, $y_2(x(s)) = y(s+1)$. Thus, in the account of (3), we have

$$p(x(s)) - \sqrt{r(x(s))} = y(s)\,, \quad p(x(s)) + \sqrt{r(x(s))} = y(s+1)$$

and

$$y(s+1) + y(s) = 2p(x(s))\,, \quad (y(s+1) - y(s))^2 = 4r(x(s))\,.$$

In the present paper we shall operate with $\mathbb{D}$ given in its general form (1). We now define other operators related to (1). Firstly, by defining $\mathbb{E}_1$ and $\mathbb{E}_2$ (see [4]), acting on arbitrary functions f as

$$(\mathbb{E}_1 f)(x) = f(y_1(x))\,, \quad (\mathbb{E}_2 f)(x) = f(y_2(x))\,, \tag{8}$$

the so-called companion operator of $\mathbb{D}$ is defined as (see [4])

$$(\mathbb{M}f)(x) = \frac{(\mathbb{E}_1 f)(x) + (\mathbb{E}_2 f)(x)}{2}. \tag{9}$$

Note that $\mathbb{M}f$ is a polynomial whenever f is a polynomial. Furthermore, if $\deg(f) = n$, then $\deg(\mathbb{M}f) = n$. Indeed, in the account of (3), one has

$$\mathbb{E}_j(x^n) = \left(p(x) + (-1)^j \sqrt{r(x)}\right)^n, \quad j = 1, 2.$$

From the binomial identity

$$(p \pm \sqrt{r})^n = \sum_{i=0}^{n} \binom{n}{i} p^i \left(\pm\sqrt{r}\right)^{n-i},$$

we get, for n odd,

$$\mathbb{D}x^n = \sum_{i=0}^{(n-1)/2} \binom{n}{2i} p^{2i} r^{(n-2i-1)/2}, \quad \mathbb{M}x^n = \sum_{i=0}^{(n-1)/2} \binom{n}{2i+1} p^{2i+1} r^{(n-2i-1)/2}, \tag{10}$$

and for n even,

$$\mathbb{D}x^n = \sum_{i=0}^{(n-2)/2} \binom{n}{2i+1} p^{2i+1} r^{(n-2i-2)/2}, \quad \mathbb{M}x^n = \sum_{i=0}^{n/2} \binom{n}{2i} p^{2i} r^{(n-2i)/2}. \tag{11}$$

In the remainder of the paper we use the following notation:

$$\mathbb{D}x^n = \sum_{k=0}^{n-1} d_{n,k} x^k, \quad \mathbb{M}x^n = \sum_{k=0}^{n} m_{n,k} x^k. \tag{12}$$

We shall consider orthogonal polynomials related to a (formal) Stieltjes function defined by

$$S(x) = \sum_{n=0}^{+\infty} u_n x^{-n-1}, \tag{13}$$

where (u_n), the sequence of moments, is such that $\det\left[u_{i+j}\right]_{i,j=0}^{n} \neq 0$, $n \geq 0$, and, without loss of generality, $u_0 = 1$. The orthogonal polynomials related to S, P_n, $n \geq 0$, are taken to be monic, and we will denote the sequence $\{P_n\}_{n\geq 0}$ by SMOP.

Monic orthogonal polynomials satisfy a three-term recurrence relation [13]

$$P_{n+1}(x) = (x - \beta_n) P_n(x) - \gamma_n P_{n-1}(x), \quad n = 0, 1, 2, \ldots, \tag{14}$$

with $P_{-1}(x) = 0$, $P_0(x) = 1$, and $\gamma_n \neq 0$, $n \geq 1$, $\gamma_0 = 1$.

The quantities β_n, γ_n are called the recurrence coefficients of $\{P_n\}_{n\geq 0}$.

Another important sequence, related to $\{P_n\}_{n\geq 0}$, is the sequence of associated polynomials of the first kind, denoted by $\{P_n^{(1)}\}_{n\geq 0}$, which satisfies the three term recurrence relation

$$P_n^{(1)}(x) = (x-\beta_n)P_{n-1}^{(1)}(x) - \gamma_n P_{n-2}^{(1)}(x)\,, \quad n = 1, 2, \ldots \tag{15}$$

with $P_{-1}^{(1)}(x) = 0,\ P_0^{(1)}(x) = 1$.

In the framework of the Hermite-Padé Approximation (see [14]), the polynomials P_n are the diagonal Padé denominators of (13), and the polynomials $P_{n-1}^{(1)}$ are the numerator polynomials, thus, also determined by the relation

$$S(x) - P_{n-1}^{(1)}(x)/P_n(x) = \mathcal{O}(x^{-2n-1})\,, \quad x\to\infty\,.$$

2.2 *Semi-classical Orthogonal Polynomials on Non-uniform Lattices: The Class One and General Difference Equations*

Semi-classical orthogonal polynomials on non-uniform lattices may be defined by a difference equation for the Stieltjes function [4, 9],

$$A\mathbb{D}S = C\mathbb{M}S + D\,, \tag{16}$$

where $A,\ C,\ D$ are irreducible polynomials (in x). In general, the polynomials $A,\ C,\ D$ in (16) satisfy, in the account of (1), (9), and (13),

$$\deg(A) \leq m+2\,,\ \deg(C) \leq m+1\,,\ \deg(D) \leq m\,, \tag{17}$$

where m is some nonnegative integer. When $m = 0$ we get the so-called classical polynomials [15, 16].

In the present paper we will study class one (see [8, Def. 8 and Th. 9]), that is, we will take the difference equation (16) for S with $m = 1$ in (17), under the following condition (in order to avoid degenerate cases):

$$\deg(A) = 3\,, 1 \leq \deg(C) \leq 2 \text{ or } \deg(A) < 3\,, \deg(C) = 2\,. \tag{18}$$

A very useful result on semi-classical orthogonal polynomials concerns the difference equations studied recently in [6, 17, 18]. In particular, SMOP related to (16) satisfy the following difference equations, for all $n \geq 0$:

$$A_{n+1}\mathbb{D}P_{n+1} = (l_n - C/2)\mathbb{M}P_{n+1} + \Theta_n\mathbb{M}P_n\,, \tag{19}$$

$$A_{n+1}\mathbb{D}P_n^{(1)} = (l_n + C/2)\mathbb{M}P_n^{(1)} + D\mathbb{M}P_{n+1} + \Theta_n\mathbb{M}P_{n-1}^{(1)}\,, \tag{20}$$

with

$$A_{n+1} = A + \frac{\Delta_y^2}{2}\pi_n\,. \tag{21}$$

Here Δ_y is defined in (5) and l_n, Θ_n, π_n are polynomials in x satisfying, for all $n \geq 0$,

$$\pi_{n+1} = -\frac{1}{2}\sum_{k=0}^{n+1}\frac{\Theta_{k-1}}{\gamma_k}\,, \tag{22}$$

$$l_{n+1} + l_n + \mathbb{M}(x - \beta_{n+1})\frac{\Theta_n}{\gamma_{n+1}} = 0\,, \tag{23}$$

$$-A + \mathbb{M}(x - \beta_{n+1})(l_{n+1} - l_n) - \frac{\Delta_y^2}{2}(\pi_{n+1} + \pi_n) + \Theta_{n+1} = \frac{\gamma_{n+1}}{\gamma_n}\Theta_{n-1}\,, \tag{24}$$

with initial conditions

$$\pi_{-1} = 0,\ \pi_0 = -D/2, \tag{25}$$

$$\Theta_{-1} = D,\ \Theta_0 = A - \frac{\Delta_y^2}{4}D - (l_0 - C/2)\mathbb{M}(x - \beta_0), \tag{26}$$

$$l_{-1} = C/2,\ l_0 = -\mathbb{M}(x - \beta_0)D - C/2\,. \tag{27}$$

3 Main Results: The Symmetric Orthogonal Polynomials of Class One

The symmetric families of class one satisfy (14) with $\beta_n = 0,\ n \geq 0$, that is,

$$P_{n+1}(x) = xP_n(x) - \gamma_n P_{n-1}(x)\,, \quad n = 0, 1, 2, \ldots\,,$$

and they are related to the Stieltjes functions such that (18) holds. Let us remark that this gives some condition on $A,\ C,\ D,\ p,\ r$. Indeed, we can find the moments u_j in terms of the coefficients of $A,\ C,\ D,\ p,\ r$. We have $u_0 = 1,\ u_1 = 0$ and by using the well-known formulas [13] for the recurrence coefficients in terms of determinants of moments, we get, for instance, $\beta_1 = 0$ provided some condition on $A,\ C,\ D,\ p,\ r$ holds.

Let us now proceed to the determination of the recurrence coefficients γ_n.

The polynomials $A_n,\ l_n,\ \Theta_n,\ \pi_n$ in the difference equations from the previous section satisfy $\deg(A_n) = 3,\ \deg(l_n) = 2,\ \deg(\Theta_n) = \deg(\pi_n) = 1$ [17]. We set

$$\begin{aligned} &A(x) = a_3x^3 + a_2x^2 + a_1x + a_0\,,\ C(x) = c_2x^2 + c_1x + c_0\,,\\ &D(x) = d_1x + d_0\,,\ A_n(x) = a_{n,3}x^3 + a_{n,2}x^2 + a_{n,1}x + a_{n,0}\,, \qquad (28)\\ &l_n(x) = \ell_{n,2}x^2 + \ell_{n,1}x + \ell_{n,0}\,,\ \Theta_n(x) = \Theta_{n,1}x + \Theta_{n,0}\,,\ \pi_n = \pi_{n,1}x + \pi_{n,0}\,,\\ &p(x) = p_1x + p_0\,,\ r(x) = r_2x^2 + r_1x + r_0\,. \qquad (29)\end{aligned}$$

In accordance with (18), we will take the case

$$|a_3| + |c_2| \neq 0\,. \tag{30}$$

In the following we shall also assume $p_1 \neq 0,\ p_1^2 - r_2 \neq 0$, that is, $\hat{a}\hat{b}\hat{c} \neq 0$ in (2).

The polynomial D is defined in terms of the polynomials $A,\ C$. Indeed, by collecting the coefficients in (26) and (27), we get

$$d_1 = \frac{-(a_3 + c_2 p_1)}{p_1^2 - r_2}\,, \quad d_0 = \frac{-a_2 - c_1 p_1 - c_2 p_0 + (r_1 - 2 p_0 p_1) d_1}{p_1^2 - r_2}\,. \tag{31}$$

Theorem 1 *Let $\{P_n\}_{n\geq 0}$ be a SMOP related to a Stieltjes function S satisfying $A(x)\mathbb{D}S(x) = C(x)\mathbb{M}S(x) + D(x)$ with $\deg(A) \leq 3$, $\deg(C) \leq 2$, $\deg(D) \leq 1$ under condition (30) in the previous notation. Let the recurrence relation*

$$P_{n+1}(x) = x P_n(x) - \gamma_n P_{n-1}(x)\,, \quad n = 0, 1, 2, \ldots$$

hold with $\gamma_0 = 1$. Under the previous notations we have, for all $n \geq 0$,

$$\gamma_{n+2} = \frac{\gamma_1 T_0 D(x_0) + \sum_{k=0}^{n} \zeta_k T_k}{T_n T_{n+1}}\,, \tag{32}$$

$$\gamma_1 = \frac{(a_1 + c_1 p_0 + c_0 p_1 + d_0(2 p_0 p_1 - r_1) + d_1(p_0^2 - r_0))(p_1^2 - r_2)}{-a_3 + c_2 p_1 + 2 d_1 (r_2 + p_1^2)}\,, \tag{33}$$

where

$$x_0 = -p_0/p_1\,, \ \zeta_n = A(x_0) + 2r(x_0)(\pi_{n+1} + \pi_n)(x_0)\,, \ T_n = \Theta_n(x_0)/\gamma_{n+1}\,.$$

The quantities ζ_n, T_n are given in terms of the polynomials $A,\ C$ as well as of p, r defined in (4). Indeed, there holds the following explicit formulae. For all $n \geq 0$,

$$\frac{\Theta_{n+1,1}}{\gamma_{n+2}} = \frac{q^{-n-1}(q+1)(q^{2n+2}-1)}{(q-1)p_1}\ell_{0,2} + \frac{q^{-n-1}(q^{2n+3}-1)}{q-1}\frac{\Theta_{0,1}}{\gamma_1}\,, \tag{34}$$

$$\pi_{n+1,1} = -\frac{d_1}{2} - \frac{\Theta_{0,1}}{2\gamma_1} - (1 - q^{-n})\left(\frac{(q+1)(q^{n+1}-1)}{2p_1(q-1)^2}\ell_{0,2} - \frac{(q^{n+2}-1)}{2(q-1)^2}\frac{\Theta_{0,1}}{\gamma_1}\right)\,, \tag{35}$$

$$\frac{\Theta_{n+1,0}}{\gamma_{n+2}} = 2(\pi_{n+1,0} - \pi_{n+2,0})\,, \tag{36}$$

$$\pi_{n+1,0} = \prod_{k=0}^{n} s_k \left(\pi_{0,0} + \sum_{k=0}^{n} \left(\prod_{j=0}^{k} s_j\right)^{-1} t_k\right)\,, \tag{37}$$

with

$$s_n = \frac{f_{n,0} + 2p_1}{2p_1 - f_{n+1,0}}, \tag{38}$$

$$t_n = \frac{f_{n+1} + f_n + f_{n+1,1}\pi_{n+1,1} + f_{n,1}\pi_{n,1} + p_0\Theta_{n,1}/\gamma_{n+1}}{2p_1 - f_{n+1,0}}, \tag{39}$$

$$f_{n,0} = \frac{2r_2 d_{n+1,n}}{m_{n+1,n+1}}, \tag{40}$$

$$f_{n,1} = \frac{1}{m_{n+1,n+1}}\left(2r_2\left(d_{n+1,n-1} - d_{n+1,n}\frac{m_{n+1,n}}{m_{n+1,n+1}}\right) + 2r_1 d_{n+1,n}\right), \tag{41}$$

$$f_n = \frac{c_1}{2} + \frac{1}{m_{n+1,n+1}}\left(a_3\left(d_{n+1,n-1} - d_{n+1,n}\frac{m_{n+1,n}}{m_{n+1,n+1}}\right) + a_2 d_{n+1,n}\right), \tag{42}$$

and the initial conditions

$$\ell_{0,2} = -p_1 d_1 - \frac{c_2}{2}, \quad \frac{\Theta_{0,1}}{\gamma_1} = \frac{-a_3 + c_2 p_1 + 2d_1(p_1^2 + r_2)}{p_1^2 - r_2}, \quad \pi_{0,1} = -\frac{d_1}{2}, \tag{43}$$

$$\pi_{0,0} = -\frac{d_0}{2}, \quad \frac{\Theta_{0,0}}{\gamma_1} = \frac{1}{p_1^2 - r_2}\Big(-a_2 + c_2 p_0 + d_1(3p_0 p_1 + 2r_1) + d_0(p_1^2 + 2r_2) + p_1(p_1 d_0 + p_0 d_1 + c_1) + \frac{\Theta_{0,1}}{\gamma_1}(r_1 - 2p_0 p_1)\Big). \tag{44}$$

Here, p_1, p_0, r_2, r_1 are the coefficients of $p(x)$ and $r(x)$ defined in (4), and q is defined by (7).

Proof First, let us we deduce (32)–(33).

Evaluating (24) at $x_0 := -p_0/p_1$ we get, as $\mathbb{M}(x) = p(x)$ (which follows from (3) and (9)),

$$-A(x_0) - 2r(x_0)(\pi_{n+1} + \pi_n)(x_0) + \gamma_{n+2}\frac{\Theta_{n+1}}{\gamma_{n+2}}(x_0) = \gamma_{n+1}\frac{\Theta_{n-1}}{\gamma_n}(x_0).$$

Thus, we write

$$\gamma_{n+2}T_{n+1} = \gamma_{n+1}T_{n-1} + \zeta_n \tag{45}$$

with $T_{n+1} = \Theta_{n+1}(x_0)/\gamma_{n+2}$, $\zeta_n = A(x_0) + 2r(x_0)(\pi_{n+1} + \pi_n)(x_0)$. By multiplying (45) by T_n we get

$$\gamma_{n+2}T_n T_{n+1} = \gamma_{n+1}T_{n-1}T_n + \zeta_n T_n, \quad n \geq 0.$$

Iterating yields

$$\gamma_{n+2}T_n T_{n+1} = \gamma_1 T_{-1}T_0 + \sum_{k=0}^{n}\zeta_k T_k, \quad n \geq 0,$$

thus we obtain (32), where we used $T_{-1} = D(x_0)$ (see (26)).

To obtain γ_1 we proceed in two steps: first, by collecting the coefficient of x in (26) and using (27), we get

$$\Theta_{0,1} = a_1 + c_1 p_0 + c_0 p_1 + d_0(2p_0 p_1 - r_1) + d_1(p_0^2 - r_0)\,.$$

Next we have (43),

$$\frac{\Theta_{0,1}}{\gamma_1} = \frac{-a_3 + c_2 p_1 + 2d_1(p_1^2 + r_2)}{p_1^2 - r_2}\,.$$

Combining the two equations above we get (33).

Now, let us we deduce the quantities $\Theta_{n,j}/\gamma_{n+1}$ and $\pi_{n,j}$, $j = 0, 1$.

To deduce (34) we start by taking $\beta_n = 0$ as well as $\mathbb{M}(x) = p(x)$ in (23), and collect the coefficients of x^2, thus getting

$$\ell_{n+1,2} = -\ell_{n,2} - p_1 \frac{\Theta_{n,1}}{\gamma_{n+1}}\,. \tag{46}$$

Also, starting with the definition of A_n in (21), and using a similar procedure as in [18, Lemma 1], we obtain $\Theta_{n+1,1}/\gamma_{n+2}$ as a linear combination of $\ell_{n,2}$ and $\Theta_{n,1}/\gamma_{n+1}$,

$$\frac{\Theta_{n+1,1}}{\gamma_{n+2}} = \frac{-4p_1}{r_2 - p_1^2}\ell_{n,2} + \left(1 - \frac{2(r_2 + p_1^2)}{r_2 - p_1^2}\right)\frac{\Theta_{n,1}}{\gamma_{n+1}}\,. \tag{47}$$

Then we write (46) and (47) in the matrix form,

$$\begin{bmatrix} \ell_{n+1,2} \\ \Theta_{n+1,1}/\gamma_{n+2} \end{bmatrix} = \mathscr{X} \begin{bmatrix} \ell_{n,2} \\ \Theta_{n,1}/\gamma_{n+1} \end{bmatrix}, \quad \mathscr{X} = \begin{bmatrix} -1 & -p_1 \\ \dfrac{-4p_1}{r_2 - p_1^2} & 1 - \dfrac{2(r_2 + p_1^2)}{r_2 - p_1^2} \end{bmatrix}. \tag{48}$$

Iterating (48) yields, for all $n \geq 0$,

$$\ell_{n+1,2} = \left(\mathscr{X}^{n+1}\right)_{(1,1)} \ell_{0,2} + \left(\mathscr{X}^{n+1}\right)_{(1,2)} \Theta_{0,1}/\gamma_1\,, \tag{49}$$

$$\Theta_{n+1,1}/\gamma_{n+2} = \left(\mathscr{X}^{n+1}\right)_{(2,1)} \ell_{0,2} + \left(\mathscr{X}^{n+1}\right)_{(2,2)} \Theta_{0,1}/\gamma_1\,. \tag{50}$$

Here, $\left(\mathscr{X}^{n+1}\right)_{(i,j)}$ denotes the element on the position (i, j) of $\mathscr{X}^{n+1}$.

The diagonalization of $\mathscr{X}$ proceeds as in [18, Lemma 1]. Indeed, the set of the eigenvalues of $\mathscr{X}$ is given by $\sigma(\mathscr{X}) = \{q, q^{-1}\}$ with q defined in (7). Therefore, we get $\mathscr{X} = \mathscr{V}\mathscr{D}\mathscr{V}^{-1}$, with $\mathscr{V}$, $\mathscr{D}$ given by

$$\mathscr{V} = \begin{bmatrix} \dfrac{-p_1}{1+q} & \dfrac{-p_1}{1+q^{-1}} \\ 1 & 1 \end{bmatrix}, \quad \mathscr{D} = \begin{bmatrix} q & 0 \\ 0 & q^{-1} \end{bmatrix}.$$

Thus, $\mathscr{X}^{n+1} = \mathscr{V}\mathscr{D}^{n+1}\mathscr{V}^{-1}$. As a consequence, from (50) we obtain (34). To get $\ell_{0,2}$ and $\pi_{0,1}$ we use (27) and (25), respectively. The quantity $\Theta_{0,1}/\gamma_1$ follows from equating the coefficients of x^3 in (20) with $n = 1$ combined with (23) with $n = 0$.

Equation (35) follows from the definition of π_n (cf. (24)) combined with (34).

Equation (36) follows from (24).

In order to deduce $\pi_{n,0}$ we start by obtaining some formulae involving $\ell_{n,1}$. By equating the coefficients of x^{n+2} in (19) we get

$$\ell_{n,1} = \frac{c_1}{2} + \frac{1}{m_{n+1,n+1}} \left(a_{n+1,3} \left(d_{n+1,n-1} - d_{n+1,n} \frac{m_{n+1,n}}{m_{n+1,n+1}} \right) + a_{n+1,2} d_{n+1,n} \right),$$

where the $m_{n+1,k}$ and $d_{n+1,k}$ are given by (12). As $a_{n+1,3} = a_3 + 2r_2\pi_{n,1}$, $a_{n+1,2} = a_2 + 2r_2\pi_{n,0} + 2r_1\pi_{n,1}$, we get

$$\ell_{n,1} = f_n + f_{n,1}\pi_{n,1} + f_{n,0}\pi_{n,0}, \tag{51}$$

with $f_{n,0}$, $f_{n,1}$, f_n given by (40)–(42). Now, to obtain $\pi_{n,0}$, we begin by collecting the coefficient of x in (23), hence,

$$\ell_{n+1,1} + \ell_{n,1} + p_1 \frac{\Theta_{n,0}}{\gamma_{n+1}} + p_0 \frac{\Theta_{n,1}}{\gamma_{n+1}} = 0. \tag{52}$$

Using (51) in (52) as well as $2(\pi_{n,0} - \pi_{n+1,0}) = \Theta_{n,0}/\gamma_{n+1}$ we get, after basic computations,

$$\pi_{n+1,0} = s_n\pi_{n,0} + t_n, \quad n \geq 0,$$

with s_n, t_n given by (38)–(39). Thus, we obtain (37).

To obtain $\pi_{0,0}$ we use (25), and to obtain $\Theta_{0,0}/\gamma_1$ we take the coefficient of x^2 in (20) with $n = 1$ combined with (23) with $n = 0$. □

Acknowledgements The authors are very grateful to the anonymous referee for her/his valuable comments.

GF acknowledges the support of the National Science Center (Poland) via grant OPUS 2017/25/B/BST1/00931. Support of the Alexander von Humboldt Foundation is also greatfully acknowledged.

The work of MNR was partially supported by the Centre for Mathematics of the University of Coimbra – UID/MAT/00324/2013, funded by the Portuguese Government through FCT/MCTES and co-funded by the European Regional Development Fund through the Partnership Agreement PT2020.

References

1. Koekoek, R., Lesky, P.A., Swarttouw, R.F.: Hypergeometric Orthogonal Polynomials and Their q-Analogues. Springer, Berlin (2010)

2. Nikiforov, A.F., Suslov, S.K., Uvarov, V.B.: Classical Orthogonal Polynomials of a Discrete Variable. Springer, Berlin (1991)
3. Nikiforov, A.F., Uvarov, V.B.: Special Functions of Mathematical Physics: A Unified Introduction with Applications. Birkhäuser, Basel (1988)
4. Magnus, A.P.: Associated Askey-Wilson polynomials as Laguerre-Hahn orthogonal polynomials. In: Alfaro, M., Dehesa, J.S., Marcellán, F.J., Rubio de Francia, J.L., Vinuesa, J. (eds.) Orthogonal Polynomials and Their Applications (Segovia, 1986). Lecture Notes in Mathematics, vol. 1329, pp. 261–278. Springer, Berlin (1988)
5. Askey, R., Wilson, J.: Some basic hypergeometric orthogonal polynomials that generalize Jacobi polynomials. Mem. AMS **54**(319). AMS, Providence (1985)
6. Magnus, A.P.: Special nonuniform lattice (snul) orthogonal polynomials on discrete dense sets of points. J. Comput. Appl. Math. **65**, 253–265 (1995)
7. Van Assche, W.: Discrete Painlevé equations for recurrence coefficients of orthogonal polynomials. In: Elaydi, S., Cushing, J., Lasser, R., Ruffing, A., Papageorgiou, V., Van Assche, W. (eds.) Difference Equations, Special Functions and Orthogonal Polynomials, pp. 687–725. World Scientific, Hackensack (2007)
8. Mboutngama, S., Foupouagnigni, M., Njionou Sadjang, P.: On the modifications of semi-classical orthogonal polynomials on nonuniform lattices. J. Math. Anal. Appl. **445**, 819–836 (2017)
9. Witte, N.S.: Semi-classical orthogonal polynomial systems on nonuniform lattices, deformations of the Askey table, and analogues of isomonodromy. Nagoya Math. J. **219**, 127–234 (2015)
10. Maroni, P., Mejri, M.: The symmetric D_{ω}-semiclassical orthogonal polynomials of class one. Numer. Algorithms **49**, 251–282 (2008)
11. Belmehdi, S.: On semi-classical linear functionals of class $s = 1$. Classif. Integral Represent. Indag. Math. **3**, 253–275 (1992)
12. Dominici, D., Marcellán, F.: Discrete semiclassical orthogonal polynomials of class one. Pac. J. Math. **268**, 389–411 (2014)
13. Szegő, G.: Orthogonal Polynomials. American Mathematical Society Colloquium Publications, vol. 23, 4th edn. American Mathematical Society, Providence (1975)
14. Magnus, A.P.: Riccati acceleration of the Jacobi continued fractions and Laguerre-Hahn polynomials. In: Werner, H., Bunger, H.T. (eds.) Padé Approximation and its Applications (Proceedings Bad Honnef 1983). Lecture Notes in Mathematics, vol. 1071, pp. 213–230. Springer, Berlin (1984)
15. Foupouagnigni, M., Kenfack Nangho, M., Mboutngam, S.: Characterization theorem for classical orthogonal polynomials on non-uniform lattices: the functional approach. Integral Transforms Spec. Funct. **22**, 739–759 (2011)
16. Nikiforov, A.F., Suslov, S.K.: Classical orthogonal polynomials of a discrete variable on non uniform lattices. Lett. Math. Phys. **11**, 27–34 (1986)
17. Branquinho, A., Rebocho, M.N.: Characterization theorem for Laguerre-Hahn orthogonal polynomials on non-uniform lattices. J. Math. Anal. Appl. **427**, 185–201 (2015)
18. Filipuk, G., Rebocho, M.N.: Orthogonal polynomials on systems of non-uniform lattices from compatibility conditions. J. Math. Anal. Appl. **456**, 1380–1396 (2017)

Determinantal Form for Ladder Operators in a Problem Concerning a Convex Linear Combination of Discrete and Continuous Measures

Carlos Hermoso, Edmundo J. Huertas and Alberto Lastra

Abstract In this contribution we complete and deepen the results in (Huertas et al, Proc Am Math Soc 142(5), 1733–1747, 2014, [1]) by introducing a determinantal form for the ladder operators concerning the infinite sequence $\{Q_n(x)\}_{n\geq 0}$ of monic polynomials orthogonal with respect to the following Laguerre–Krall inner product

$$\langle f, g\rangle_\nu = \int_0^{+\infty} f(x)g(x)x^\alpha e^{-x}dx + \sum_{j=1}^{m} a_j f(c_j)g(c_j),$$

where $c_j \in \mathbb{R}_- \cup \{0\}$. We obtain for the first time explicit formulas for these ladder (creation and annihilation) operators, and we use them to obtain several algebraic properties satisfied by $Q_n(x)$. As an application example, based on the structure of the above inner product $\langle f, g\rangle_\nu$, we consider a convex linear combination of continuous and discrete measures that leads to establish an interesting research line concerning the Laguerre–Krall polynomials and several open problems.

Keywords Second order differential equations · Orthogonal polynomials · Ladder operators · Combination of measures

MSC Primary 33C45 · Secondary 33C47

C. Hermoso · A. Lastra
Departamento de Física y Matemáticas, Universidad de Alcalá, Ctra. Madrid-Barcelona, Km. 33.600, 28871 Alcalá de Henares, Madrid, Spain
e-mail: carlos.hermoso@uah.es

A. Lastra
e-mail: alberto.lastra@uah.es

E. J. Huertas (✉)
Departamento de Ingeniería Civil: Hidráulica y Ordenación del Territorio, E.T.S. de Ingeniería Civil, Universidad Politécnica de Madrid, C/ Alfonso XII, 3 y 5, 28014 Madrid, Spain
e-mail: ej.huertas.cejudo@upm.es

G. Filipuk et al. (eds.), *Formal and Analytic Solutions of Diff. Equations*, Springer Proceedings in Mathematics & Statistics 256, https://doi.org/10.1007/978-3-319-99148-1_16

1 Introduction

Let $\{L_n^\alpha\}_{n\geq 0}$ be the sequence of monic classical Laguerre polynomials, orthogonal with respect to the inner product $\langle p, q\rangle_\alpha : \mathbb{P}\times\mathbb{P}\to\mathbb{R}$, with

$$\langle p, q\rangle_\alpha = \int_0^\infty p(x)q(x)x^\alpha e^{-x}dx,\ \alpha > -1,\ p, q \in \mathbb{P}, \tag{1}$$

where $\mathbb{P}$ is the linear space of polynomials with real coefficients, and the corresponding norm is given by $\|p\|_\alpha^2 = \int_0^\infty |p(x)|^2 x^\alpha e^{-x}dx$, $\alpha > -1$. They can be explicitly defined through the so called Rodrigues formula

$$L_n^\alpha(x) = (-1)^n\, x^{-\alpha} e^x \frac{d^n}{dx^n}\left(e^{-x}x^{n+\alpha}\right) = (-1)^n\, x^{-\alpha}\left(\frac{d}{dx} - 1\right)^n x^{n+\alpha},\quad n \geq 0. \tag{2}$$

Among any other nice properties (see, for example, [2, 3]), they satisfy the following three term recurrence relation (TTRR)

$$xL_n^\alpha(x) = L_{n+1}^\alpha(x) + \beta_n L_n^\alpha(x) + \gamma_n L_{n-1}^\alpha(x),\quad n \geq 1,$$

with $L_0^\alpha(x) = 1,\ L_1^\alpha(x) = x - (\alpha+1)$, $\beta_n = 2n + \alpha + 1$ and $\gamma_n = n\,(n+\alpha)$.

On the other hand, let $\{Q_n\}_{n\geq 0}$ be the sequence of monic orthogonal polynomials (SMOP in short) associated with the modified Laguerre measure supported in the whole real line

$$d\mu_M = x^\alpha e^{-x}dx + M\,\delta(x-c)$$

with $M \in \mathbb{R}_+$, $\delta(x-c)$ the Dirac delta function at $x = c$, and $c \in \mathbb{R}_- \cup \{0\}$. This is known as an Uvarov canonical spectral transformation (or perturbation) of the Laguerre measure (see [4]). The polynomials $\{Q_n\}_{n\geq 0}$ are called Laguerre–Krall (or Laguerre-type) orthogonal polynomials, and they are orthogonal with respect to the inner product

$$\langle p, q\rangle_M = \int_0^\infty p(x)q(x)x^\alpha e^{-x}dx + Mp(c)q(c).$$

We can iterate the above canonical transformation of the Laguerre measure applying an integer number m of mass points (or Dirac deltas) located in the negative real semiaxis, i.e.

$$d\nu = x^\alpha e^{-x}dx + \sum_{j=1}^m a_j\,\delta(x-c_j), \tag{3}$$

with $a_j \in \mathbb{R}_+$, $c_j \in \mathbb{R}_- \cup \{0\}$, and therefore, the polynomials $\{Q_n\}_{n\geq 0}$ are now orthogonal with respect to the inner product

$$\langle p, q\rangle_{\nu} = \int_{0}^{+\infty} p(x)q(x)x^{\alpha}e^{-x}dx + \sum_{j=1}^{m} a_j\, p(c_j)q(c_j). \tag{4}$$

In recent years, increasing attention has been paid to the notion of spectral transformations of measures. They have been analyzed from different points of view by several authors. From the point of view of adding mass points at the end points of the support of the classical Jacobi measure we can find in the literature the pioneer work of T. H. Koornwinder [5]. Not much later, the works [6, 7] studied several analytic properties of orthogonal polynomials with respect to a perturbation of the Laguerre weight when a mass is added at $x = 0$. In this same case, an electrostatic interpretation of the zeros as equilibrium points with respect to a logarithmic potential, under the action of an external field, has been done in [8, 9]. Other analytic properties, as well as the outer relative asymptotics, Mehler–Heine, and Plancherel–Rotach formulas for these polynomials have been obtained in [10].

When the mass point is located outside the support of the standard Laguerre measure, that is $c \in \mathbb{R}_{-}$ and $M > 0$ in (1), a first approach was done in [11, 12], where, among other results, a representation of these polynomials in terms of standard Laguerre polynomials, and a hypergeometric representation was obtained.

The next natural step, which is consider more than one discrete mass points at $\mathbb{R}_{-}$ (i.e., consider the SMOP with respect to (4) was analyzed in [1]. There, among other results, was presented a second order linear differential equation (also known as holonomic equation) that the corresponding Laguerre–Krall polynomials satisfy, and the authors provide a complete electrostatic description of the zeros of these polynomials in terms of a logarithmic potential interaction under the action of an external field.

In this contribution we focus our attention on some important mathematical operators called *ladder operators*, also known as raising and lowering operators, or creation and annihilation operators in quantum mechanics (see, among others [13, Ch. 3]). The origin of the ladder operators find its roots in the representation theory of Lie groups and Lie algebras (see [14]), but given its particular importance, the concept soon spread out beyond that framework to other fields of mathematics and physics. We present explicit formulas for the ladder operators of the Laguerre–Krall SMOP, which were not provided in [1]. Moreover, among other results, we recover the aforementioned second order differential equation for Laguerre–Krall polynomials obtained in [1] by means of tedious and cumbersome computations, but using the elegant formalism provided by the ladder operators, which reduces considerably the computations.

The structure of the manuscript is as follows. In Sect. 2 we present without proof four connection Lemmas previously developed in [1], necessary to find our main result, namely the explicit formulas of the aforementioned ladder operators for the SMOP $\{Q_n\}_{n\geq 0}$, which are presented in Sect. 3. The structure of these operators can be expressed in a simple and compact way as 2×2 determinants with polynomial entries. To show its usefulness and usability, we obtain in Sect. 3 a Rodrigues type formula for $\{Q_n\}_{n\geq 0}$ using these ladder operators. As a second example, we also recover the holonomic equation satisfied by these polynomials, previously obtained in [1], but using the framework of the ladder operators. In Sect. 4, as an application of the Laguerre–Krall SMOP, we consider (4) when the weights $a_j \geq 0$ are given by the well known binomial distribution of probability theory

$$\binom{m}{j} p^j (1-p)^{m-j} \quad \text{at } j = 0, 1, 2, \ldots, m,$$

and $c_j = -j$, with $j = 0, 1, 2, \ldots, m$. Technical difficulties appear when one considers this situation, and therefore several open problems are pointed out.

2 Connection Lemmas Between Q_n and L_n^α

A first step to describe the new Laguerre–Krall polynomials of the SMOP $\{Q_n\}_{n\geq 0}$ is to express them in terms of the classical Laguerre orthogonal polynomials. This problem was already solved in [1] . For the convenience of the reader we repeat some relevant material from [1] without proofs, thus making our exposition self-contained.

From now on, $\rho_m(x)$ denotes the mth degree polynomial

$$\rho_m(x) = \prod_{j=1}^{m} \left(x - c_j\right). \tag{5}$$

Lemma 1 *([1, Lemma 4.1]) For the sequences of polynomials $\{Q_n\}_{n\geq 0}$ and $\{L_n^\alpha\}_{n\geq 0}$, we get*

$$\begin{aligned} \rho_m(x) Q_n(x) &= A_1(x;n) L_n^\alpha(x) + B_1(x;n) L_{n-1}^\alpha(x), \\ x\left[\rho_m(x) Q_n(x)\right]' &= C_1(x;n) L_n^\alpha(x) + D_1(x;n) L_{n-1}^\alpha(x), \end{aligned}$$

where $\rho_m(x)$ is given in (5), and

$$\begin{aligned} A_1(x;n) &= \rho_m(x) - \sum_{j=1}^{m} \left(\frac{a_j\, L_{n-1}^\alpha(c_j)\, Q_n(c_j)}{(n-1)!\Gamma(n+\alpha)} \right) \rho_{m,j}(x), \\ B_1(x;n) &= \sum_{j=1}^{m} \left(\frac{a_j\, L_n(c_j)\, Q_n(c_j)}{(n-1)!\Gamma(n+\alpha)} \right) \rho_{m,j}(x), \\ \rho_{m,k}(x) &= \prod_{j=1,\, j\neq k}^{m} \left(x - c_j\right), \\ C_1(x;n) &= nA_1(x;n) - B_1(x;n) + xA_1'(x;n), \\ D_1(x;n) &= n(n+\alpha)A_1(x;n) + (x-(n+\alpha))B_1(x;n) + xB_1'(x;n). \end{aligned}$$

Lemma 2 *([1, Lemma 4.2]) The sequences of monic polynomials* $\{Q_n\}_{n\geq 0}$ *and* $\{L_n^{\alpha}\}_{n\geq 0}$ *are also related by*

$$\begin{aligned}\rho_m(x)Q_{n-1}(x) &= A_2(x;n)L_n^{\alpha}(x) + B_2(x;n)L_{n-1}^{\alpha}(x),\\ x\left[\rho_m(x)Q_{n-1}(x)\right]' &= C_2(x;n)L_n^{\alpha}(x) + D_2(x;n)L_{n-1}^{\alpha}(x),\end{aligned}$$

where

$$\begin{aligned}A_2(x;n) &= \frac{-1}{(n-1+\alpha)(n-1)}B_1(x;n-1),\\ B_2(x;n) &= A_1(x;n-1) + \frac{(x+1-2n-\alpha)}{(n-1+\alpha)(n-1)}B_1(x;n-1),\\ C_2(x;n) &= \frac{-1}{(n-1+\alpha)(n-1)}D_1(x;n-1),\\ D_2(x;n) &= C_1(x;n-1) + \frac{(x+1-2n-\alpha)}{(n-1+\alpha)(n-1)}D_1(x;n-1).\end{aligned}$$

Lemma 3 *([1, Lemma 4.3]) The classical Laguerre polynomials can be expressed in terms of the Laguerre–Krall polynomials in the following way*

$$\begin{aligned}L_n^{\alpha}(x) &= \frac{\rho_m(x)}{\Delta(x;n)}\left(B_2(x;n)Q_n(x) - B_1(x;n)Q_{n-1}(x)\right),\\ L_{n-1}^{\alpha}(x) &= \frac{\rho_m(x)}{\Delta(x;n)}\left(-A_2(x;n)Q_n(x) + A_1(x;n)Q_{n-1}(x)\right).\end{aligned}$$

where

$$\Delta(x;n) = A_1(x;n)B_2(x;n) - B_1(x;n)A_2(x;n), \quad \deg \Delta(x;n) = 2m.$$

Lemma 4 *([1, Lemma 4.4]) The following ladder equations follows from the above three connection lemmas*

$$G(x;n)Q_n(x) + F(x;n)[Q_n]'(x) = H(x;n)Q_{n-1}(x), \tag{6}$$
$$J(x;n)Q_{n-1}(x) + F(x;n)[Q_{n-1}]'(x) = K(x;n)Q_n(x), \tag{7}$$

where

$$\begin{aligned}F(x;n) &= x\Delta(x;n)\rho_m(x),\\ G(x;n) &= x\Delta(x;n)\rho_m'(x) + \rho_m(x)\left[D_1(x;n)A_2(x;n) - C_1(x;n)B_2(x;n)\right],\\ H(x;n) &= \rho_m(x)[D_1(x;n)A_1(x;n) - C_1(x;n)B_1(x;n)],\\ J(x;n) &= x\Delta(x;n)\rho_m'(x) + \rho_m(x)\left[C_2(x;n)B_1(x;n) - D_2(x;n)A_1(x;n)\right],\\ K(x;n) &= \rho_m(x)[C_2(x;n)B_2(x;n) - D_2(x;n)A_2(x;n)].\end{aligned} \tag{8}$$

It is easy to realise that all the above polynomials in Lemma 4 can be expressed as 2×2 determinants with polynomial entries. Thus, we have

$$\begin{aligned}
\Delta(x;n) &= A_1(x;n)B_2(x;n) - B_1(x;n)A_2(x;n) = \begin{vmatrix} A_1(x;n) & B_1(x;n) \\ A_2(x;n) & B_2(x;n) \end{vmatrix}, \\
F(x;n) &= x\Delta(x;n)\rho_m(x) = x\rho_m(x)\begin{vmatrix} A_1(x;n) & B_1(x;n) \\ A_2(x;n) & B_2(x;n) \end{vmatrix}, \\
G(x;n) &= x\Delta(x;n)\rho_m'(x) + \rho_m(x)\left[D_1(x;n)A_2(x;n) - C_1(x;n)B_2(x;n)\right] \\
&= \begin{vmatrix} A_2(x;n) & \rho_m(x)C_1(x;n) - x\rho_m' A_1(x;n) \\ B_2(x;n) & \rho_m(x)D_1(x;n) - x\rho_m' B_1(x;n) \end{vmatrix}, \\
J(x;n) &= x\Delta(x;n)\rho_m'(x) + \rho_m(x)\left[C_2(x;n)B_1(x;n) - D_2(x;n)A_1(x;n)\right] \\
&= \begin{vmatrix} \rho_m(x)C_2(x;n) - x\rho_m' A_2(x;n) & A_1(x;n) \\ \rho_m(x)D_2(x;n) - x\rho_m' B_2(x;n) & B_1(x;n) \end{vmatrix},
\end{aligned}$$

and

$$H(x;n) = \rho_m(x)\begin{vmatrix} A_1(x;n) & C_1(x;n) \\ B_1(x;n) & D_1(x;n) \end{vmatrix}, \quad K(x;n) = \rho_m(x)\begin{vmatrix} C_2(x;n) & A_2(x;n) \\ D_2(x;n) & B_2(x;n) \end{vmatrix}.$$

3 Ladder Operators for the Laguerre–Krall SMOP

Being I the identity operator, we can rewrite the ladder equations (6) and (7) as

$$\left(\frac{G(x;n)}{H(x;n)} I + \frac{F(x;n)}{H(x;n)}\frac{d}{dx}\right) Q_n(x) = Q_{n-1}(x), \tag{9}$$

$$\left(\frac{J(x;n)}{K(x;n)} I + \frac{F(x;n)}{K(x;n)}\frac{d}{dx}\right) Q_{n-1}(x) = Q_n(x). \tag{10}$$

Observe that the action of the bracketed structures in the left hand sides of the above equations modify the degree in the Laguerre–Krall polynomials.

Proposition 1 (ladder operators) *We have*

$$b_n := \frac{G(x;n)}{H(x;n)} I + \frac{F(x;n)}{H(x;n)}\frac{d}{dx} =$$

$$\frac{1}{H(x;n)}\begin{vmatrix} A_2(x;n) & \left\{\rho_m(x)C_1(x;n) - x\rho_m' A_1(x;n)\right\} I - x\rho_m(x)A_1(x;n)\dfrac{d}{dx} \\ B_2(x;n) & \left\{\rho_m(x)D_1(x;n) - x\rho_m' B_1(x;n)\right\} I - x\rho_m(x)B_1(x;n)\dfrac{d}{dx} \end{vmatrix},$$

$$b_n^\dagger := \frac{J(x;n)}{K(x;n)} I + \frac{F(x;n)}{K(x;n)}\frac{d}{dx} =$$

$$\frac{1}{K(x;n)}\begin{vmatrix} A_1(x;n)\ \{-\rho_m(x)C_2(x;n)+x\rho'_m A_2(x;n)\}\ I + x\rho_m(x)A_2(x;n)\dfrac{d}{dx} \\ B_1(x;n)\ \{-\rho_m(x)D_2(x;n)+x\rho'_m B_2(x;n)\}\ I + x\rho_m(x)B_2(x;n)\dfrac{d}{dx} \end{vmatrix}$$

satisfying $b_n Q_n(x) = Q_{n-1}(x)$, *and* $b_n^{\dagger} Q_{n-1}(x) = Q_n(x)$ *respectively.*

Proof From the ladder equations (9) and (10), and the expression of $G(x;n)$, $J(x;n)$ and $F(x;n)$ in (8), it's straightforward to check that the above determinantal expressions on the right hand side, lead to the expressions of the ladder operators b_n and $b_n^{\dagger}$ on the left hand side.

The framework of the ladder operators is very powerful and allows us to obtain new properties of Laguerre–Krall polynomials. For the special cases of Legendre, Hermite, Laguerre or Jacobi sequences of orthogonal polynomials, there is a formula which is referred to as the Rodrigues formula, that allows to express the nth degree polynomial of the sequence as the repeated action (n times) of certain differential operator on the first polynomial of the sequence (i.e. the polynomial of zero degree). See for example (2) for the Laguerre case. Thus, we can obtain a Rodrigues type formula for the Laguerre–Krall SMOP as follows.

Theorem 1 *The nth degree Laguerre–Krall polynomial of the SMOP, can be given by*

$$Q_n(x) = \left(b_n^{\dagger} b_{n-1}^{\dagger} b_{n-2}^{\dagger} \cdots b_1^{\dagger}\right) Q_0(x),$$

where $Q_0(x) = 1$.

Proof Using (10) for $n = 1$, the statement is true. Next, the expression for $Q_n(x)$ is a straightforward consequence of the definition of the raising operator.

As a second example, we obtain the same second order differential (holonomic) equation for $Q_n(x)$ as in [1] but using the framework of ladder operators. The usual technique consists in applying the raising operator $b_n^{\dagger}$ to both sides of expression $b_n Q_n(x) = Q_{n-1}(x)$. Therefore, we first have

$$b_n Q_n(x) = \left(\frac{G(x;n)}{H(x;n)} I + \frac{F(x;n)}{H(x;n)} \frac{d}{dx}\right) Q_n(x) = Q_{n-1}(x)$$

and then we apply $b_n^{\dagger}$ to both sides of the above equation, obtaining

$$\left(\frac{J(x;n)}{K(x;n)} I + \frac{F(x;n)}{K(x;n)} \frac{d}{dx}\right) \left(\frac{G(x;n)}{H(x;n)} I + \frac{F(x;n)}{H(x;n)} \frac{d}{dx}\right) Q_n(x)$$
$$= b_n^{\dagger} Q_{n-1}(x) = Q_n(x).$$

Next, we can rewrite the above expression as

$$
\begin{aligned}
&\frac{F^2(x;n)}{K(x;n)H(x;n)} Q_n''(x) \\
&+ \left(\frac{J(x;n)F(x;n) + F(x;n)G(x;n)}{K(x;n)H(x;n)} + \frac{F(x;n)}{K(x;n)} \frac{F'(x;n)H(x;n) - H'(x;n)F(x;n)}{H^2(x;n)} \right) Q_n'(x) \\
&+ \left(\frac{J(x;n)G(x;n)}{K(x;n)H(x;n)} - 1 + \frac{F(x;n)}{K(x;n)} \frac{G'(x;n)H(x;n) - H'(x;n)G(x;n)}{H^2(x;n)} \right) Q_n(x) = 0,
\end{aligned}
$$

and dividing all the above equation by the coefficient of $Q_n''(x)$, we obtain exactly the same coefficients $\mathscr{A}(x;n)$ and $\mathscr{B}(x;n)$ given in [1, Theorem 4.5]. Thus

Theorem 2 (The holonomic equation, *[1, Theorem 4.5]*) *The nth monic orthogonal polynomial with respect to the inner product (4) is a polynomial solution of the second order linear differential equation with rational functions as coefficients*

$$
[Q_n]''(x) + \mathscr{A}(x;n)[Q_n]'(x) + \mathscr{B}(x;n)Q_n(x) = 0,
$$

with

$$
\begin{aligned}
\mathscr{A}(x;n) &= -[\ln H(x;n)]' + [\ln F(x;n)]' + \frac{G(x;n) + J(x;n)}{F(x;n)} \\
&= \frac{-u_{2m}'(x;n)}{u_{2m}(x;n)} + 2\frac{\rho_m'(x)}{\rho_m(x)} + \frac{\alpha+1}{x} - 1, \\
\mathscr{B}(x;n) &= \frac{J(x;n)G(x;n) - K(x;n)H(x;n)}{F^2(x;n)} + \frac{G'(x;n)H(x;n) - H'(x;n)G(x;n)}{F(x;n)H(x;n)},
\end{aligned}
$$

where $u_{2m}(x;n) = \frac{H(x;n)}{\rho_m(x)} = A_1(x;n)D_1(x;n) - C_1(x;n)B_1(x;n)$ *is a polynomial of degree* $2m$.

4 A Convex Linear Combination of Measures

Let $\{\mathscr{K}_n^{p,m}\}_{n\geq 0}$ be the sequence of monic Krawtchouk polynomials, orthogonal with respect to the following inner product on the the linear space of polynomials with real coefficients (see [13])

$$
\langle f, g\rangle_{p,m} = \int_0^\infty f(x)\,g(x)\,d\psi^{p,m}(x), \tag{11}
$$

where $0 < p < 1$, and m is an integer number. Here $\psi^{p,m}$ is the binomial distribution of probability theory

$$\psi^{p,m}(x) = \binom{m}{x} p^x (1-p)^{m-x} \quad , \text{at } x = 0, 1, 2, \ldots, m.$$

The above measure can be seen as $m + 1$ Dirac deltas with weight

$$\binom{m}{x} p^x (1-p)^{m-x} \quad \text{at points } x = 0, 1, 2, \ldots, m.$$

Therefore, one can express the inner product (11) as

$$\langle f, g \rangle_{p,m} = \sum_{j=0}^{m} f(c_j) g(c_j) \binom{m}{j} p^j (1-p)^{m-j}$$

where $c_j = 0, 1, 2, \ldots, m$.

Next, we use the aforementioned results concerning a ladder operators in a particularization of the inner product under study (4), introducing the following inner product induced by a convex linear combination of Laguerre and Krawtchouk measures

$$\langle p, q \rangle_\lambda = \lambda \int_0^{+\infty} p(x) q(x) x^\alpha e^{-x} dx + (1-\lambda) \sum_{j=0}^{m} \binom{m}{|c_j|} p^{|c_j|} (1-p)^{m-|c_j|} p(c_j) q(c_j), \tag{12}$$

where the weights a_j in (4) have been replaced by $\binom{m}{|c_j|} p^{|c_j|} (1-p)^{m-|c_j|}$, and $c_j = -j = 0, -1, -2, \ldots, -m$. From (3), one could say that the above inner product is associated with the measure

$$d\nu^{(\lambda)} = \lambda x^\alpha e^{-x} dx + (1-\lambda) \sum_{j=0}^{m} \binom{m}{|c_j|} p^{|c_j|} (1-p)^{m-|c_j|} \delta(x - c_j), \tag{13}$$

which is also a linear combination of an absolutely continuous part, and a pure discrete part.

The monic polynomials orthogonal with respect to the discrete part of the above inner product (12) are the monic Krawtchouk polynomials defined on the negative real semiaxis, that is $T_n^{p,m}(x) := \mathcal{K}_n^{p,m}(-x)$. It means that the above Laguerre–Krall polynomials $Q_n^\lambda(x)$ are now a “mixture” , governed by the parameter $\lambda \in [0, 1]$, of classical monic Laguerre polynomials $L_n^\alpha(x)$ and the discrete monic Krawtchouk polynomials defined in the negative real semiaxis $T_n^{p,m}(x)$.

To illustrate the above linear combination, we present here two graphical examples. Figure 1 shows above the weight function of the measure ν in (3) without parameter λ, $m = 15$ (16 mass points), $p = 0.3$ for the Krawtchouk measure and $\alpha = 1$ for the classical Laguerre measure. Below, to show the overall reducing effect of the introduction of λ, it appears the weight function of the measure (13) for the same parameter values m, p, and α, but with $\lambda = 0.3$.

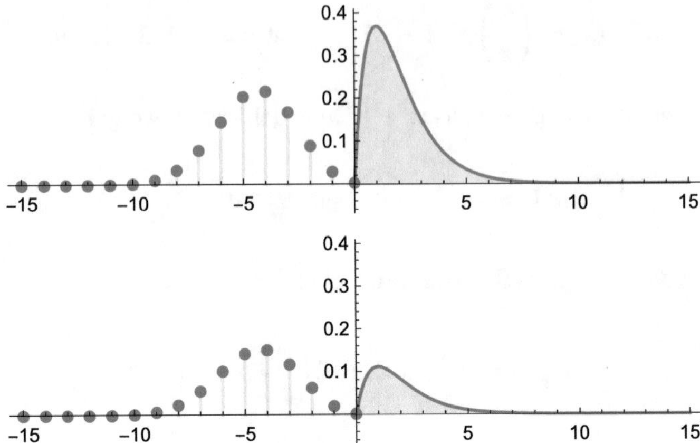

Fig. 1 Parameters $m = 15$, $p = 0.3$, $\alpha = 1$, and $\lambda = 0.3$ below

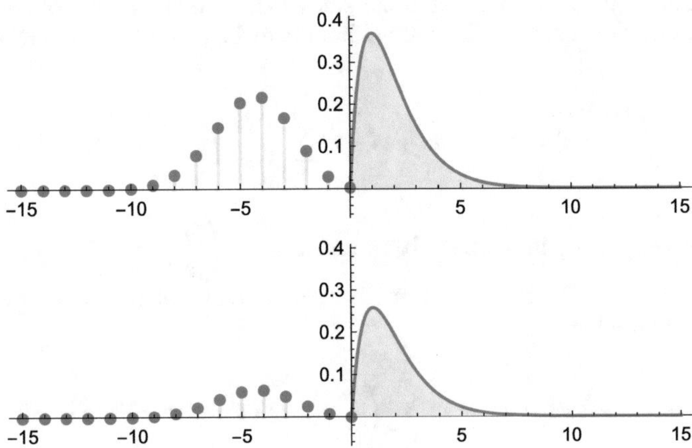

Fig. 2 Parameters $m = 15$, $p = 0.3$, $\alpha = 1$, and $\lambda = 0.7$ below

Figure 2 shows an analogous situation but giving more relevance to the continuous part using parameter $\lambda = 0.7$. It is clear that the above part of both Figures is the same, and we observe that the weight of the continuous and discrete parts are reduced in factors corresponding to the value of λ in each case.

The Laguerre–Krall orthogonal polynomials $Q_n(x)$ become $Q_n^\lambda(x)$ when the parameter λ comes into play. The connection formula becomes

$$\rho_m(x)Q_n^\lambda(x) = A_1(x; n, \lambda)L_n^\alpha(x) + B_1(x; n, \lambda)L_{n-1}^\alpha(x), \tag{14}$$

where the coefficients $A_1(x; n, \lambda)$ and $B_1(x; n, \lambda)$ show now an explicit dependence of the parameter λ as follows

$$A_1(x;n,\lambda) = \rho_m(x) - \frac{1-\lambda}{\lambda}\sum_{j=0}^{m}\left(\binom{m}{|c_j|}p^{|c_j|}(1-p)^{m-|c_j|}\frac{L_{n-1}^{\alpha}(c_j)\,Q_n^{\lambda}(c_j)}{(n-1)!\Gamma(n+\alpha)}\right)\rho_{m,j}(x),$$

$$B_1(x;n,\lambda) = \frac{1-\lambda}{\lambda}\sum_{j=0}^{m}\left(\binom{m}{|c_j|}p^{|c_j|}(1-p)^{m-|c_j|}\frac{L_n(c_j)\,Q_n^{\lambda}(c_j)}{(n-1)!\Gamma(n+\alpha)}\right)\rho_{m,j}(x),$$

with $c_j = 0, -1, -2, \ldots, -m$, and

$$\rho_m(x) = \prod_{j=0}^{m}(x+|c_j|), \quad \text{and } \rho_{m,k}(x) = \prod_{j=0,\, j\neq k}^{m}\left(x+|c_j|\right).$$

Remark 1 All the other coefficients in the connection Lemmas $A_2(x;n)$, $B_2(x;n)$, $C_1(x;n)$, $D_1(x;n)$, $C_2(x;n)$, and $D_2(x;n)$, depend on these two "main" coefficients $A_1(x;n)$, and $B_1(x;n)$, therefore all of them will show now an explicit dependence on λ as $A_2(x;n,\lambda)$, $B_2(x;n,\lambda)$, $C_1(x;n,\lambda)$, $D_1(x;n,\lambda)$, $C_2(x;n,\lambda)$, and $D_2(x;n,\lambda)$. The same situation arises with polynomials F, G, H, J, and K, (8) which become $F(x;n,\lambda)$, $G(x;n,\lambda)$, $H(x;n,\lambda)$, $J(x;n,\lambda)$, and $K(x;n,\lambda)$.

Hence, all the Connection Lemmas, theorems and results above have an automatic and trivial extension to the case with the parameter λ. It is straightforward to write the second order differential equation and the Rodrigues formula for this SMOP $\{Q_n^{\lambda}\}_{n\geq 0}$. The presence of the discrete (Krawtchouk) or continuous (Laguerre) features in $\{Q_n^{\lambda}\}$ is controlled by λ, so this parameter plays a key role to connect Krawtchouk and Laguerre measures in this framework.

At this point, it is not difficult to see that $A_1(x;n,\lambda)$ and $B_1(x;n,\lambda)$ in (14) become $A_1(x;n)$ and $B_1(x;n)$ when $\lambda = 1$, the discrete part of (13) vanishes, and then $Q_n^{\lambda}(x) \equiv L_n^{\alpha}(x)$. On the contrary, when $\lambda = 0$, is the continuous part of (13) which vanishes, but here the factor $(1-\lambda)/\lambda$ in $A_1(x;n,\lambda)$ and $B_1(x;n,\lambda)$ diverges, and we do **not** have $Q_n^{\lambda}(x) \equiv T_n^{p,m}(x)$.

In this framework, the following open problems appear which will be interesting to address in future research.

Problem 1 Based on the aforementioned divergence of $A_1(x;n,\lambda)$ and $B_1(x;n,\lambda)$ when $\lambda = 0$, it would be interesting to find out if there exists a connection formula which provides $Q_n^{\lambda}(x) \equiv T_n^{p,m}(x)$ when $\lambda = 0$.

Problem 2 It would be also very interesting to explore the limit when $m \to \infty$. In this case, we have Charlier polynomials instead of Krawtchouk polynomials in the negative real semiaxis. Using λ we expect to discover new properties of Charlier polynomials from those of Laguerre polynomials via the Laguerre–Krall polynomials $Q_n^{\lambda}(x)$.

Acknowledgements The first author (CH) wishes to thank the Dpto. de Física y Matemáticas de la Universidad de Alcalá for its support. The work of the second author (EJH) was partially supported by Dirección General de Investigación Científica y Técnica, Ministerio de Economía y Competitividad of Spain, under grant MTM2015-65888-C4-2-P. The work of the third author (AL) was partially supported by the Spanish Ministerio de Economía y Competitividad under the Project MTM2016-77642-C2-1-P.

References

1. Huertas, E.J., Marcellán, F., Pijeira, H.: An electrostatic model for zeros of perturbed Laguerre polynomials. Proc. Am. Math. Soc. **142**(5), 1733–1747 (2014)
2. Chihara, T.S.: An Introduction to Orthogonal Polynomials. Gordon and Breach, New York (1978)
3. Szegő, G.: Orthogonal Polynomials. American Mathematical Society Colloquium Publication Series, vol. 23, 4th edn. American Mathematical Society, Providence (1975)
4. Zhedanov, A.: Rational spectral transformations and orthogonal polynomials. J. Comput. Appl. Math. **85**, 67–83 (1997)
5. Koornwinder, T.H.: Orthogonal polynomials with weight function $(1-x)^{\alpha}(1+x)^{\beta} + M\delta(x+1) + N\delta(x-1)$. Can. Math. Bull. **27**, 205–214 (1984)
6. Koekoek, R.: Generalizations of classical Laguerre polynomials and some q-analogues. Doctoral Dissertation, Technical University of Delft, The Netherlands (1990)
7. Marcellán, F., Ronveaux, A.: Differential equations for classical type orthogonal polynomials. Canad. Math. Bull. **32**, 404–411 (1989)
8. Dueñas, H., Marcellán, F.: Laguerre-type orthogonal polynomials. Electrost. Interpret. Int. J. Pure Appl. Math. **38**, 345–358 (2007)
9. Ismail, M.E.H.: More on electrostatic models for zeros of orthogonal polynomials. Numer. Funct. Anal. Optim. **21**, 191–204 (2000)
10. Álvarez-Nodarse, R., Moreno-Balcázar, J.J.: Asymptotic properties of generalized Laguerre orthogonal polynomials. Indag. Math. N. S. **15**, 151–165 (2004)
11. Dueñas, H., Huertas, E.J., Marcellán, F.: Analytic properties of Laguerre-type orthogonal polynomials. Integral Transforms Spec. Funct. **22**, 107–122 (2011)
12. Fejzullahu, B.X., Zejnullahu, R.X.: Orthogonal polynomials with respect to the Laguerre measure perturbed by the canonical transformations. Integral Transforms Spec. Funct. **17**, 569–580 (2010)
13. Ismail, M.E.H.: Classical and quantum orthogonal polynomials in one variable. Encyclopedia of Mathematics and its Applications, vol. 98. Cambridge University Press, Cambridge (2005)
14. Fuchs, J.: Affine Lie Algebras and Quantum Groups. Cambridge University Press, Cambridge (1992)

Printed in the United States
By Bookmasters